Sustainable Tourism Futures

Routledge Advances in Tourism

EDITED BY PROFESSOR STEPHEN J. PAGE,
University of Stirling, Scotland

Sustainable Tourism Futures

Perspectives on Systems, Restructuring and Innovations

Edited by Stefan Gössling, C. Michael Hall, and David B. Weaver

Routledge
Taylor & Francis Group

New York   London

First published 2009
by Routledge
711 Third Avenue, New York, NY 10017

Simultaneously published in the UK
by Routledge
2 Park Square, Milton Park, Abingdon, Oxfordshire OX14 4RN

Routledge is an imprint of the Taylor & Francis Group, an informa business

First issued in paperback 2012

© 2009 Taylor & Francis

Typeset in Sabon by IBT Global.

Library of Congress Cataloging in Publication Data

Sustainable tourism futures : perspectives on systems, restructuring, and innovations / edited by Stefan Gössling, C. Michael Hall & David B. Weaver.
    p. cm. — (Routledge advances in tourism)
    Includes bibliographical references.
    1. Ecotourism—Congresses. I. Gössling, Stefan. II. Hall, Colin Michael, 1961–
III. Weaver, David B. (David Bruce)
    G156.5.E26S875 2009
    910.68'4 — dc22
    2008029065

ISBN13: 978-0-415-99619-8 (hbk)
ISBN13: 978-0-415-54225-8 (pbk)
ISBN13: 978-0-203-88425-6 (ebk)

Contents

PART III
Innovation: Sustainable Tourism Futures

Figures

Tables

Abbreviations

AAU	Assigned Amount Unit
BAU	Business As Usual
CDM	Clean Development Mechanism
CER	Certified Emissions Reduction
CCX	Chicago Climate Exchange
COP	Conference of the Parties
DNA	Designated National Authority
DOE	Designated Operational Entity
EB	Executive Board
ERU	Emissions Reductions Unit, credited via JI
EUA	European Union Allowance
EU ETS	European Union Emissions Trading Scheme
GHG	Greenhouse Gas
GWP	Global Warming Potential
IETA	International Emissions Trading Association
JI	Joint Implementation
LULUCF	Land Use, Land Use Change and Forestry
IPCC	Intergovernmental Panel on Climate Change
ITL	International Transaction Log

Mt	Mega-tonne, 10^6 metric tonnes
NAP	National Allocation Plan
NGO	Non-governmental Organisation
NO_x	Any and All Oxides of Nitrogen
PDD	Project Design Document
RFI	Radiative Forcing Index
UNEP	United Nations Environment Programme
UNFCCC	United Nations' Framework Convention on Climate Change
VER	Voluntary (or Verified) Emissions Reduction
WBCSD	World Business Council for Sustainable Development

Preface

This book is in great part the result of an expert meeting on sustainable tourism, which took place in Helsingborg, Sweden in September 2007, bringing together a range of acknowledged tourism sustainability scientists from a range of disciplines. The meeting had the goal of summarising the state-of-the-art in sustainable tourism development, to present new research results, and to discuss avenues to establish sustainable tourism on a broader basis. Some of the results from the three-day workshop are presented in this book, which sets out with a collection of theoretical papers, a reconsideration of the sustainable tourism paradigm from various perspectives, particularly in the light of new developments associated with climate change. It also provides practical examples of tourism sectors and enterprises that have managed to establish sustainable operations, and discusses the preconditions for sustainable futures, with a focus on innovation- and restructuring needs.

The book originated out of the notion that "sustainable tourism" has, rhetoric aside, largely remained an academic debate although, as several cases in this book demonstrate, some businesses have made substantial progress in terms of operating sustainably. Given the rapid growth in international tourist arrivals, the expansion of tourism infrastructure, aviation and individual motorised transport, or the development of huge remote areas for tourism as currently planned in Cambodia and Cyprus, there is little evidence that the tourism production- and consumption system in its entity is moving towards greater sustainability. Particularly with regard to climate change, it becomes ever more obvious that tourism is an emerging conflict. A recent study by UNWTO-UNEP-WMO (2008) suggests that emissions from global tourism will grow by a factor 2.5 by 2035. With regard to tourism's economic-environmental performance, calculated as the turnover associated with tourism in comparison to the sector's contribution to emissions of greenhouse gases, tourism needs to be seen as one of the least sustainable economic sectors in the world (e.g., Becken 2008; Gössling et al. 2005, Gössling and Hall 2008).

While supranational tourism organisations such as the UN World Tourism Organization and the World Travel and Tourism Council (WTTC) have started to realise that tourism is not the "smokeless" or "white" industry it has come to be associated with over several decades, it is unclear of whether such insights will be followed by concrete, measurable action. In the context of climate change, there are already calls by UNWTO and other industry organisations not to focus mitigation action on aviation, the most important tourism sub-sector in terms of emissions, in order to not potentially deprive developing countries of their opportunities to economic growth (e.g., Caribbean Hotel Association and Caribbean Tourism Organisation 2007). The aviation sector in particular has become a test case for sustainable transport developments. With moderate to low profitability (IATA 2007), it is dependent on volume growth to increase gains, while often lacking the financial resources to replace aging fleets. Scandinavian Airlines, often perceived as a pro-environmental airline, is a case in point: in its latest annual report (SAS 2008), the airline outlines two options, replacing part of its fleet rather soon in favour of moderately more energy efficient aircraft (10–15% greater efficiency), or to wait for another decade to buy considerably more energy efficient aircraft (30–35% greater efficiency). This illustrates the dilemma of the aviation sector more generally, where even high fuel prices do not seem to make it economically attractive to renew fleets, while the lifetimes of new aircraft are now calculated at 30 years (SAS 2008). At such renewal rates and small financial margins, it is clear that the sector will necessarily become increasingly unsustainable and contribute to growing emission levels.

In other tourism sectors, however, a few sustainability pioneers have made considerable progress (see Simpson et al. 2008 for case studies). Scandic Hotels, with its 17,000 hotel rooms, has moved almost all of its hotels through a certification process (Nordic Swan). A large share of the chain's food is organic and local, and all coffee is organic and/or fair trade. By 2025, the chain's energy will entirely be sourced from renewables. German operator Avanti Bus Travel has systematically developed bus travel as a convenient means of transport by offering comfortable travel in long-distance buses with low emission levels. Aspen Ski Company in the USA was the first resort operator to join the Chicago Climate Exchange, and to thereby legally commit itself to annual reduction in greenhouse gas emissions. All of its electricity is now derived from renewable sources. In central Europe, seventeen Alpine towns in five nations (Austria, France, Germany, Italy and Switzerland) founded the Alpine Pearls Association in 2006, providing customers with sustainable public transport options both to and within the towns. These and other examples show that there is ample room to move towards sustainable tourism, but the scale at which such changes take place is, from a global point of view, negligible. This book analyses and discusses this situation, and provides new perspectives on sustainable tourism.

The editors would like to thank the Helsingborg City Council for showing leadership by hosting and sponsoring the meeting of international experts, and Sparbankstiftelsen Skåne, Rica Hotels, and The Swedish Research

Council Formas for providing financial support that collectively made the conference possible. Benjamin Holtzman of Routledge is recognised for his efforts as Commissioning Editor, and two students of Stefan Gössling—Nadine Heck and Ajana Sadicovic—for their excellent work in formatting and processing the manuscripts. Last but not least, the editors are grateful for the participation of the thirty-five international experts who attended the Helsingborg conference and whose collective expertise and wisdom helped to advance the state of knowledge in the field of sustainable tourism. In particular we thank those participants who contributed to this collection.

Helsingborg, Richmond and the Gold Coast
Stefan Gössling, C. Michael Hall, and David B. Weaver

REFERENCES

Becken, S. (2008) 'Developing indicators for managing tourism in the face of peak oil'. *Tourism Management* 29(4): 695–705.

Caribbean Hotel Association and Caribbean Tourism Organization (2007) 'CHA-CTO Position Paper of Global Climate Change and the Caribbean Tourism Industry'. Available at: http://www.caribbeanhotels.org/ClimateChangePosition0307.pdf (18 February 2008).

Gössling, S. and Hall, C.M. (2008) 'Swedish tourism and climate change mitigation: An emerging conflict?'. *Scandinavian Journal of Hospitality and Tourism* 8(2): 141–158.

Gössling, S., Peeters, P.M., Ceron, J.-P., Dubois, G., Patterson, T., and Richardson, R.B. (2005) 'The eco-efficiency of tourism'. *Ecological Economics* 54(4): 417–434.

IATA (2007) '*New IATA Financial Forecast Predicts 2008 Downturn*'. 12 December 2007. Available at: http://www.iata.org/pressroom/pr/2007-12-12-01 (14 February 2008).

Scandinavian Airlines (2008) '*Your Choice for Air Travel. SAS Group Annual Report and Sustainability Report 2007*'. Stockholm, Sweden: SAS AB.

Simpson, M.C., Gössling, S., Scott, D., Hall, C.M., and Gladin, E. (2008) 'Climate Change Adaptation and Mitigation in the Tourism Sector: Frameworks, Tools and Practice'. *United Nations Environment Programme (UNEP), University of Oxford, United Nations World Tourism Organization (UNWTO), World Meteorological Organization (WMO), Paris, France*

UNWTO-UNEP-WMO (*United Nations World Tourism Organization, United Nations Environment Programme, World Meteorological Organization*) (2008) 'Climate Change and Tourism: Responding to Global Challenges'. [Scott, D., Amelung, B., Becken, S., Ceron, J.-P., Dubois, G., Gössling, S., Peeters, P., Simpson, M.] *United Nations World Tourism Organization (UNWTO), United Nations Environmental Programme (UNDP) and World Meteorological Organization (WMO). UNWTO, Madrid, Spain*.

1 Sustainable Tourism Futures

Perspectives on Systems, Restructuring and Innovations

Stefan Gössling, C. Michael Hall, and David B. Weaver

TOURISM AND SUSTAINABILITY: KEY ISSUES

Although often thought of as a recent phenomenon, the concept of sustainable development has been part of the resource conservation and management debate for almost one hundred and fifty years. George Perkins Marsh's book *Man and Nature Or, Physical Geography as Modified by Human Action* (1965), originally published in 1864, had enormous impact on conservation debates, and the questions that Marsh poses as to what is the best economic use of our natural resources still reverberates in debates with respect to sustainable development to the present day.

The concept of sustained yield had become well established in forest conservation practices by the early part of the twentieth century including the setting aside of land for timber and other forest values such as recreation (Nash 1968). However, the concept arguably become part of broader concerns over the carrying capacity of land for human use from the late 1950s on, with issues of biophysical and social carrying capacity becoming a significant focus for tourism research as a result of the influence of outdoor recreation studies (Mitchell 1989). Moreover, at the global scale the concept of carrying capacity also became a significant part of individual studies of the resource limits to economic growth at a global scale (e.g., Meadows et al. 1972), that were extremely influential in the development of the UN biosphere and habitat programmes that are the intellectual and administrative antecedents for the current UN programmes on sustainable development.

The concept of sustainability first came to public attention with the publication of the World Conservation Strategy (WCS) in March, 1980 (IUCN 1980). The WCS was prepared by the International Union for Conservation of Nature and Natural Resources (IUCN) with the assistance of the United Nations Environment Education Programme (UNEP) the World Wildlife Fund (WWF), the Food and Agricultural Organisation of the United Nations (FAO) and the United Nations Educational, Scientific and Cultural Organisation (UNESCO). The WCS was a strategy for the conservation of the earth's biological resources in the face of international environmental problems such as deforestation, desertification, ecosystem degradation and destruction, species extinction and loss of genetic diversity, loss of cropland, pollution, and soil erosion.

The WCS defined conservation as 'the management of human use of the biosphere so that it may yield the greatest sustainable benefit to present generations while maintaining its potential to meet the needs and aspirations of future generations' (IUCN 1980, *s.1.6*). The WCS is important in historical terms because it highlighted the global nature of environmental problems, emphasised the significance of the environmental–economic development relationship in the relationship between the developed and less developed countries (the north–south debate), and provided a basis for some government and private sector response, albeit limited, to the problems and issues identified in the report (Hall and Lew 1998). The WCS was also significant in that it represented the halfway mark between the 1972 United Nations Stockholm Conference on the Human Environment (which established the UNEP) and the 1992 UN Conference on Environment and Development (UNCED) in Rio de Janeiro (usually referred to as 'the Rio Summit'). However, it should be noted that tourism only received extremely limited coverage in the WCS.

In addition to assisting in the development and promotion of the WCS, the UNEP promoted the idea of the creation of a World Commission on Environment and Development (WCED) at its ten-year review conference in 1982. In 1983 an independent commission reporting directly to the United Nations Assembly was established with Gro Harlem Brundtland, then Parliamentary Leader of the Norwegian Labour Party appointed as its chair. Although the term *sustainability* was also used in a 1981 book by Lester Brown of the Worldwatch Institute, *Building a Sustainable Society*, by Myers and Gaia Ltd. (1984) and Clark and Munn (1986), *Ecologically Sustainable Development of the Biosphere*, it was the publication of the report of the WCED in 1987 entitled *Our Common Future*, commonly referred to as the Brundtland report that the term 'sustainable development' became popularised. According to the WCED (1987, 43) sustainable development is development which 'meets the needs of the present without compromising the ability of future generations to meet their own needs'. However, although the concept of sustainable development has been incredibly influential in tourism research, it is notable that tourism was only briefly mentioned in the Brundtland report.

Although formal interest in sustainability on the part of the tourism industry and tourism academics dates from the late 1980s, earlier developments in the sector were critical in setting the stage for this involvement. The decades following World War II, specifically, were characterised by a virtually uncritical embrace of mass tourism, representing what Jafari (1989, 2001) has described as the 'advocacy platform'. Inevitably, market-driven expansion fostered ecological, sociocultural, and economic problems in many locations, and within the sea-sand-sun destinations of the

Third World pleasure periphery in particular (Turner and Ash 1975). Various 'cautionary platform' advocates chronicled the perils of unregulated mass tourism, culminating in Butler's (1980) well known destination life cycle model, which suggested laissez-faire tourism's tendency under conditions of sustained demand to expand beyond the carrying capacities of individual destinations.

During the early 1980s, small-scale, locally controlled modes of 'alternative tourism' (that is, alternative to mass tourism) were proffered as being more adapted to the needs and conditions of most destinations. However, the limitations of this 'adaptancy platform' quickly became apparent in the realisation that such limited forms of tourism activity could never substitute for, or replace, the established mass tourism industry that was by then implicating hundreds of millions of tourist trips per year. Many developing countries and communities, moreover, were still compelled by the prospect of large-scale tourism activity generating development-inducing inputs of revenue and employment. The idea that a destination could have mass tourism *and* positive economic, sociocultural, and ecological outcomes was stimulated by Brundtland's popularisation of 'sustainable development' in the late 1980s (WCED 1987). Building on the momentum that this concept generated in broad-based fora such as the 1992 Rio Earth Summit, 'sustainable tourism' quickly emerged as both a principle and objective of tourism organisations, businesses, and academics. Its subsequent institutionalisation at the highest levels is evidenced by the establishment of a dedicated sustainable tourism entity within the UNWTO, and also within the private sector by the World Travel and Tourism Council's *Blueprint for New Tourism* document (WTTC 2003), although the latter makes no mention of the potential implications of climate change. During the early 2000s, additional momentum for this institutional embrace has been provided by growing concern about global climate change and its substantial apparent linkages with tourism systems.

The embrace of sustainable tourism is a core premise of the 'knowledge-based platform', which Jafari (1989, 2001) additionally characterises as a less ideologically constrained scientific approach that regards both mass and alternative tourism as legitimate and beneficial models of tourism development, depending on the local circumstances, background research, and management/planning considerations that pertain to any given destination. Yet, although something approaching consensus has been attained with respect to the basic principle of sustainable tourism, implementation in an era of climate change concern is a far more confounding matter, reflecting sustainable development's status as an 'essentially contested concept' susceptible to multiple interpretations depending on the 'values and ideologies of various stakeholders' (Hall 1998, 13). 'Sustainable tourism', as such, can be equated with virtually any type of activity, the term at times being cynically employed to gain added ethical standing leading to what is sometimes referred to as 'greenwash'. On the other hand, Hunter (1997) and others regard this flexibility (or malleability?) as a strength which recognises that its implementation in a densely urban environment such as Hong Kong or the Gold Coast of Australia must be predicated on entirely different assumptions from those which pertain to wilderness settings such as Antarctica or northern Siberia.

The dilemma of flexibility permeates to all aspects of implementation, including the selection and weighting of relevant indicators as well as the identification of appropriate monitoring protocols, thresholds, and benchmarks for these indicators. Such challenges are evident in many of the case studies described in this book, as are the inherent complexities of tourism systems, which are characterised by fuzzy boundaries and often unclear or unanticipated indirect and induced impacts that arise from unpredictable cause and effect relationships (Weaver 2006). Should, for example, a hotel which is internally stellar in terms of its operational sustainability still be regarded as 'sustainable' if management does not mitigate the carbon emissions generated by its guests, or if substantial habitat clearance results from the nearby settlement of new employees and their families? Ultimately, the issue of sustainable tourism is moot if the external systems which interact with tourism are unsustainable or otherwise inimical to sustainable outcomes. The devastating tsunami of 2004 is but one dramatic illustration of this imperative of inter-dependency, though less dramatic but more pervasive practices such as dynamite and cyanide fishing may be just as harmful to the ideal of sustainable tourism in the long term. Also daunting are escalating costs of energy and other inputs, which on one hand might stimulate the pursuit of cost-saving measures such as energy conservation and recycling, but on the other hand might dissuade tourism authorities and businesses from addressing long-term concerns just as or more important for the attainment of long-term sustainability goals.

TOURISM AND CLIMATE CHANGE

One of the most relevant, if not *the* single most relevant issue for environmentally sustainable tourism is climate change, both because tourism is affected by climate change and because the sector is a considerable force of climate change (Hall and Higham 2005; Gössling and Hall 2006; Becken and Hay 2007; UNWTO-UNEP-WMO 2008). According to the Intergovernmental Panel on Climate Change (2007), temperature increases due to growing concentrations of greenhouse gases in the atmosphere will lead to globally averaged surface temperature rises by 1.8–4.0°C by the end of the twenty-first century. Changes in temperatures and other climate parameters will not be linear, rather they will be accompanied by heat waves, heavy precipitation events, or storms. Decreases in snow cover are expected to continue. These are but some examples of the consequences of global climate and environmental change for tourism (for the most recent and comprehensive review see UNWTO-UNEP-WMO 2008).

Tourism is itself an important contributor to emissions, as shown in various studies (e.g., Becken 2002; Gössling 2002; Gössling and Hall 2008). The sector's contribution to climate change has usually been calculated by including three sub-sectors, transport to and from the destination, accommodation, and activities. A recent study (UNWTO-UNEP-WMO 2008) has quantified tourism's share as 5 per cent of global CO_2 emissions. Although this may appear minor, it deserves mention that tourism also causes significant emissions of noncarbon greenhouse gasses, mostly by air transport (Sausen et al. 2005). Transport is responsible for about 75 per cent of the CO_2 emissions contributed by tourism, and up to 90 per cent of its contribution to radiative forcing. A focus on transports, and within transports aviation (40 per cent of CO_2 emissions), is thus of great importance in addressing emissions from this sector. Further insight for mitigation priorities can be gained from the distribution of emissions across tourism sectors and within individual trips. For instance, trips by rail and bus/coach account for about 34 per cent of all tourist journeys, causing 13 per cent of all CO_2 emissions. In contrast, long-haul trips by air account for 2.7 per cent of all tourist trips, but are responsible for 17 per cent of emissions. Individual trips can cause emissions with a factor 1,000 difference: from close to zero (a holiday by bicycle and tent) to more than 10 t of CO_2 (a journey to Antarctica) (UNWTO-UNEP-WMO 2008).

In the future, tourism is very likely to contribute to increasing emissions. A current estimate is that emissions of CO_2 will grow by a factor 2.5 in the period 2005–2035, mostly as a result of increasing air travel (UNWTO-UNEP-WMO 2008). This will be in contrast to global emission reduction needs as agreed upon in the Kyoto Agreement, and should thus be seen as a call for strong governmental action: the IPCC (2007) suggests to cut emissions to about 1 t CO_2 per capita per year by the end of the century, and countries like Sweden or the UK already discuss emission cuts by 60–80 per cent by 2050. However, at the time of writing of this book (May 2008), very few governments seem to have taken any action to address emissions from aviation or tourism more generally.

INNOVATION AND SUSTAINABILITY

For many people innovation is often primarily understood in terms of technological change. However, this is only part of what constitutes innovation. Innovation also includes ideas and knowledge with respect to 'ways of doing'. Sustainability itself was, and arguably still is, an innovative idea. One that has profound implications for the stewardship of natural resources, environmental conservation, our understanding of the connectivity between natural capital and economic capital, and the distribution of material benefits. This is not of course to suggest that there is unanimity in adoption of the concept of the idea of sustainability. Indeed, as this book suggests there is a substantial implementation gap between the theory and practice of sustainable development and sustainable tourism in particular.

The role of theory per se should not be blamed for the lack of implementation of sustainable tourism practices. Campbell and Fainstein (2003, 3) observed that with respect to planning, 'Theory can inform practice. Planning theory is not just some idle chattering at the margins of the field. If done poorly, it discourages and stifles; but if done well, it defines the field and drives it forward'. Indeed, within the public policy tradition it is important to realise that government policies with respect to environmental conservation and sustainable development imply theories (Hall 2008). Tourism planning and policy making reflects assumptions about the manner in which people, organisations and, in some cases, the environment, will act given an authoritative decision or set of decisions. As Pressman and Wildavsky (1973, xv) stated:

> Whether stated explicitly or not, policies point to a chain of causation between initial conditions and future consequences. If X, then Y. Policies become programs when, by authoritative action, the initial conditions are created, X now exists. Programs make the theories operational by forging the first link in the causal chain connecting actions to objectives. Given X, we act to obtain Y.

Similarly, Majone (1980, 178) argued that 'policies may be viewed as theories from two different but related perspectives'. First, 'they can be seen as an analyst's rational reconstruction of a complex sequence of events'. Second, 'they can be seen from the point of view of actions, giving them stability and internal coherence' (Majone 1980, 178). The theory and planning and implementation of sustainable tourism therefore go hand-in-hand. Indeed, the implementation of sustainable tourism initiatives arguably has as much to do with the barriers to adoption and diffusion of ideas and practices as it does with the intrinsic characteristics of a tourism innovation. And limits to adoption can arise as a result of a wide variety of factors ranging from hard barriers such as trade restrictions, government regulation, and cost, as well as soft barriers such as culture, knowledge, political structures, and risk perception. Therefore, consideration of sustainable tourism futures requires the development of an understanding not only of adaptive practices but also the systems into which they are embedded.

FROM RELATIVE TO ABSOLUTE
MEASURES OF SUSTAINABILITY?

Over the past two decades, a wide range of indicators for sustainable tourism has come into existence (e.g., Choi and Sirakaya 2006). Virtually none of these indicators, however, considers differences between relative and absolute sustainability, or what is here defined as a relatively better standard of environmental performance as compared to the overall or absolute sustainability of the activity. To illustrate this, a hotel might substantially reduce its use of water, thereby achieving more efficient and sustainable operations, but its overall water use may, in a theoretical situation where water is scarce and abstracted from limited, non-renewable sources, still remain unsustainable. In another example, a luxurious, eco-friendly, and certified hotel may still use more resources per guest night than a simple, non-certified accommodation. In this situation, the certification is an indicator of relative sustainability within one type of accommodation, and not making any statement of sustainability in a comparative assessment. Yet another example is air travel. Even though airlines become more energy efficient every year, i.e., using less energy to transport one person from location A to location B, overall energy use and associated climate change-relevant emissions are increasing. It is anticipated that within the European Union, emissions will double in the period 2005–2020 (European Commission 2006). The paradoxical situation is thus that if measured using relative indicators, air travel could be seen as continuously moving towards sustainability, identified as declining emissions per passenger kilometre travelled. In absolute terms, however, aviation becomes continuously less sustainable. Based on these examples, it is clear the consideration of relative and absolute measures of sustainability in tourism is meaningful, and should receive more attention by industry and academics alike.

In recent years, the number of tour operators, hotel chains, and travel companies engaged in proenvironmental activities within their respective fields of operation has grown in absolute numbers, though probably not in relative terms, i.e., as a share of all tourism businesses. This process is mirrored in the emergence of an increasing number of ecolabels, codes of conduct, sustainability reporting schemes, awards, benchmarking programmes, and best practice approaches (e.g., Bodhanowicz and Martinac 2006; Font and Harris 2004). Skinner, Font, and Sanabria (2004), for instance, counted over sixty certification programmes setting standards, with an average of about fifty certified tourism firms per programme. Most of the certification programmes are regional or national, and are linked to specific destinations with locally relevant standards. Most focus on environmental issues (Font and Harris 2004), which is a generally complex task, as pointed out by Weaver (2005, 26): 'indicator-based sustainable tourism strategies are complicated by the actual process of selecting, measuring, monitoring, and evaluating a viable set of relevant variables'.

The importance of indicators in the development of more sustainable tourism notwithstanding, there has so far not been any distinction of relative and absolute perspectives on sustainability. This may largely be a result of the notion that sustainability is a process without an ultimate goal, or as stated by Farrell and Twinning-Ward (2004, 275): '[. . .] sustainability must be conceived as a transition, journey or path, rather than an end point or an achievable goal'. Indeed, this would seem to apply to all socio-cultural aspects of tourism, as it is impossible to define one single sustainable 'end state' society from a socio-cultural point of view, particularly in the absence of any knowledge on future generations' perspectives on sustainability. Even with regard to environmental aspects, it is clear that sustainability currently refers to the transition from a resource-intense, non-renewable society to one that is dematerialising, re-using resources, or using renewable resources. However, it may also be argued that the notion of sustainability as a transitory process is one of the greatest inherent weaknesses of the sustainability concept. Criticism of the concept began as soon as the World Commission on Environment and Development presented its report 'Our Common Future' (WCED 1987), which brought the term *sustainable development* to a global audience. One of the key problems with the report is the interpretability of what sustainability is, implicitly referring to the problem of a missing, measurable goal. Almost two decades ago, Lele (1991, 608) remarked:

> Sustainable development is a 'metafix' that will unite everybody from the profit-minded industrialist and risk-minimizing subsistence farmer to the equity-seeking social worker, the pollution-concerned or wild-life-loving First Worlder, the growth-maximizing policy maker, the goal-oriented bureaucrat and, therefore, the vote-counting politician.

Given that the definition of sustainability has been up to each and everyone, again reflecting the notion that it is a *process of change* and not an *ideal end state*, could thus be made responsible for the past two decades of rhetorical (but hardly real) progress towards achieving sustainability. This is also evident in tourism. For instance, Butler's (1993) often-cited definition of sustainable tourism refers to change (and the limits to change):

> [. . .] developed and maintained in an area (community, environment) in such a manner and at such a scale that it remains viable over an indefinite period and does not degrade or alter the environment (human and physical) in which it exists to such a degree that it prohibits the successful development and well-being of other activities and processes.

As such definitions are largely nonoperationable, in the sense that they do specify the areas where change is needed, over which timescales change should occur, and how fundamental the change should be, it should not come as a surprise that progress towards sustainability has been slow in tourism—indeed, it could be argued that tourism is still becoming less sustainable on the global scale, at least in the environmental domain. Given this situation, there may indeed be a need to define 'an ideal end-state' with regard to some parameters of sustainability. This appears feasible within timeframes of up to one hundred years (cf. e.g., IPCC 2007), and for selected aspects. Climate change may be one such aspect; the scientific consensus being that greenhouse gas emissions should not exceed levels causing 'dangerous interference with the climate system' (e.g., Schellnhuber et al. 2006). Dangerous climate change is anticipated to occur if global average temperatures increase in excess of 2°C, which in turn translates into maximum carbon dioxide (CO_2) concentrations of around 450 ppm in the atmosphere by 2100 (IPCC 2007). This definition of climate sustainability may be acceptable even for future generations, as it ensures that living conditions on Earth remain viable, while simultaneously representing an absolute sustainability goal considering scale, location, and time. Maximum concentrations of CO_2 are known and can be broken down to annual global emission scenarios, where a share of sustainable emissions is allocated to each country—the Kyoto protocol is essentially doing just this. Each country can in turn allocate these emissions to different economic sectors, considering sector-specific reduction or growth scenarios. Emissions are thus a very suitable indicator for sustainability, and some indicator systems are now emerging focusing on this parameter, even though it deserves mention that Choi and Sirakaya (2006) do not list even a single climate indicator in their Delphi-based collection of 125 indicators (energy use is included, though). With regard to emerging carbon indicators, Becken (2008) suggests no less than ten indicators to better understand emissions caused by tourism.

In summary, indicators as currently used within tourism may usually show whether a product or service is less environmentally harmful, thus reflecting aspects of *relative* sustainability, but they do not indicate whether overall levels of consumption are sustainable (in terms of *absolute* sustainability). Resource use standards based on maximum consumption levels (see, for instance, water and energy use; Chapter 7, this volume) may be seen as absolute standards, but these are not currently based on systems' perspectives of sustainability, as for instance suggested by Peeters et al. (Chapter 13, this volume) in the form of per capita per day emission calculations. The global challenge is that a rapidly growing number of tourism businesses need to become relatively more resource efficient at a pace that leads to absolute reductions in resource use, as exemplified by emissions of greenhouse gases. There is thus a need to rethink indicators in general, to consider aspects of scale (local/regional, national, and global), location, and whether an indicator informs about relative or absolute sustainability.

SYSTEMS, SCALE, RESTRUCTURING

Systems concepts and thinking provides an important integrative set of ideas for tourism, innovation, and sustainability. For example, the use of tourism for sustainable development purposes can be expressed in systems terms as the process of meeting the needs of current and future generations via tourism production and consumption without undermining the resilience of the life-supporting properties or the integrity and cohesion of natural and social systems. While sustainable tourism development is the process of meeting the needs of current and future generations of tourism producers and consumers without undermining the resilience of tourism firms and the life-supporting properties or the integrity of natural and social systems on which they rely.

Systems thinking is a means of abstracting from a reality in a manner that makes it more understandable. A system is an object of study and comprises a set of elements; the set of relationships between those elements and a set of relationships between the elements and the environment in which they are placed. However, it is important to note that one of the characteristics of understanding systems is the problem of scale. Systems are embedded within systems. What may be regarded as an element of a system at one level of analysis (a subsystem) may itself constitute a system at a lower level of analysis. For example, in tourism research the movement of tourists within an international tourism system is often studied by analysing the flows of tourists between different countries, which constitute the elements of such a system. However, if we change our resolution we may then examine the flows of tourists within a country by looking at the intra-regional flows of tourists. In the latter example it is the country which is the system and the regions the elements. How we define an element therefore depends on the scale at which we conceive the system, otherwise referred to as the level of resolution (Hall 2008).

Three basic questions arise with respect to scale (Haggett 1965; Harvey 1969):

1. Scale coverage—do we have regular and comprehensive monitoring of the world at all relevant scales?
2. Scale standardisation—do we have comparable data from equivalent sampling frames?
3. Scale linkage—what connections can be identified between the various scale levels?
   (a) same level—which refers to a comparative relationship
   (b) high to low level—which is a contextual relationship
   (c) low to high level—which is an aggregative relationship

Such a discussion may sound extremely abstract to some but it is actually crucial to understanding sustainability because, as noted in the previous section, a great challenge with the selection of indicators is ensuring that they apply appropriately across a scale and can therefore be effectively utilised for purposes of measurement and benchmarking. Substantial inferential problems may also arise

> because generalisations we make at one level may not hold for another. Indeed, the idea of emergence, i.e. that the whole is greater than the

sum of the parts, makes this virtually a certainty. Such a situation creates substantial difficulties for explanation in tourism studies which the field has not addressed, especially as most tourism analysis often does not acknowledge the scale at which work is being undertaken, or the contexts of that scale, and the capacity to generalise from one scale to another. (Hall 2008, 81)

The study of systems has been well established with respect to ecology and natural systems since the 1930s but arguably its greatest influence was not felt until the 1950s on as systems theory revolutionised not only the study of natural history but was also influential in the social sciences, including management and politics. The concept of a tourism system had been utilised in tourism research since the late 1960s with the idea becoming popularised in tourism texts from the 1980s on. However, in tourism the notion of a system is often used as a metaphor or analogue as opposed to a set of relationships between elements and their environment that can be expressed mathematically. Nevertheless, even analogue approaches have proven to be useful conceptual tools with respect to understanding the role of collaboration and networks in tourism development (Farrell and Twining-Ward 2004; Hall 2008). For example, the issue of relationships between scales is extremely important for tourism planning and policy and the solution of policy issues, such as the environment, that cross over different scales of governance (Gössling and Hall 2006).

The systems concept is also important for understanding change. As Mill and Morrison (1985, xix) commented with respect to the concept of a tourism system as used in their text: 'The system is like a spider's web—touch one part of it and reverberations will be felt throughout'. Tourism systems, as well as innovation systems, are examples of social-ecological systems that are linked systems of people and nature. However, they are both subsets of broader social-ecological systems. The purpose of a tourism system is to satisfy the needs of tourists in such a way as to provide appropriate returns to producers and maintain the system, while the purpose of an innovation system is to generate innovations and appropriately transfer them. Tourism and innovation is appropriate for social-ecological systems to the extent to which they enhance a system's capacity to deal with change and continue to develop. This is also referred to as resilience, which is a term that is increasingly used interchangeably with sustainability.

Scale therefore is clearly an important issue for understanding sustainability. Members of the tourism system, who see the wellbeing of that system as being synonymous with the health of the wider social-ecological system, whether on a local or a global scale, are sadly mistaken. The outcomes of sustainable tourism are not the same as those of utilising tourism as a means of promoting sustainable development. However, ultimately it is the broader social-ecological system on which we depend, even though tourism may have an important contribution to make towards encouraging ecosystem and socio-economic resilience.

As Chapter 15 by Hall argues, the capacity of tourism businesses to survive and grow is a measure of their individual resilience as well as a potential indicator of system resilience at a destination level. Furthermore, firm survival is recognised as being linked to the innovative capacity of firms (Hall and Williams 2008) often via various adaptations to their environment. The capacity of firms to adapt and innovate is a product of both their system characteristics and their embeddedness within innovation systems (Edquist 2005). From a system-wide perspective, the resilience of a social-ecological system, such as a destination, will be affected by the response diversity of the tourism firms within it. This means that the different sensitivities of firms to specific disturbances to the system, such as economic and social change and restructuring, will decrease the vulnerability of that destination to such disturbances. Nevertheless, change does occur. The fundamental issue being not so much change per se but the magnitude of change in relation to a systems capacity to adapt and develop without the loss of the ecosystem services essential to social, economic, and political wellbeing.

CLIMATE CHANGE MITIGATION AND ADAPTATION

Climate change mitigation relates to technological, economic and social changes, and substitutions that lead to emissions reductions (IPCC 2007). Current perspectives on mitigation are largely technical in character, with a focus on innovation and market mechanisms such as emissions trading. In tourism, it is increasingly clear that absolute reductions in greenhouse gas emissions can only be achieved through behavioural change (UNWTO-UNEP-WMO 2008). This is primarily because annual reductions in energy use and emissions are outpaced by the rapidly growing number of humans participating in tourism, as well as more energy-intense travel patterns in already mature tourism markets. Mitigation and adaptation, i.e., the adjustment of human or natural systems to climate change, can be complementary. For instance, climate policy in the European Union, which is likely to be implemented in other important tourism countries as well, is likely to contribute to considerable increases in mobility costs. For tourism-dependent countries, it should thus be a priority to restructure their tourism industries towards low-carbon alternatives (Gössling et al. 2008).

While mitigation is a necessity to reduce greenhouse gas emissions to levels that will not constitute a dangerous interference with the climate system, adaptation will still be necessary to deal with the changes of the climate that will be resulting from current and historic rises in greenhouse gas emissions. Such changes are already causing economic and social interference, and are anticipated to cause considerably greater interferences in the future (Stern Review 2006). As unabated climate change would increase risks for tourism, particularly in tropical destinations, mitigation and adaptation should be seen as a combined process in which destinations, businesses, tourism organisations, and scientists can work together.

REFERENCES

Becken, S. (2008) 'Developing indicators for managing tourism in the face of peak oil'. *Tourism Management* 29(4): 695–705.

Becken, S. (2002) 'Analysing international tourist flows to estimate energy use associated with air travel'. *Journal of Sustainable Tourism* 10(2): 114–131.

Becken, S. and Hay, J.E. (2007) *Tourism and Climate Change, Risks and Opportunities*. Clevedon: Channel View.

Bohdanowicz P. and Martinac I. (2007) 'Determinants and benchmarking of resource consumption in hotels—case study of Hilton International and Scandic in Europe'. *Energy and Buildings* 39: 82–95.

Brown, L. (1981) *Building a Sustainable Society*. New York: W.W. Norton.

Butler, R.W. (1980) 'The concept of a tourist area cycle of evolution: implications for management of resources'. *Canadian Geographer* 24(1): 5–12.

Butler, R. (1993) 'Tourism—an evolutionary perspective'. In Nelson, J.G., Butler, R., and Wall, G. (eds.) *Tourism and Sustainable Development: Monitoring, Planning, Managing*, Publication Series no. 37, 27–43. Waterloo: University of Waterloo, Department of Geography.

Campbell, S. and Fainstein, S. (2003) 'Introduction: The structure and debates of planning theory'. In Campbell, S. and Fainstein, S. (eds.) *Readings in Planning Theory*, 2nd ed., 1–16. Oxford: Blackwell.

Choi, H.C. and Sirakaya, E. (2006) 'Sustainability indicators for managing community tourism'. *Tourism Management* 27: 1274–1289.

Clark, W. and Munn, R.E. (eds.) (1986) *Ecologically Sustainable Development of the Biosphere*. New York: Cambridge University Press.

Edquist, C. (2005) 'Systems of innovation: Perspectives and challenges'. In Fagerberg, J., Mowery, D., and Nelson, R.R. (eds.) *The Oxford Handbook of Innovation*, 181–208. Norfolk: Oxford University Press.

European Commission (2006) *Summary of the Impact Assessment: Inclusion of Aviation in the EU Greenhouse Gas Emissions Trading Scheme* (EU ETS), Commission staff working document. Brussel: European Commision.

Farrell, B.H. and Twining-Ward, L. (2004) 'Reconceptualizing tourism'. *Annals of Tourism Research* 31(2): 274–295.

Font, X. and Harris, C. (2004) 'Rethinking standards from green to sustainable'. *Annals of Tourism Research* 31(4): 986–1007.

Gössling, S. (2002) 'Global environmental consequences of tourism'. *Global Environmental Change* 12(4): 283–302.

Gössling, S. and Hall, C.M. (eds.) (2006) *Tourism and Global Environmental Change. Ecological, Social, Economic and Political Interrelationships*. London: Routledge.

Gössling, S. and Hall, C.M. (2008) 'Swedish tourism and climate change mitigation: An emerging conflict?'. *Scandinavian Journal of Hospitality and Tourism* 8(2): 141–158.

Gössling, S., Peeters, P., and Scott, D. (2008) 'Consequences of climate policy for international tourist arrivals in developing countries'. *Third World Quarterly* 29(5): 869–897.

Haggett, P. (1965) 'Scale components in geographical problems'. In Chorley, R.J. and Haggett, P. (eds.) *Frontiers in Geographical Teaching*, 164–185. London: Methuen.

Hall, C.M. (2008) *Tourism Planning*, 2nd ed. Harlow: Pearson Education.

Hall, C.M. and Higham, J. (eds.) (2005) *Tourism, Recreation and Climate Change*. Clevedon: Channelview Publications.

Hall, C.M. and Lew, A. (1998) 'The geography of sustainable tourism development: an introduction'. In Hall, C.M. and Lew, A.A. (eds.) *Sustainable Tourism Development: Geographical Perspectives*, 1–12. London: Addison-Wesley Longman.

Hall, C.M. and Williams, A. (2008) *Tourism and Innovation*. London: Routledge.

Harvey, D. (1969) *Explanation in Geography*. London: Edward Arnold.

Hunter, C. (1997) 'Sustainable tourism as an adaptive paradigm'. *Annals of Tourism Research* 24(4): 850–867.

International Union for the Conservation of Nature and Natural Resources (IUCN) (1980) *World Conservation Strategy*. Morges: The IUCN with the advice, cooperation and financial assistance of the United Nations Environment Education Program and the World Wildlife Fund and in collaboration with the Food and Agricultural Organization of the United Nations and the United Nations Educational, Scientific and Cultural Organization, IUCN.

Intergovernmental Panel on Climate Change (IPCC) (2007) *Summary for Policymakers. Intergovernmental Panel on Climate Change Fourth Assessment Report Climate Change 2007: Synthesis Report*. Cambridge, United Kingdom and New York, NY, USA: Cambridge University Press.

Jafari, J. (1989) 'An English language literature review'. In Bystrzanowski, J. (ed.) *Tourism as a Factor of Change: A Sociocultural Study*. Centre for Research and Documentation in Social Sciences: Vienna.

Jafari, J. (2001) 'The scientification of tourism'. In Smith, V.L. (ed.), *Hosts and Guests Revisited: Tourism Issues of the 21st Century*, 28–41. New York: Cognizant.

Lele, S.M. (1991) 'Sustainable development: a critical review'. *World Development* 19(6): 607–620.

Majone, G. (1980a) 'The uses of policy analysis'. In Raven, B.H. (ed.) *Policy Studies Annual Review*, Vol. 4, 161–180. Beverly Hills: Sage.

Marsh, G.P. (1965) *Man and Nature or Physical Geography as Modified by Human Action*, orig. Lowenthal, D. (1864) (ed.). Cambridge: The Belknap Press of Harvard University Press.

Meadows, D.H., Meadows, D.L., Randers, J., and Behrens III, W.W. (1972) *The Limits to Growth. A Report for The Club of Rome's Project on the Predicament of Mankind*. London: Pan Books.

Mill, R.C. and Morrison, A.M. (1985) *The Tourism System: An Introductory Text*. Englewood Cliffs: Prentice-Hall International.

Mitchell, B. (1989) *Geography and Resource Analysis*, 2nd ed. Harlow: Longman Scientific and Technical.

Myers, N. and Gaia Ltd. staff (1984) *Gaia: An Atlas of Planet Management*. New York: Anchor/Doubleday.

Nash, R. (ed.) (1968) *The American Environment, Readings in the History of Conservation*. Reading: Addison-Wesley Publishing.

Pressman, J.L. and Wildavsky, A.B. (1973) *Implementation: How Great Expectations in Washington are Dashed in Oakland; or, Why it's Amazing that Federal Programs Work at All*, 2nd ed. Berkeley: University of California Press.

Sausen, R., Isaksen I., Grewe, V., Hauglustaine, D., Lee, D.S., Myhre, G., Köhler, M.O., Pitari, G., Schumann, U., Stordal, F., and Zerefos, C. (2005) 'Aviation radiative forcing in 2000: an update on the IPCC report 1999'. *Meteorologische Zeitschrift* 14: 555–561.

Schellnhuber, J., Cramer, W., Nakicenovic, N., Wigley, T., and Yohe, G. (eds.) (2006) *Avoiding Dangerous Climate Change*. Cambridge: Cambridge University Press.

Skinner, E., Font, X., and Sanabria, R. (2004) 'Does stewardship travel well? Benchmarking, accreditation and certification'. *Corporate Social Responsibility and Environmental Management* 11: 121–132.

Stern, N. (2006) *The Economics of Climate Change: The Stern Review*. Cambridge: Cambridge University Press.

Turner, L. and Ash, J. (1975) *The Golden Hordes*. London: Constable.

UNWTO-UNEP-WMO (United Nations World Tourism Organization, United Nations Environment Programme, World Meteorological Organization) (2008) *Climate Change and Tourism: Responding to Global Challenges*. United Nations World Tourism Organization (UNWTO), United Nations Environmental Programme (UNDP) and World Meteorological Organization (WMO): UNWTO, Madrid, Spain.

Weaver, D. (2005) *Sustainable Tourism*. Amsterdam: Elsevier.

Weaver, D. (2006) *Sustainable Tourism: Principles and Practices*. London: Butterworth-Heinemann.

World Commission on Environment and Development (WCED) (the Brundtland Report) (1987) *Our Common Future*. London: Oxford University Press.

World Travel and Tourism Council (2003) *Blueprint for New Tourism*. London: World Travel and Tourism Council.

Part I

Theoretical Foundations

Re-thinking the Tourism System

2 Thirty Years of Sustainable Tourism
Drivers, Progress, Problems—and the Future

Bernard Lane

Tourism has had a long history. Some commentators place its origins in mediaeval pilgrimages, some in the Grand Tours of the eighteenth and nineteenth century, and others in the railway age world of the spa and seaside resort. But the real rise of tourism as a major pursuit and as a major industry begins in the post war period, especially after that key year in the UN World Tourism Organization statistics, 1950, when 25 million international travellers were recorded. Then the meteoric rise of the tourism industry began in earnest. Average year-on-year growth rates of 6.5 per cent have been achieved over the period 1950–2006. The year 2006 saw 846 million international arrivals world wide. UNWTO looks forward to 1.6 billion arrivals by 2020 (www.unwto.org). No one working in the industry today can personally recall the pregrowth era. Growth—in numbers—in geographical impacts—in product terms—is regarded as an ongoing and given norm.

This chapter discusses how the linear growth of tourism was first challenged by the concept of sustainable tourism, how the challenge went largely unheeded, and how only now, thirty years on, is the industry beginning to fear an unsustainable future.

Back in the exciting days of the 1950s and 1960s tourism was hailed as a clean industry that was revolutionising the world. The tyranny of distance was being conquered. By 1970, the WTO was able to report that 166 million international travellers arrived at their destinations. Fuelled by increasing levels of disposable income, by cheap oil, by changing fashions and life styles, by tour operating companies, by technological change, by rising levels of education and by media commentaries, the tourism boom had been established. The travel drug began to addict more and more people. Most addictions are a form of escapism; travel is the ultimate escapist activity, taking body as well as soul and dreams along the way.

Few people then understood the problems that the tourism boom would create. They believed that mobility was a good thing, that travel broadened the mind, spread happiness, prevented war. But in the German-speaking— and to some extent, the French-speaking—academic world, criticism of tourism grew in the 1970s. Jost Krippendorf, from the University of Bern, was a leading critic from that era. Describing and analysing the environmental and social consequences of tourism, he became an advocate of *Sanfter Tourismus*—Soft Tourism—as an alternative (Müller and Lane 2003). *Sanfter Tourismus* was the forerunner of sustainable tourism.

In 1975 Jost Krippendorf had published *Die Landschaftsfresser* (The Landscape Eaters), describing the impact of tourism on the Alpine landscape. His 1984 book *Die Ferienmenschen* took the discussion much further; in 1987 his book was translated into English as *The Holiday Makers: Understanding the Impact of Leisure and Travel*. It sought not regulation but a change in life style and behaviour from all parties. It sought a more human form of tourism. That form of tourism would use more informed and responsible marketing; more thoughtful and better trained tourism personnel of all kinds; host populations would be prepared to manage tourism; holidaymakers would be better informed; research would be undertaken into the drivers and issues of a more sustainable tourism. Re-reading *The Holiday Makers* twenty years later reveals that Jost Krippendorf was remarkably far sighted—except about the threat of climate change, an almost unknown in 1984. Jost Krippendorf was not alone in his work; his compilation *Für einen andern Tourismus* (Towards an Alternative Tourism; Krippendorf, Zimmer, and Glauber 1988), brought together the thoughts of no less than seventeen of the leading sustainable tourism protagonists of that time from Austria, France, Germany, Italy, and Switzerland.

In the mid- to late- 1980s the concept of *Sanfter Tourismus* broke out of its German speaking heartland and spread across the wider world. After a period of discussion, the term *sustainable tourism* was established. A concise definition of sustainable tourism can be found in Bramwell and Lane (1993). By 1991 discussions about launching a *Journal of Sustainable Tourism* had begun. Its first volume was published in 1993. It now ranks fourth in the peer reviewed tourism journal hierarchy (McKercher, Law, and Lam, 2006).

There was much dispute about the concept of sustainable tourism in those early days. It was looked at with apprehension and rejection by the industry, which was unhappy about any limits to growth and found it an ivory tower concept, failing to fulfil the needs of the market. It was also felt to be intellectually arrogant, expensive, elitist, and unnecessary. Conventional tourism worked well, and used a proven and successful business model.

Apprehension and rejection was not confined to the industry. Governments toyed with the idea of sustainable tourism, but the existing tourism business model was as attractive to them as to the industry. Conventional tourism created jobs and transferred wealth from richer to poorer regions. It was politically popular: Holidays were like the astrological Jupiter in Gustav Holst's *Planet Suite*, "the bringer of Jollity". The electorate liked jollity. The media—the hidden link between the industry and the market—also toyed with the idea of sustainable tourism but soon found that it was easier, and much more fun, to deride it than to explain it.

The concept was also derided by some tourism academics. It was described as 'wishful thinking' and as 'an impossible dream'. One well-known commentator wrote that the concept of managing tourism's style and growth was like asking a woman to become just a little bit pregnant. Another frequently referenced academic felt passionately that the whole idea was intrinsically impossible and a greenwash trick (see, for example, Butler 1990; Wheeler 1992, 1993). Many of their comments were actually not far short of the truth; at that stage the concept had not been researched and its techniques were neither implemented nor evaluated.

EARLY DAYS

Sustainable tourism grew out of emerging unease about the impacts of tourism. While tourism was a capable economic development tool, it had strong negative environmental, cultural, and social impacts. It was not planned or operated holistically. It tended to operate on a short-term growth and decline cycle. It tended to retain its profits in visitor source market areas rather than in destination areas. It was rarely used as a tool for conservation or for sustainable development. It went along with the fundamentally selfish, short term, and hedonistic approach common to many holidaymaking pursuits.

Sustainable tourism began as a purely reactive concept to the above issues, trying to stop negative change. Early outlines simply listed the negative impacts down the left side of the page and then had a wish list of their opposites, presumed to be positive outcomes, down the right side of the page. To be fair to their authors, there were no research findings or exemplars of successful sustainable tourism to draw on. Only gradually did sustainable tourism become proactive, trying to create positive change. Many commentators—professional as well as amateur—enjoy criticising tourism. The key to achieving sustainable tourism is, however, to carry out analytical review and criticism, then implement effective management techniques, and then carry on a rolling review, criticism, and management process.

WHAT HAS BEEN ACHIEVED SO FAR

Progress in sustainable tourism to date has concentrated on:

- discussions and definitions, and devising basic assessment/evaluation programmes for small scale sites.
- testing a range of individual management techniques, notably a range of visitor management programmes, especially those for protected areas, more sustainable accommodation provision, transport centred research, and the creation of partnership programmes.

- local and individual projects, often innovative, many very short term.
- local sustainable tourism strategies, usually written by or for local governments.
- a number of certification programmes of varying types and varying quality, largely voluntary membership programmes with all the inherent problems that membership programmes bring with them. Such programmes are essentially prisoners of their members, succeeding with the success of their members, failing if their members either dilute their aims or leave the programmes.
- discussion and trialling of a range of indicators designed to show progress (or lack of progress) in implementing sustainable tourism.
- the thinking through of the ethics and key concepts of the 'subject'— one of the most important examples of this has been work by authors such as Bob McKercher, Bryan Farrell, Louise Twining Ward, and John Shultis, which introduced uncertainty, risk, chaos, and organic change into the previously linear, inevitable progression development scenario.
- research and case study work: A wealth of knowledge now exists on some issues. We understand, for example, much more about the role of information provision and interpretation in implementing sustainable tourism. Much research remains to be done, even more remains to be implemented.
- the emergence of a 'first generation' of academics who have worked on sustainable tourism. Many members of that first generation are now beginning to reach retirement or to take senior posts that make active research and authorship difficult.

BUT WHY HAS SUSTAINABLE TOURISM BEEN SO SLOW TO DEVELOP?

Despite the hopeful start thirty years ago, real progress in sustainable tourism—especially in implementation—has been remarkably slow. Why?

We can note a number of overlapping problems across many stakeholders:

The Industry:

- *No driver or imperative:* The tourism industry has not been driven, either by government or market forces, to achieve a more sustainable form of tourism. The industry has successfully opposed attempts to regulate its impacts by invoking the idea of self regulation as being the best way forward. And the regulators have been delighted to avoid the real and complex problems that would otherwise be involved. The market for tourism remains strongly driven by price and fashion factors, and both the market and the industry remain very conservative.

- *Growth* is a key belief within society and especially within the tourism industry. Sustainable development questions some forms of growth, and of any growth at all in some places. The industry has not come to terms with those issues: after sixty years of growth it is addicted to growth. It fears and tries to avoid—for many good reasons—any 'downturn'.
- *Few short term benefits* accrue to the development of more sustainable forms of tourism: In the long term, as the London docker famously told the economist John Maynard Keynes, 'we are all dead'. Tourism is strangely like farming: One sows one's crop annually (product and marketing) and it is then dependent on the weather and the market for success. A year is a long time in tourism; failed experiments can be disastrous.
- *Denial.* The industry denies some of its impacts, and denies the need for a changing approach. That is a common approach to problems across most established sectors of our economy. It is an easy approach to take in a sector like tourism where there has been, until now, a ready supply of alternative destinations, and a ready demand for holidays with no real market interest in new approaches.
- *The concept of Social Marketing*, of using marketing techniques to encourage behavioural change, rather than increased consumption of existing products, is in its infancy—and little understood by tourism marketing agencies, or the media. The media's role in the sustainable development/sustainable tourism story is an especially central but a rarely researched one. The media's close financial relationship with the industry (through advertising revenues) also makes for uncritical treatment by the press.
- *Partnership building* is in its infancy—and partnership work is essential to get the greatest benefits from the holistic sustainable tourism approach. For example, few in the public transport sector understand the concept of sustainable tourism—even though they would benefit from it; few in the heritage or farming sectors understand and try to implement the concept. Governments speak of partnerships but do not understand their requirements, implications, and limitations.
- *A leaderless industry.* The diverse and fragmented nature of the tourism industry has helped to prevent strong and far seeing leadership from emerging—regionally, nationally, or internationally—to promote thought or long term change. The well known rhetorical question from the world of politics—'Who are the leaders, and who are the led?'—is peculiarly apposite to tourism. In theory the industry leads; in practice the market leads, but it is leaderless.
- *The industry does not work closely with academic researchers.* Sustainable tourism originated through commentators, critics, and thinkers—not through the industry. Effective links between industry

and academic research are rare. Many academics do not understand the needs of the industry; much of the industry is afraid of researchers who may uncover commercial knowledge.

- *The concept of Social and Environmental Responsibility* remained new or even unknown across most industries—not just tourism—until very recently.
- *Ecotourism* became a development trap for some sustainable tourism advocates. It was relatively easy to develop and assess small scale sustainable tourism projects in rural areas. These projects appealed to the 'small is beautiful' beliefs common amongst many and avoided the problems of contact with the mainstream tourist industry. Many commentators seemed to assume that it would be impossible to make 'mass' tourism sustainable. Valuable time was lost before the dangers inherent in some forms of ecotourism became apparent, and before the pressing need to work with mass tourism was realised.

Governance and the Regulators

Governments have been shy to encourage or require change in the tourism sector beyond basic safety regulations. Governments have traditionally practised boosterism towards tourism. In the new privatism that dominates governance, regulation is not welcome. Instead many governments have published advisory documents urging others to do their work of regulation, rather than supporting the creation of a more sustainable tourism industry. Governments are keen to create jobs, are keen to support economic growth; sustainable development generally has a poor image on those issues. Further, one of the problems within the democratic process is that it encourages short-termism, because of the typical three- to five-year electoral cycle. Sustainable tourism is a long term approach. And regulation is not an easy process, and especially difficult in an internationally competitive area like tourism.

The obvious places to try out regulatory systems, the urban and rural protected areas, are typically weak in tourism management skills, funds, political support, and the new ethos required by the sustainable tourism approach (see Eagles 2002).

Few politicians understand tourism; even fewer understand sustainable tourism; very, very few seek to actively implement sustainable tourism. It tends to be a slogan rather than practical politics.

The Market, Tourists, and Society:

- *Society* generally has not understood the need for sustainable development of most kinds. Sustainable development requires thought, change, and investment: All are difficult to achieve. Sustainable living needs behavioural change by all stakeholders. Behavioural change is very hard to bring about. It is seen by many as unnecessary and painful.

- *Denial.* Most tourists—perhaps more so than the industry—are in denial about their environmental and cultural impacts. Recent research by Becken (2007) shows that the majority of air travellers, for example, sincerely believe that others should refrain from travel—not them. And research by Shaw and Thomas (2006) also shows that most young people have come to believe that travel is a fundamental right, not a privilege. Serious questions about freedom, political problems, and the functions of the market economy are raised.

- Sustainable tourism is *a very Reithian concept.* Lord Reith was the first Chief Executive of the BBC (from 1922–1938) and worked to use Radio and TV to educate society for its own good. His stamp on the work of the BBC, from its domestic to its World services, remained strong for many years. The annual Reith lectures survive. But the concept of educating society through well meaning media finds limited acceptance in a consumer society.

- *The Nature of Holiday Making.* Many years ago (1990), the author was asked to address the main board of Thomson Travel about Sustainable Tourism. Thompson was, at that time, the largest tour operator in the UK, with a market share of the outbound holiday market in excess of 40 per cent. The request was one that could not be refused—a major challenge. The address was made. The Board's reply can be summarised as: 'nice idea, but the future is bright, the future is Euro-Disney, we do not need a more sustainable product'. A discussion ensued, followed by lunch. I was taken aside by a wise and experienced member of the board, who said, very gently but firmly, that I had to understand that holidays were the two weeks of the year when selfishness and thoughtless consumption were possible for everyone, when caution could be relaxed. She was the director of marketing. She was, in the real world of that time, correct.

The Research Community

By this stage, readers will be awaiting the arrival of the white knight, riding to the rescue. Is the white knight the research community, dedicated seekers of the truth and sustainable progress? The answer is a very, very, qualified yes. Sustainable tourism has been pushed slowly along, largely by academics. But there have been a whole series of problems which have not assisted progress.

Working with the industry has been rare. Very few academic researchers have worked inside the tourism industry, and they remain outsiders, neither understanding the pressures and the drivers within the industry, nor how to work with the industry. Equally, the industry has not been keen to work with academics because of the industry's essentially utilitarian, typically short term approach. There is an ongoing tension here. It emerges when the value of engaging tourism graduates is discussed in business circles. It emerges when discussing the tensions within Australia's Sustainable Tourism Co-operative Research Centre (STCRC), the industry-backed Australian government initiative to carry out research using University research skills. It emerges when researchers seek material that is 'commercially sensitive'.

This is a fascinating but frustrating area. The industry does not really wish to change. It does not like being asked to use bright new ideas that may not increase profitability. It does not like to take unnecessary risks. No one likes criticism. Academic researchers are trained to analyse and criticise, and to put forward ideas and suggest change. It requires great skill to understand the industry and to analyse the personalities involved. It requires both a thick *and* a sensitive skin. Most academics fight shy of the tensions involved.

The *Journal of Sustainable Tourism* has a story to tell here. When founded, it, perhaps naively, hoped to be read and have contributors from both sides of the industry/academic divide. The Editors quickly learned that the content requirements and the research and writing abilities of the two sides were incompatible.

Understanding the market has been another problem. Most academics do not understand or research the fundamental nature of the markets that drive tourism. The classic work on the subject by Stanley Plog (1991) never achieved star rating amongst the ivory tower critics. Only now is research being carried out to begin this process (Dolnicar, Crouch, and Long, 2008).

Understanding marketing is also a problem. Most tourism academics do not come from a business background. Many were geographers in their formative years. There can be a distrust of Business Schools and their perceived narrow view of society, human behaviour, and research techniques, and sometimes unhelpful use of management speak. Tourism academics have very usefully discussed de-marketing, but the whole area of Social Marketing, of how to promote behavioural change, seems to be a blank for sustainable tourism researchers. Names like Alan Andreasen (1995) or even Philip Kotler (cf. Kotler, Roberto, and Lee, 2002), or journals such as the *Journal of Social Marketing*, do not occur in sustainable tourism research paper reference lists.

Researching the mass media's role in the tourism industry has also been slow to develop. Yet the media shapes the travel addictions of both the mass and niche markets, and has developed a special form of infotainment which masks their hidden persuasions. Perhaps just one hundred travel editors and TV/Radio programme commissioners control the key to the addictions of hundreds of millions of travellers. Who are these people, what are their techniques, how can they be influenced, what are their rewards?

Popularising sustainable tourism is difficult, and not something that the research community indulges in. It brings few rewards. Set against that the activity can lead to colleague condemnation, as if it were some form of peculiar diversion from 'real' work.

The narrow approach of the research community is a final problem to be noted. The tourism industry is a mongrel, an assembly industry of many talents, with widespread drivers and impacts. One of the reasons that the *Journal of Sustainable Tourism* has two Editors is the breadth of different skills and subject areas required. But too often tourism research is relatively narrow. It rarely looks at green building techniques, it rarely engages in the science of ecology, it rarely works on political analysis, it rarely tackles the issues raised by many commentators in the social sciences—such as Alan Andreason, previously mentioned, or the prolific writings of Ulrich Beck, the German author of key works such as *Risk Society* (1992), or *Ecological Politics in an Age of Risk* (1995). Beck notes that post industrial society brings both new freedoms and new risks, that those risks cross political and social frontiers, and have deep behavioural and political consequences. On the ecological front, 'hazards can only be minimized by technological means, never ruled out', and they can 'cause irreversible damage and destruction that may have a determinable beginning but no foreseeable end'. Those words are especially relevant to the new world of climate change. For a longer discussion of related themes to these topics, read the Editorials by Bramwell and Lane in *Journal of Sustainable Tourism*, 13/1, 2005 and 14/1, 2006.

SO—WHY HAS SUSTAINABLE TOURISM MOVED UP THE PUBLIC AGENDA NOW?

The fundamental reason is the recognition that climate change is happening, that its consequences could be seriously damaging, and that a series of changes are required in our existing life styles. Those changes could affect us all. And they could impact very strongly upon tourism and its growth. For the first time since 1950, tourism's growth rates are being threatened; for some regions the very existence of the tourism industry is threatened. The media is displaying new interest in green issues, and new anti-travel, anti-tourism pressure groups have developed. Air travel is being scapegoated. And it must also be recognised that wider pronature, proheritage interests are growing in many (but not all) societies and parts of society, with strong implications for nonsustainable tourism.

It is, however, an ill wind that blows no one any good, and the winds of climate change have the potential to give new life to the concept of sustainable tourism. Suddenly governments, regulators, the media, the industry, and even a few travellers are questioning the survival of the status quo. They are less scornful of sustainable development. A powerful driver has emerged. Greed has, to a small extent, been replaced by fear.

Fear is, however, a simplistic driver. Fear leads to panic. Panic stops the painful but necessary process of thought. A new form of short-termism usually emerges from fear and panic. But the great fear for many of the aficionados of the sustainable tourism world is that the holistic concept of sustainable tourism may be overwhelmed by a new form of single issue politics—the politics of the carbon footprint. The wider requirements and opportunities of sustainable development may be lost in the rush to claim smaller carbon footprints. At the same time opponents of sustainable development are quick to seize the many contradictions within aspects of the carbon offset industry (see Gössling et al. 2007), and the ways in which governments are caught in dilemmas about their approaches to climate change. We are living in interesting times.

WHERE NEXT—IS THERE A FUTURE FOR SUSTAINABLE TOURISM—CAN IT BE ACHIEVED?

The Research Community

Back in the 1970s, when Sustainable tourism could be summed up on three OHP slides, its protagonists posed the failings of 'conventional' tourism against their dreams for sustainable tourism. Repeating that approach now, it is easy to say that to achieve sustainable tourism researchers need to:

- engage and work with the industry.
- engage the problems of governance and regulation.
- research market beliefs and new forms of marketing.
- research the role of the media industry in opinion forming.
- use the ideas developed in the social sciences to explore decision making and social trends.

But beyond those perhaps obvious and mechanical needs, researchers need to go further. Ulrich Beck can write almost impenetrable sociological texts, but he can also, and is proud to, write regularly, for quality newspapers such as *Die Zeit* or the *Frankfurter Allgemeine Zeitung*, in a style accessible to most. He sees that as the duty of a professional thinker, and the media see it as their professional duty to publish such pieces. Jost Krippendorf was also a skilled communicator on a range of levels. Academics working in the field of sustainable tourism need to develop the skills of accessible writing for multilayered audiences, ranging from the public through to nonspecialist decision makers, and on to professionals in the industry, in planning and regulation, and in consultancy.

There are, inevitably, implications to the aforementioned. The world within the ivory tower now requires a thick but sensitive skin. The world outside can be ten times more abrasive. First, counselling may be needed to survive some aspects of working with the media. Second, academics need to understand the politics of change a little better. Sustainable tourism is as much an art as a science. They need to reflect on, use, and analyse skills from both the arts and the sciences. Third, researchers need to press home the fact that sustainable tourism is a positive approach, and tourism can be a powerful tool for supporting conservation aims, cultural goals, rural and urban regeneration, and a range of public transport systems. All those areas need careful analytical research. There is hope: A new generation of highly skilled academics are now researching, publishing, and building on the work of the first generation with new insights and vigour.

The Tourism Industry

The industry has an especially difficult role to play. Always a relatively short term player, it needs to retain its adaptability while developing a longer term approach. And it needs to realise that there is no magic bullet that will transform its future. The future is complex. It will also need a partnership approach—working with stakeholders and with competitors. Partnership approaches are fraught with problems but necessary. The survival of the industry is at stake; in crises all parties survive by working together (Bramwell and Lane 2000, 2004).

One of the greatest problems is that, within the industry, there is a powerful belief that technical and managerial changes can quickly solve the problems, including those of climate change. Technical changes can help but there is little evidence that they can help on the scale required to stave off climate change, which is recognised to be the global challenge of the next thirty years (see Viner 2006). And within the tourism industry, technical developments in transport are especially important but very difficult to achieve (see Becken 2007). Parallel to technical and managerial change comes the requirement to encourage life style changes, using new marketing techniques and product development. The industry is uniquely placed to work on this broad but risk-laden and difficult issue. Like an animal approaching winter, some form of activity reduction process may need to be thought out and undertaken to ensure survival. New value systems across management are urgently needed.

The Market, Tourists, and Society

The research community can think and evaluate and do so independently. The industry can work to implement new products, technologies, and marketing systems. But the market—the whole of society—will be responsible for embracing and taking to heart the work of researchers and industry. Success will require life style changes to be accepted *and* enjoyed.

This chapter has referred to the market's need for selfish escapism into hedonism and consumption. Together all parties have to develop and enjoy new forms of tourism activity: the equivalent of nonalcoholic beer and better. The task is made more difficult because the market for tourism is fast taking in a huge new world of tourists from the developing world, notably India and China, keen to enjoy the pleasures that the developed world has revelled in for the last sixty years.

And Governments?

Until recently the role of governments in tourism has largely been one of arm's length boosterism. Command economies have found it difficult to develop and control the many nuances and qualities required for success in the tourism market place. The role of governments in achieving sustainable tourism will most likely, therefore, also be one of boosterism, boosting the chances of achieving sustainable tourism, offsetting market failures, assisting innovation and change, managing planning permissions, creating, improving, and retaining sustainable infrastructures of many types including public transport and protected areas. And it must lead where neither researchers, nor the market, nor the industry can lead, by bringing stakeholders together to create holistic and implementable solutions to the sustainable tourism paradigm. That leadership must come through helping finance thought and change, and that leadership will require political as well as financial support. In January 2008, the Norwegian government announced an important initiative in the field: six million krone (1.2 million US$) has been awarded to develop a sustainable tourism strategy for Norway that will be carbon neutral by 2030 (Gössling, 2009).

CONCLUSION

The achievement of sustainable tourism will be a necessary but lengthy and most likely contested procedure. Sustainable tourism began life by being contested; contestation is a feature of change. And change takes time.

All long journeys begin with first steps. The world and its peoples cannot change in just a few years. Change comes first from small projects (and we should learn from their mistakes as well as successes). There is a time scale. There is also a geographic scale. One size rarely fits all. It is important to recognise the different scales at which sustainable tourism can operate: local, regional, national, continental, and global scales, and scales of rising difficulty. Small may be beautiful because it can be manageable. And it is also usually at the local scale that some of the most immediately damaging impacts from tourism take place. Do not forget the global impact however. Ulrich Beck noted the new global risks to society and environment nearly twenty years ago. Our new driver, climate change, does not respect boundaries.

ACKNOWLEDGMENT

The author thanks his colleague and co-editor of the *Journal of Sustainable Tourism*, Bill Bramwell, for his assistance and encouragement in developing this chapter.

REFERENCES

Andreasen, A.R. (1995) *Marketing Social Change: Changing Behavior to Promote Health, Social Development, and the Environment.* San Francisco: Jossey-Bass.

Beck, U. (1992) *Risk Society: Towards a New Modernity* (translated by Mark Ritter). London: Sage.

Beck, U. (1995) *Ecological Politics in an Age of Risk* (translated by Amos Weisz). Cambridge: Polity Press.

Becken, S. (2006) 'Editorial: Tourism and transport: The sustainability dilemma'. *Journal of Sustainable Tourism* 14(2): 113–115.

Becken, S. (2007) 'Tourists' perception of international air travel's impact on the global climate and potential climate change policies'. *Journal of Sustainable Tourism* 15(4): 351–368.

Bramwell B. and Lane, B. (1993) 'Sustainable tourism: An evolving global approach'. *Journal of Sustainable Tourism* 1(1): 1–5.

Bramwell, B. and Lane, B. (eds.) (2000/2004) *Tourism Partnerships and Collaboration: Politics, Practice and Sustainability.* Clevedon: Channel View Publications.

Bramwell, B. and Lane, B. (2005a) 'From niche to general relevance? Sustainable tourism, research and the role of tourism journals'. *Journal of Tourism Studies* 16(2): 52–62.

Bramwell, B. and Lane, B. (2005b) 'Editorial: Sustainable tourism research and the importance of societal and social science trends'. *Journal of Sustainable Tourism* 13(1): 1–3.

Bramwell, B. and Lane, B. (2006) 'Editorial: Policy relevance and sustainable tourism research: Liberal, radical and post-structuralist perspectives'. *Journal of Sustainable Tourism* 14(1): 1–5.

Butler, R. (1990) 'Alternative tourism: Pious hope or Trojan horse'. *Journal of Travel Research* 28(3): 40–45.

Dolnicar, S., Crouch, G., and Long P. (2008) 'Environmentally friendly tourists: What do we really know about them?'. *Journal of Sustainable Tourism* 16(2): 197–210.

Eagles, P.J.F. (2002) 'Trends in park tourism: Economics, finance and management'. *Journal of Sustainable Tourism* 10(2): 132–153.

Gössling, S. (2009) 'Carbon neutral destinations: A conceptual analysis'. *Journal of Sustainable Tourism* 17(1).

Gössling, S., Broderick, J., Upham, P., Ceron, J.P., Dubois, G., Peeters, P., Strasdas, W. (2007) 'Voluntary carbon offsetting schemes for aviation: Efficiency, credibility and sustainable tourism'. *Journal of Sustainable Tourism* 15(3): 223–248.

Kotler, P., Roberto, N., and Lee, N. (2002) *Social Marketing: Improving the Quality of Life.* London: Sage.

Krippendorf, J. (1975) *Die Landschaftsfresser* (The Landscape Eaters). Bern/Stuttgart: Hallwag Verlag.

Krippendorf, J. (1984) *Die Ferienmenschen, Zürich/Schwäbisch Hall.* Orrel Füssli Verlag.

Krippendorf, J. (1987) *The Holiday Makers: Understanding the Impact of Leisure and Travel.* London: Heinemann.

Krippendorf, J., Zimmer, P., and Glauber, H. (1988) *Für einen andern Tourismus* (Towards an Alternative Tourism). Frankfurt am Main: Fischer Taschenbuch Verlag.

McKercher, B., Law, R., and Lam, T. (2006) 'Rating tourism and hospitality journals'. *Tourism Management* 27: 1235–1252.

Müller, H.-R. and Lane B. (2003) 'Obituary: Jost Krippendorf'. *Journal of Sustainable Tourism* 11(1): 3–4.

Plog, S.C. (1991) *Leisure Travel: Making it a Growth Market . . . Again!* New York: John Wiley.

Shaw, S. and Thomas, C. (2006) 'Social and cultural dimensions of air travel demand: hyper-mobility in the UK?' *Journal of Sustainable Tourism* 14(2): 209–215.

Viner, D. (guest editor) (2006) 'Tourism and its interactions with climate change—special issue'. *Journal of Sustainable Tourism* 14(4): 317–415.

Wheeller, B. (1992) 'Alternative tourism— a deceptive ploy'. *Progress in Tourism, Recreation and Hospitality Management* 4: 140–145.

Wheeller, B. (1993) 'Sustaining the ego'. *Journal of Sustainable Tourism* 1(2): 121–130.

3 Reflections on Sustainable Tourism and Paradigm Change

David B. Weaver

INTRODUCTION: PARADIGM SHIFT OR PARADIGM NUDGE?

Weaver (2007) describes the current ubiquitous interest in, and involvement with, sustainable tourism as an example of 'paradigm nudge' rather than 'paradigm shift'. The latter, in the Kuhnian sense (Kuhn 1970), describes fundamental changes in the underlying cosmology of a society, as most notably occurred in the West from the 1400s to the 1600s when the scientific worldview gradually supplanted a prevailing theological worldview increasingly incapable of explaining and accommodating contradictions such as the heliocentric solar system. It has been argued by Knill (1991) and others that a new shift is currently underway as this scientific worldview and its attendant anthropocentric emphasis on human separation from and superiority to nature is giving way to a 'green' or 'ecological' paradigm predicated on equality and integration within nature, soft technologies, and equilibrium rather than linear growth. Emergent contradictions supposedly facilitating this transition include the 'Frankenstein' effects of new genetic, medical, and nuclear technologies, and ecological disruptions both gradual (e.g., climate change and ozone depletion) and sensational (e.g., Chernobyl, Bhopal, Katrina) stemming directly or indirectly from relentless development and unrestrained capitalism.

Within the tourism sector, the comparable contradiction/crisis of the dominant paradigm is embodied in the well-known destination life cycle sequence of Butler (1980), which posits that escalating tourism development in destinations with laissez-faire markets and sustained levels of demand eventually gives way to stagnation and degradation as ecological and sociocultural carrying capacity thresholds are inevitably exceeded. New and overlapping manifestations of tourism that purport to resolve this contradiction by allegedly adhering to the precepts of the emerging green paradigm include alternative tourism (Boxill 2004; Dernoi 1981), geotourism (Buckley 2003; Pralong 2006), responsible tourism (Harrison and Husbands 1996), pro-poor tourism (Hall 2007; Roe, Goodwin, and Ashley 2004) and, of course, sustainable tourism. Of these, the latter to date has attained the most inclusive and widest currency; given that the term was coined only as recently as the early 1990s, the formal contemporary adoption of its rhetoric by most tourism-related entities is a remarkable example of accelerated institutionalisation.

Widespread rhetorical adoption, however, does not necessarily indicate paradigm shift, with Weaver (2007) going on to describe the tourism sector's adoption of sustainability *practices* as neither deep nor broad. The lack of breadth is revealed by the small number of conventional tourism companies generally regarded as being exceptionally engaged in things 'sustainable'. The accommodation and airline sectors are relatively well represented in this regard by major companies such as Marriott, Starwood, Grecotel, and Scandic for the former, and by British Airways and American Airlines for the latter. Tour operators and travel agencies, by contrast, have few high profile champions (the German tour operator TUI is a notable exception). An important indicator of breadth is the degree to which relevant enterprises participate in certification-based ecolabels utilising third party verification procedures. The Green Globe ecolabel stands out as the most ambitious of these due to its global coverage of virtually all tourism sectors (Parsons and Grant 2007), yet just 211 products (individual hotels, etc.) were participating in this scheme as of late 2007 according to the Green Globe website, down from 372 in mid-2004 (Weaver 2006). This miniscule rate of penetration is barely elevated when other still-active or new ecolabels are taken into account.

The concomitant lack of depth is revealed in the nature of the purported sustainability practices, which in the case of many hotels do not extend beyond signage that encourages guests to reuse towels before relegating them to the laundry basket, or, on the social side, donations to charity (Holcomb, Upchurch, and Okumus 2007). Sector champions do go beyond such superficial initiatives through involvement in recycling, energy cogeneration, energy-efficient design, etc. (Enz and Sigauw 2003), but in almost all cases even these activities are implemented mainly because they yield, in addition to their environmental and social benefits, a high return in positive public relations and/or profit. Practices that move meaningfully beyond the imperatives of profit and growth, such as a freeze on new development in growing resort areas, or major environmental and social measures that cut deeply into earnings, are conspicuous by their absence within the conventional tourism sector.

Revisiting the idea of paradigm shift, the assessment of the contemporary tourism sector just discussed could still be interpreted as early-stage evidence of a movement toward the green paradigm, especially considering that sector participation in such activities was extremely limited as recently as the early 1990s (Weaver 2006). Supporters of paradigm shift, with reasonable confidence, could then predict substantial broadening and deepening of sustainable tourism over the next two or three decades simply by extrapolating the rate over which such practices have been developed and disseminated over the past fifteen years. The probability of such a transition would be increased if more events such as Hurricane Katrina were to occur and be linked decisively to the dynamics of climate change.

However, another interpretation is possible in lieu of catastrophe-induced shift. Kuhn (1970) suggests that proponents of the existing dominant paradigm consciously or subconsciously adapt and strengthen the latter by selectively incorporating elements of the challenging paradigm that help it to resolve emergent contradictions and crises without undermining its core principles. In the current clash of paradigms, it may therefore be that proponents of the scientific/capitalist paradigm are opportunistically assimilating perspectives and practices associated with the green paradigm that increase its popular appeal in the light of growing environmental public sensitivities, mask its darker social Darwinian elements and effects, and (ironically) enhance its capacity to generate profit and growth without incurring additional regulation from the public sector—hence a 'paradigm nudge' that diabolically reinforces the incumbent worldview. This may also help to account for the widespread adoption by industry of the mother term 'sustainable development', whose noun embodies a core attribute of the growth paradigm and its adjective a desirable but nonthreatening (if not contradictory) qualifier from the green paradigm. The concomitant appeal of 'sustainable tourism' to interests in that sector is similarly based on the possibility of 'strong' and 'weak' manifestations that respectively recognise the legitimacy of alternative and conventional mass tourism development in presumably suitable destination settings (Hunter 1997).

MIRRORING VENEER ENVIRONMENTALISM IN THE MARKETPLACE

The adaptations of the anthropocentric growth paradigm proponents are often perceived by opponents and critics as a cynical exercise in greenwashing that runs counter to the public interest and changing public will. Examples of egregious greenwashing no doubt can be cited in the tourism sector (Duffy 2002), but it can be counter argued that this criticism fails to recognise the realities and influence on industry of the 'public will' if not the public interest. It may be, as Weaver (2007) contends, that industry (and government) is engaging in a 'veneer sustainability' version of paradigm nudge primarily as a response to the veneer environmentalism exhibited by the general public itself, in high per capita GNP societies at least. Conditionally environmentalist consumers, for example, are consistently revealed in early twenty-first century US public opinion polls. These polls, in the first instance, express strong pro-environmentalist attitudes. Typical is a Gallup poll of 1,000 adults taken in March 2006, in which 85 per cent generally favored increased federal government spending to develop alternative sources of fuel for automobiles. Similarly strong was support for stronger enforcement of federal environmental regulations (79 per cent), increased government spending on the development of solar and wind power (77 per cent), setting higher emissions and pollution standards for business (77 per cent), imposing mandatory controls on CO_2 emissions and other greenhouse gases (75 per cent) and setting higher emissions standards for automobiles

(73 per cent). An ABC/Washington Post/Stanford University poll of 1,002 adults in April 2007 found 70 per cent of respondents agreeing that the federal government should do more to deal with global warming, with only 17 per cent citing the latter as a 'not too important' or 'not important at all' personal issue. However, as is also typical, only 20 per cent went on to favor an increased tax on electricity to reduce use, while 32 per cent favored increased taxes on gasoline (PollingReport 2007).

Thus, the rhetoric of environmentalism, and relatively abstract and impersonal pro-environment government efforts, are strongly supported, while concrete and personal measures focused on the same outcome are strongly opposed. Weaver (2007) suggests that about one half of consumers in the United States and similar societies are 'veneer environmentalists' who sympathise with the attendant concerns but are unwilling to make major lifestyle changes to redress those concerns. Another quarter are more or less 'environmentalist' in their willingness to make such changes, while the remaining quarter are 'nonenvironmentalists' generally unconcerned about environmental issues and therefore unwilling to make any sacrifices. Major public opinion polls generally do not extend their environmentalism-related questions to the realm of tourism attitudes and behaviour, but several focused surveys support this pattern (Chafe 2007). In a 2003 Travel Industry Association survey of US adult travelers, for example, 74 per cent of respondents regarded it as important for their personal travel not to damage the natural environment. Minorities of the sample, however, indicated a willingness to pay more (amount not stipulated) to use tourism businesses that try to avoid damage to the environment (38 per cent), or an interest to assist in protecting the environment without altering their existing lifestyle (40 per cent) (TIA 2003). Negligible industry participation in ecolabels such as Green Globe, accordingly, is explained mainly by the minute minority of consumers who are aware of such schemes and consciously seek out participating tourism products.

The respective attitudes of industry and the general public are therefore remarkably similar, from strong accordance with the rhetoric of sustainability to an equally strong unwillingness to engage in personal sacrifice to attain the ends espoused by that rhetoric. Given the current lack of incentives—for example, imminent global collapse of ecosystems, imposition of draconian government regulations, or, most relevant to this discussion, a major public shift from conditional to unconditional environmentalism—it is unrealistic to expect that industry will initiate any substantive effort that moves beyond paradigm nudge in the tourism (or any other) sector; this simply is not currently being demanded by most consumers. However, industry presumably must respond at some point to a hypothetical shift in public behavioural intentions, or risk the loss of its markets and/or drastic regulations imposed by government as a result of public pressure. The remainder of this brief chapter considers how such a shift (which may or may not constitute a paradigm shift as such) might be facilitated based on a strategy of raising public awareness and subsequently inducing widespread public activism with regard to tourism-related purchasing and behaviour.

TOWARD GREATER PUBLIC AWARENESS AND ACTION

Factors that impede the dissemination of public awareness about tourism-related issues include the relative infrequency of tourism-related travel, which impedes the development of purchasing habits that take environmental and sociocultural considerations into account (Hjalager 1999). Consumers may also not be able to draw a clear and direct connection between their tourism activity—and tourism overall as a benign 'smokeless' industry—and negative environmental and social consequences, in the same way that is done for other products. Organisations that attempt to promote a more sustainable mode of travel, in addition, are either obscure and poorly funded entities such as Tourism Concern that focus more on the small-scale, alternative side of the tourism spectrum (Turner, Miller, and Gilbert 2001), or ecolabels that are similarly obscure (see earlier references to Green Globe) and subject to serious limitations in terms of criteria used to award and monitor participating products (Black and Crabtree 2007; Schott 2006). It may also be that the travelling consumers who are most likely to be aware of the issues and to advocate for change are already involved with the alternative tourism sector and less so or not at all with the conventional mass tourism sector. Another possibility worthy of investigation is that consumers who are otherwise conscientious about environmental and social issues in their everyday lives may suspend these attitudes and behaviour during times of conventional tourism travel, which researchers have long perceived as a time when normal behaviour patterns are suspended (Gottlieb 1982; Snepenger et al. 2006). Finally, few if any objective and reliable sources are available that identify qualifying hotels, attractions, tour operators, etc. for individuals who do want to patronise the more active practitioners of responsible conventional tourism.

One possible way of overcoming these obstacles to mass awareness and trigger change is to encourage the publication of a book for the mass market that clearly (i.e., in nonacademic language) and comprehensively describes the environmental and social problems associated with conventional mass tourism, and the measures that the travelling public can take to encourage greater sustainability. Research that estimates the carbon emissions effects of international and domestic tourism (see Chapters 9 and 10, this volume), for example, could be incorporated into this popular publication so that readers could better appreciate the links between climate change and tourism, and calculate their own actual and potential 'carbon footprint'. Preferably, the author of such a book would be a charismatic individual in the manner (if not substance) of a Suzuki or Bellamy who might be fortunate enough to appear on highly influential television programmes such as the US-based Oprah Winfrey Show. This female-centric show is especially relevant given the crucial role of women in making household leisure travel decisions (Koc 2004; Mottiar and Quinn 2004; Zalatan 1998).

It is to be hoped that this proposed book would do for tourism what Rachel Carson's *Silent Spring* did for the environmental movement in the 1960s, although the potential effect may be even greater given the contemporary sophistication and diversity of facilitating mass media. More specifically, the dual goal would be to mobilise those already disposed to pro-environmental behaviour, whilst moving the 'veneer environmentalist' one-half of the population from sympathy for the rhetoric of tourism sustainability to concrete action, thereby forcing industry to react accordingly to the resulting critical mass of activist consumers. What kinds of action should be encouraged? Ideally, receptive consumers would be encouraged to patronise tourism products that meet given minimal standards of sustainability, while boycotting those that do not. The problem, however, is that no such list of products exists. Consumers must rely on the unverified, self-serving claims of individual companies (usually as described on their websites), or refer to the small lists of products that are 'certified' under the questionable criteria employed by various tourism-related ecolabels. A more fruitful starting point might be for the aforementioned book to equip concerned tourists with a list of questions that could be presented to potential providers (e.g., hotels, tour operators, attractions) as the basis for deciding whether or not to patronise those providers. An expectation would be that the some kind of verification is provided. Over time, through the cumulative effects of bloggers, customer experience, etc., a ranking of providers (sorted by sector) would emerge that would more readily facilitate the selection process of concerned consumers and provide competitive advantage to those products yielding higher rankings. The emergence of a credible and widely subscribed universal certification-based ecolabel is an equally desirable alternative, though such seems unlikely in the foreseeable future.

FINAL THOUGHTS

The four core contentions of this chapter may be reiterated as follows:

- The adoption of sustainability-related practices within the tourism-related industries is neither broad nor deep, nor commensurate with the pro-sustainability rhetoric that these industries espouse.
- Existing practices enhance public relations and profitability, and do not reflect any fundamental changes in underlying assumptions about the desirability of growth and profit. Hence, they indicate 'paradigm nudge', or opportunistic adaptations to the dominant paradigm, rather than paradigm shift.

- Paradigm nudge is a reaction to the conditional environmentalism dominant in major tourist-generating markets. This entails sympathetic attitudes toward sustainability but concomitant unwillingness to enable the rhetoric through major personal sacrifice. Most consumers, accordingly, do not demand pro-sustainability structural changes within tourism.
- Changes in consumer attitudes and behaviour are necessary to effect change in the tourism industry, and the former can best be achieved through a well-written and well-publicised mass market book that describes the links between tourism and sustainability and recommends practices that consumers should demand from tourism providers.

The purpose of this chapter is to stimulate discussion—and action—as to the interrelationship between tourism, sustainability, and paradigm change. A confounding aspect is the ambiguous nature of 'paradigm shift', with an argument being possible that a nudge is already a type of shift. The pertinent issue, then, may be the degree of shift that is necessary to 'save the planet'. Some may argue for a radical replacement shift on the assumption that the premises of the existing dominant paradigm are ultimately inimical to the survival of global life support systems. A counter argument is that paradigm nudge is adequate, though it probably needs to be taken to a greater extent than has occurred thus far. A third possibility, perhaps most in line with Kuhn (1970), is the gradual emergence of a new synthesis-based green paradigm that incorporates elements of its predecessor, including some semblance of capitalism, anthropomorphism, etc. The boundary between the last two options may be impossible to discern, but it is within that spectrum that one might hope that short-to-medium term change may occur as a result of the consumer-driven approach advocated here, especially if the issue of climate change continues to gain traction in the public imagination.

REFERENCES

Black, R. and Crabtree, A. (2007) (eds.) *Quality Assurance and Certification in Ecotourism.* Wallingford, UK: CABI.

Boxill, I. (2004) 'Towards an alternative tourism for Jamaica'. *International Journal of Contemporary Hospitality Management* 16: 269–272.

Buckley, R. (2003) 'Environmental inputs and outputs in ecotourism: Geotourism with a positive triple bottom line?' *Journal of Ecotourism* 2: 76–82.

Butler, R. (1980) 'The concept of a tourist area cycle of evolution: Implications for management of resource'. *Canadian Geographer* 24: 5–12.

Carson, R. (1963) *Silent Spring.* Boston: Houghton Mifflin.

Chafe, Z. (2007) 'Consumer demand for quality in ecotourism'. In Black, R. and Crabtree, A. (eds.) *Quality Assurance and Certification in Ecotourism.* Wallingford, UK: CABI.

Dernoi, L. (1981) 'Alternative tourism: Towards a new style in north–south relations'. *International Journal of Tourism Management* 2: 253–264.

Duffy, R. (2002) *A Trip Too Far: Ecotourism, Politics and Exploitation*. London: Earthscan.

Enz, C. and Sigauw, J. (2003) 'Revisiting the best of the best: Innovations in hotel practice'. *Cornell Hotel and Restaurant Administration Quarterly* 44(5/6): 115–123.

Hall, C.M. (ed.) (2007) *Pro-Poor Tourism: Who Benefits? Perspectives on Tourism and Poverty Reduction*. Clevedon, UK: Channel View.

Harrison, L. and Husbands, W. (eds.) (1996) *Practicing Responsible Tourism: International Case Studies in Tourism Planning, Policy and Development*. New York: Wiley.

Hjalager, A. (1999) 'Consumerism and sustainable tourism'. *Journal of Travel and Tourism Marketing* 8(3): 1–20.

Holcomb, J., Upchurch, R., and Okumus, F. (2007) 'Corporate social responsibility: What are top hotel companies reporting?' *International Journal of Contemporary Hospitality Management* 19: 461–475.

Hunter, C. (1997) 'Sustainable tourism as an adaptive paradigm'. *Annals of Tourism Research* 24: 850–867.

Knill, G. (1991) 'Towards the green paradigm'. *South African Geographical Journal* 73: 52–59.

Koc, E. (2004) 'The role of family members in the family holiday purchase decision-making process'. *International Journal of Hospitality and Tourism Administration* 5: 85–102.

Kuhn, T. (1970) *The Structure of Scientific Revolutions*, 2nd ed. Chicago: University of Chicago Press.

Mottiar, Z. and Quinn, D. (2004) 'Couple dynamics in household tourism decision making: Women as the gatekeepers?' *Journal of Vacation Marketing* 10: 149–160.

Parsons, C. and Grant, J. (2007) 'Green Globe 21: A global environmental certification program for travel and tourism'. In Black, R. and Crabtree, A. (eds.) *Quality Assurance and Certification in Ecotourism*. Wallingford, UK: CABI.

Pralong, J. (2006) 'Geotourism: A new form of tourism utilizing natural landscapes and based on imagination and emotion'. *Tourism Review* 61(3): 20–25.

Roe, D., Goodwin, H., and Ashley, C. (2004) 'Pro-poor tourism: Benefiting the poor'. In Singh, T. (ed.) *New Horizons in Tourism: Strange Experiences and Stranger Practices*, 147–161. Wallingford, UK: CABI.

Schott, C. (2006) 'Proactive crises management tools: Ecolabel and Green Globe 21 experiences from New Zealand'. *Tourism Review International* 10: 81–90.

TIA (2003) *Geotourism: The New Trend in Travel*. Washington, DC: Travel Industry Association of America.

Turner, R., Miller, G., and Gilbert, D. (2001) 'The role of UK charities and the tourism industry'. *Tourism Management* 22: 463–472.

Weaver, D. (2006) *Sustainable Tourism: Theory and Practice*. London: Routledge.

Weaver, D. (2007) 'Towards sustainable mass tourism: Paradigm shift or paradigm nudge?' *Tourism Recreation Research* 32: 65–69.

Zalatan, A. (1998) 'Wives' involvement in tourism decision processes'. *Annals of Tourism Research* 25: 890–903.

Part II

Restructuring the Tourism System

Practical Examples

4 Inaction More than Action

Barriers to the Implementation of Sustainable Tourism Policies

Rachel Dodds and Richard W. Butler

INTRODUCTION

Since the report of the Brundtland Commission (World Commission for Economic Development [WECD] 1987) there has been a substantial body of literature emerge on the concept of sustainable development. Not surprisingly, a part of this literature has dealt with the related concept of sustainable tourism, and has proved so appealing that a journal on this topic has been published since 1993 (*Journal of Sustainable Tourism*). Despite the literary and other outpourings of academics, media writers, nongovernmental organisations (NGOs), and public sector agencies, one might argue that the realisation of the establishment of sustainable tourism on the ground and the application of sustainable principles in tourism in the real world have yet to be seen except in isolated circumstances (Wheeller 2005) and is often difficult to define (Butler 1999). The reason for this apparent failure to achieve success when so much good will appears to exist towards the concept of sustainable development is inevitably complicated and the intended goals themselves are not always clear, but it does seem clear that the host of good intentions manifested in consultant reports, academic writings, and official planning and policy documents has not achieved the goals that have been defined for sustainable tourism development (see Dodds 2008; Bianchi 2004; Tosun 2001; Singh and Singh 1999; Dovers and Handmer 1993 in MacLellan 1997). This chapter examines why so many policies that espouse sustainable tourism principles have not been successful in establishing sustainable tourism in their respective destinations by exploring the barriers to the success of those policies. It is based on a multi-method research study, incorporating a detailed review of relevant literature and appropriate findings from field research conducted in two destinations which have endeavoured to implement sustainable tourism related policies.

Despite a great deal of research about tourism policy and a plethora of information on sustainable tourism in general, research on the implementation of tourism policy in general is weak and on sustainable tourism policy is even less (see the few exceptions such as Hall 2004, 2008; Hall and Jenkins 2005). We have seen far more examples of unsustainable developments and destinations than sustainable ones, and one conclusion has been that appropriate policy and its implementation are what are needed to make tourism more sustainable (Asher 1984; Edgell 1999). However, 'Management decisions are not worth the paper they are written on unless the policies and decisions are implemented' (Elliot 1997, 97), and herein lies the rub. Even though tourism is important from an economic point of view, and several authors (Caffyn and Jobbins 2003; Elliott 1997; Hall 1994; Hall and Jenkins 1995) have investigated levels of power, control, and ownership of tourism, and how political systems have influenced decision making, tourism itself still remains relatively neglected as an important policy issue. There are relatively few studies of tourism policy implementation (Hall 2008; Coles and Church 2007; Hall 1994). Others (Hall 2008; Williams and Shaw 1998; Elliot 1997; Hall and Jenkins 1995; Gunn 1994; Hall 1994; Dye 1992; Johnson and Thomas 1992; Choy 1991; Younis 1990; Richter 1989; Jenkins 1980) have noted their scepticism of government and the intended consequences and impact of government policies. The evaluation of tourism policy is rare and most studies of policy within the frame of tourism have been prescriptive studies of what governments should do rather than evaluations of what has happened and why. 'Unfortunately, the contemporary discussion of tourism policy development has failed to illustrate the political dimensions of tourism policy action or implementation of policies in the form of specific tourism developments' (Hall 1994, 47).

ISSUES

Policy Definitions

One definition of public policy is that of Dye (1992, 2, in Hall 2000) who declares it 'is whatever governments chose to do or not do'. Healey (1975) suggests that policy may be regarded as the strategy by which objectives are pursued (in Hall and Jenkins 1995), while Anderson (1984, 3, in Elliot 1997) defines policy as 'a purposive course of action followed by an actor or set of actions in dealing with a problem or matter of concern'. Goeldner, Ritchie, and McIntosh (2000, 1) define policy for a destination as 'a set of regulations, rules, guidelines, directives, and development/promotion objectives and strategies that provide a framework within which the collective and individual decisions directly affecting tourism development and the daily activities within a destination are taken'. Policy with respect to tourism is generally linked to economic factors such as regional development and is created at different levels in different jurisdictions. The growth and formulation of policies relating to tourism can be tracked over time to show this trend.

Resource Management

Hardin (1968), in his well known article entitled *The Tragedy of the Commons*, developed the nineteenth century ideas of Lloyd in the context of population growth and its effects on the earth's finite resources. Hardin related the Tragedy of the Commons concept to other environmental concerns, such as green/public space and pollution. Resources owned collectively will be destroyed by each individual overusing that resource, ignoring the group's collective interests in favour of their own. Because all users tend to behave in this manner, the resource is ultimately doomed. 'Ruin is the destination toward which all men rush, each pursuing his own best interest in a society that believes in the freedom of the commons' (Hardin 1968, 1245).

In a tourism context, Healy (1994, 596) suggested that scenic and historic landscapes and other elements such as beaches are often susceptible to overuse because of the lack of 'incentive for productivity-enhancing investment' and because spaces such as beaches usually have no limits on use and thus may be overused by individuals concerned only with their own enjoyment. The experience one may hope for when visiting a beach or park may be diminished by vast numbers of other visitors, each hoping for their own unique experience (Brougham 1982). Relatively few tourism destinations have established policies aimed at preventing such overuse, and those that have taken such a step have frequently found that policy implementation is more difficult than policy creation.

Implementation of Policy

Implementation, according to Inskeep (1991) should be considered throughout the planning process and requires taking into consideration what is realistic from multiple perspectives including the financial and political ones. Who, if anyone, implements policy depends on market forces and also what type of government is in power, and as Hall (2008) notes, implementation is often more likely and effective if the implementing agency is the same as the policy development agency. Policy implementation in the tourism field faces various difficulties including the different definitions of tourism, the source of policy formulation versus the implementation, often unreliable tourism growth predictions and the short-term view of operators within the tourism industry. 'Most good policy formulation requires considerable research and inputs from those who are implementing policy at the grass roots or impact level' (Elliot 1997, 101). This is especially true in tourism because of the diversity which exists within the private and public sector.

Crosby (1996) argues that constituency building, resource accumulation, and mobilisation of resources and actors are also imperative. Walker, Rahman, and Cave (2001) claim that implementation must have defensive or corrective actions to identify the conditions that must be met for the policy to succeed and Pigram (1990, 5) notes:

Policy must, first, be perceived by policy makers as conceptually robust, defensible and amenable to implementation. Secondly, the various interests involved in the implementation process must be convinced that the net outcome will be positive, or at least benign, in the longer term.

Implementation Responsibility

One of the key issues inevitably is who, or what level of decision-making should implement and control such policies. United Nations Environment Programme/International Council of Local Environmental Initiatives (UNEP/ICLEI 2003) agree that local authorities are the best placed to manage tourism in a destination for the following reasons:

- They have democratic legitimacy.
- They have relative permanence and the ability to take a long term view.
- They are responsible for a range of functions that can influence tourism development, including special planning, development control, environmental management, and community service (UNEP/ICLEI 2003, 8).

UNEP/ICLEI also declares that good tourism and policy management should:

- be coordinated by a single authority.
- be a progressive process.
- involve all relevant government bodies, research institutions, NGOs, and the local population.
- establish management strategies that rely on a systems approach recognising connections among different activities (UNEP 1996).

They point out that a key attribute of an effective policy is that the desire to begin the process must come from officers of the local authority or the interest of the elected council must be engaged. In addition, a vision for the future and an outline of principles to follow should be agreed by all stakeholders.

> Sustainability will not be easily achieved. It is not an absolute fixed position which could be attained and then forgotten about. It is a position of dynamic equilibrium which will require constant adjustment of a multitude of parameters to maintain.
>
> (ICLEI 1995, 34).

Lickorish (1991) and Krippendorf (1992) suggest a more integrated role is needed for tourism policy, and multiple authors (Aynsley 1997; Briassoulis 2002; Crosby 1996; Eber 1992; Hall 1994; Inskeep 1991; Jackson and Morpeth 1999; Krippendorf 1992; Vera and Rippin 1996) have also suggested that the key to successful policy implementation is more emphasis on local planning participation. Therefore, local involvement is viewed as fundamental to the successful planning and management of destinations (Coccossis 1996; Meetham 1998; Middleton and Hawkins 1998; Ryan 2002). The focus of policies at the international and national levels will change as they are reinterpreted and implemented at a local level and each country or destination should establish an operational definition for sustainable development so a bottom-up and top-down consensus approach can be achieved.

BARRIERS TO SUSTAINABLE TOURISM POLICY IMPLEMENTATION

An examination of multiple destinations including Calviá, Spain, and Malta (Dodds 2008, 2007a, 2007b); Tenerife, Spain (McNutt and Oreja-Rodriguez 1996), the Caribbean (Weaver 2001; Wilkinson 1989, 1997); Goa, India (Singh and Singh 1999); Pattaya, Thailand (Wong 1998); Kuta, Bali (Wong 1998); Cyprus (Godfrey 1996, Ioannides 1996, Sharpley 2000); Turkey (Tosun 2001); Tunisia (Poierer 1995); and Torremolinos and Mallorca, Spain (Bruce and Cantallops 1996; Vera and Rippin 1996) reveals a number of perceived barriers to policy implementation. Overall sustainable policy implementation faces problems from many barriers, including both private and public sectors. A number of themes can be identified in the literature, ranging from power clashes between political parties at a national level to lack of stakeholder involvement and accountability at the local level.

Power and Politics

Power struggles arise in all areas and have impeded policy implementation by all facets of government and industry and across many other sectors as well as tourism. 'Tourism development is not created exclusively by private commercial enterprise, but an adversarial attitude often inhibits tourism progress' (Gunn 1994, 435). Different stakeholders have different agendas and there is often a disconnect between ideal policy goals and achievable outcomes. In addition, the local communities who vote political parties into power are also partly responsible for power struggles over sustainability. Planning is rarely exclusively devoted to tourism *per se* but instead is a mix of economic, social, and environmental considerations which reflect all factors that influence tourism development (Hall 1994). In addition, in many destinations such as Calviá (Dodds 2008; Dodds 2007a), Malta (Dodds 2007b), Canary Islands (McNutt and Oreja-Rodriguez 1996), and Cyprus (Ioannides 1996, Sharpley 2000), much development happened before sustainability was considered as important and often sustainability means working with what exists to improve it rather than starting with a blank slate. Stated succinctly, sustainability has questioned the 'assumption of a continuous, linear, and more or less harmonious development for societies along a given track' (Becker et al. 1997, in Pollacco 2003, 359). Power is the underlying element of politics and this discussion reviews specific details to try to clarify issues resulting from this state of affairs.

Economic Priority over Social and Environmental Aspects

The most obvious barrier identified is economic priority over social and environmental concerns. This is inextricably linked with political governance's short term focus and multiple other barriers arise out of this. A focus on short term objectives creates a negative feedback loop with economic priority—the shorter the political term, the more attention is focused on job creation and development for growth and other immediate results, which leads to economics being given priority over environmental and social concerns. A four- to five-year political term is simply not long enough to achieve ST policy objectives, which by definition are long term. The majority of initiatives which have been undertaken in destinations (e.g., Mallorca) have tended to be ones that were very visible to the community and to businesses, so that there were tangible examples of what had changed. The private sector mentality also supports this negative loop as its main considerations are most often focused on return on investment and the economic bottom line for understandable reasons.

Many destinations examined also showed their past and future short term focus through their development patterns. Some destinations developed new product offerings or exploited resources solely because their competitors had done so and they feared a loss of competitiveness. This approach has not changed since the tourism boom of the 1960 and 1970s and continuous attempts to make a product competitive with that of other destinations. Destination managers justify this approach by arguing that new projects are necessary to maintain competitiveness and prevent decline. Seeking a more up-market clientele often only results in up-scaling the consumption patterns of their visitors. As Bianchi (2004) suggests, aspects of sustainability have been framed in a way that do not challenge the core pillars of free markets and profit-maximisation.

Most destinations focus on numbers of tourists rather than yield, and measures of the effectiveness and success of tourism policies invariably relate to the numbers of tourists that arrive at destinations rather than the net benefits that accrue to a destination. This suggests that there needs to be a change in the role of governments from promotion to protection (Hall 1994; Hall and Jenkins 1995; Elliot 1997). This focus is also a function of choice and markets. Conflicts in policy objectives often arise as job creation might harm the environment and society may have difficulty expressing its preferences. Election campaigns generally involve a complex system with multiple elements (e.g., taxation, services, health, defence, education.) which gives politicians considerable opportunity to interpret the 'public interest'.

Lack of Integration

Many frameworks for policy development are for new or developing destinations (United Nations 2003; Weaver 2001; McHardy 2000; Hall 1999; Craik 1990) rather than for developed or mature destinations, and have been based on the assumption that planning for tourism can incorporate issues of carrying capacity, social and cultural concerns, and environmental issues. However, the destinations which attract the greatest number of tourists are mature mass tourism destinations and decision makers in these locations often tend to view sustainability as a way to regenerate and rejuvenate stagnant or declining tourism numbers (Dodds 2008; Bianchi 2004), i.e., a means to an end rather than the end itself. In the literature it is often stated that it is at the local level where sustainability measures are achieved (Sharpley 2003, UN/ICLEI 2003). However at the local level in Calviá, Spain for example, field research revealed that many policy implementers believed that policy aims could not successfully be achieved without support and coordination from higher level governments and that without such support, policy plans could not be effective because sustainability extends much more widely than the local level (Dodds, 2007a). For example, economic growth and prosperity often hides growing social problems. One problem in Calviá was that of low education standards and high drop out rates from school as the skill set needed for jobs in the mass tourism sector (waiting tables, housekeeping, and bartending) is low. Although straightforward in theory, a change in education usually requires the involvement of higher level government than the municipality level.

Other examples of a lack of integration include areas such as transportation and waste management. Transportation is dependant on a regional plan including such elements as public bus routes and trains to make sure all public transport systems link together. Although the literature (Elliot 1997; Godfrey 1998; Meetham 1998;UNEP/ICLEI 2003) suggests that local level policy implementation is more effective as local governments have more specific control over the issues of sustainability within their areas or planning jurisdictions, there is clearly a need to have an overarching framework and principles in place and operating effectively at an international or national level to provide guidance if local level policy implementation is to be successful, particularly in the context of environmental issues.

Ecosystems rarely conform to the boundaries of cities or town, counties, states, election districts, or even nations. This means that no single government jurisdiction may possess the authority to deal completely with a particular environmental problem or to achieve sustainability results. Clearly, larger, more encompassing jurisdictions have advantages in terms of fewer externalities, but there may not be the political will to address sustainability (Portney 2003, 14–15).

A potential explanation for the lack of integration of policy initiatives is that tourism is not regarded as important by many government sectors and there is a general lack of recognition of tourism on political agendas (Richter 1989; Hall 1994).

Lack of Coordination with Other Sectors

Singh and Singh (1999) and Dodds (2008) also note the problem of lack of coordination between government bodies. It is suggested that politics and programmes of governments at different levels are poorly coordinated as there is often a problem with integrating tourism policy, and actions and policies of one level of government may contradict policies at another level or with little consultation between levels or departments. Williams and Shaw (1998) suggest that there is often conflict between a local desire to limit tourism's impacts and the generally expansionist economic interests at the regional or national level. Sustainable tourism requires close coordination between many sectors. It is often the case that a policy is subjected to change during the process of implementation (Younis 1990) and every sector needs to be aware of the others and communicate their needs and concerns in order to achieve policy implementation.

Stakeholder Participation and Accountability

Participation by stakeholders such as the local community, private sector, NGOs, and different levels of government is often regarded as imperative, but some, such as NGOs, are often excluded from policy development and implementation because they have tended to showcase environmental and social concerns and have argued that economic priorities may be detrimental to achieving sustainability.

Hall (1999), discussing policy issues regarding tourism states that:

> often, the problems of developing coordinated approaches towards tourism planning and policy problems, such as the metaproblem of sustainability, are identified in organisational terms, e.g. the creation of new organisations or the allocation of new responsibilities to existing ones. However, such a response does not by itself solve the problem of bringing various stakeholders and interests together which is an issue of establishing collaborative processes
>
> (Hall 1999, 278).

One possible problem to local sustainable policy being achieved is communitarian. Community characteristics create an enormous challenge as there is a clash between traditional economic development and a different, more sustainable path. For some projects such as public transportation initiatives, dominant social values turn out to be more resistant to change than anticipated. Another possibility is that as long as people (political and business leaders as well as the general public) are willing to accept the status quo, little progress towards sustainability is possible. 'The lack of political will to pursue sustainability prevents all those professionals and technical experts from doing their part' (Portney 2003, 128). This is eloquently stated by Parlato:

An altruistic attitude towards the environment, resulting in behavioural change, is more likely to occur if motivation to do so is on a personal and individual level, through one's own beliefs and value system rather than if it were enforced legally or simply viewed as a social or political ideal.

(Parlato, 2004, 57)

Community involvement is very popular in the academic and policy literature, however, equitable and fair multi-stakeholder consultation and cooperation is difficult to achieve, and in the end relies heavily on the power distribution arrangement in a community, going back to earlier discussion. A rampant sense of individualism can also be to blame for problems. Portney argues (2003, 135). 'the problem is that there is usually no incentive for individuals, acting purely in pursuit of the short-term, self interested bargain to use less air or water. To the contrary, in the absence of aggressive regulation, the incentives usually motivate the depletion of such common goods'. This 'tragedy of the commons' or 'rampant individualism' is where individuals are free to act on what they believe to be their own immediate self interest—a mismatch between what is good for society or the community and what individual people see as good for them personally. Such a 'communitarian view suggests that what is good for the community in aggregate is not always the simple sum total of what is good for each of the individuals in that community' (Portney 2003, 130).

Political Priorities

The constant priority given by elected representatives to attaining votes is widely discussed in the literature. Boehmer-Christiansen (2002, 357) noted that bureaucracies often have to implement and regulate for unpopular issues and therefore they act more on a NIMTO ('not in my term of office') principle. He concludes his argument by wondering whether sustainability

will turn out to be 'reactionary' or progressive [and] is of course a matter of judgment reflecting the values of the judge. When judged from the perspective of human equality, the evidence suggests that seeking "sustainability" is as yet little more than a political tactic that de facto serves capital accumulation.

(Boehmer-Christiansen 2002, 368)

Lack of Understanding/Awareness of Sustainable Tourism

Another explanation for policy failure has been an overall lack of awareness and understanding of sustainable tourism. The literature suggests that one finding problem with sustainability is that it is hard to define (Butler 1999). Dodds (2007a, 2007b) partly refute this claim as all interviewees in Calvia and Malta claimed to have had a clear understanding of what was meant by sustainability. A related potential problem is that those who influence policy often have a poor understanding of why sustainability is needed or fail to see all aspects of the triple bottom line, instead focusing exclusively on economic priorities as noted earlier. Society in a particular area usually expresses its preference for environmental and social issues through voting, however in both local and national elections, tourism is only one, usually a minor issue in any election compared to taxation, health care, security and job creation, if it is an issue at all. One might argue that while there is some confusion over sustainability in the context of tourism (Smith 1973), there is even less appreciation of the overall importance of the concept.

Avoidance of Sustainability Issues in Policy

Another potential reason sustainable tourism policy is often ineffective is because such policy may not adequately address specific sustainability issues. For example, the problem of second home development occurred in Calviá after the sustainable tourism policy was adopted and this was not adapted to mitigate that specific problem. The declassification law stopped the building of hotels but did not control the conversion and building of second homes, which are perceived to have created great pressure on resources. The policy addressed only the issue of beds proffered by officially registered tourism providers rather than the total number of beds available to tourists. In Goa policy did not address overbuilding and overuse of ground water by tourism units (Singh and Singh 1999).

Policy has often focused on attracting new markets and reducing seasonality, thus addressing economic priorities rather than environmental and social concerns. Bianchi (2004, 497) notes 'public authorities continue to promote growth centred policies based on new 'tourism products' such as marinas and golf courses, which fail to adequately address the more pressing aspects of ecological degradation and the broader social dimensions of sustainability'. He concludes by saying,

> Merely restricting the quantitative growth of tourism whilst upgrading the 'quality' of the provision, does little to ensure that the aggressive pursuit of profitability by tourism corporations and the resultant shift of capital into 'new products' will not eventually lead to the same cycle of oversupply and concentration of tourism' means of production into fewer hands.
> (Bianchi 2004, 520)

CONCLUSIONS: MOVING FROM POLICY CREATION TO IMPLEMENTATION

> If moves toward a sustainable tourism development pattern are to be successful, attention will need to be paid to institution building in the spheres of policy management and implementation.
> (de Kadt 1992, 66)

Figure 4.1 outlines the findings of this research review regarding barriers to policy implementation. The figure portrays the various elements which were considered to hinder or block the achievement of successful sustainable tourism policy implementation. The seven inner boxes contain the impediments to successful implementation while the outer boxes provide examples to explain the barriers. In summary, the principle findings of this review conclude that there is often more than one barrier to implementation and that many factors overlap or influence each other.

The process of policy and planning never ends, as any decision or action eventually needs implementation and revision, however, it can be concluded that the push for economic growth resulting in economic factors having priority over social and environmental concerns is the major causal factor affecting policy non-implementation, a view supported by other writers (Bianchi 2004; Elliot 1997; Fayos-Sola 1996; Hashimoto 1999).

Sustainability has been endorsed and adapted in many areas and has been seen as the best way forward for all sectors (economic, social, and environmental), however the execution of sustainability initiatives prove difficult and

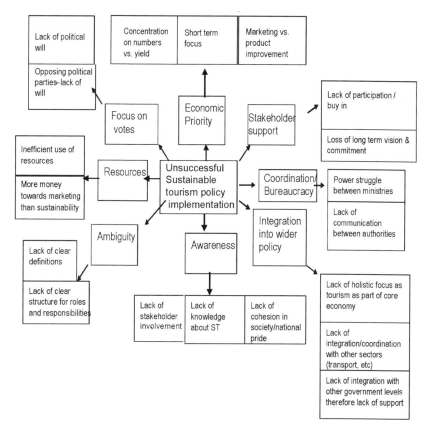

Figure 4.1   Barriers to achieving successful sustainable tourism policy.

many goals are not reached despite the impacts of existing forms of tourism being clear. It is likely, therefore, that the problem with achieving sustainability lies not in definition, but in implementation. It may be that policymakers believe that achieving sustainable tourism development requires little more than a shift away from the traditional three S's (sun, sea, sand) mode of tourism towards a niche product focus and quality initiatives to attract a more up-market tourist. As discussed earlier, this is rarely a move towards sustainability as resource requirements of such visitors are normally greater on a per capita basis than other tourists.

The problem of policy implementation cannot be resolved solely by technical means as the answer lies in political, cultural, economic, social, and psychological change. Collective action, regime, and adaptive management theories have been put forward along with long term and holistic thinking as ways to address these barriers. Tourism policy by its very nature is complex and the management of a policy in its formulation, adoption, and implementation makes the operational and functional element a part that policy itself does not control. Hall (2001) notes:

> There is no perfect planning or policy process. . . . Nevertheless, through an improved understanding of the policy processes and institutional arrangements by which coastal and ocean areas are managed, better integration of tourism development within coastal communities and ecosystems without undue negative impacts may be achieved.
>
> (Hall 2001, 610)

Although Hall's (2001) comment above is made in a marine and coastal context, the policy implications suggest a wider applicability. It is clearly a question of those in control of tourism and tourist destinations being willing not only to 'talk the talk' to create policy but also to 'walk the walk' to achieve sustainable tourism goals through policy implementation, a much harder but ultimately necessary task.

REFERENCES

Anderson, O.J. (2004) 'Public-private partnerships: Organizational hybrids as a channel for local mobilization and participation?' *Scandinavian Political Studies* 27(1): 1–21.

Asher, A. (1984) 'Barriers to international travel and tourism'. *Journal of Travel Research* 22(3): 2–16.

Aynsley, K. (1997) 'Problems in international environmental governance or a policy analysis looks at the world'. *Australian Journal of Public Administration* 56(2): 54–66.

Bianchi, R.V. (2004) 'Tourism restructuring and the politics of sustainability: A critical view from the European periphery (the Canary Islands)'. *Journal of Sustainable Tourism* 12(6): 495–529.

Boehmer-Christiansen, S. (2002) 'The geo-politics of sustainable development: Bureaucracies and politicians in search of the holy grail'. *Geoforum* 33(3): 351–365.

Bramwell, B.W., Henry, I., Jackson, G., Prat, A.G., Richards, G., and van der Straaten, J. (eds.) (1996) *Sustainable Tourism Management: Principles and Practice*. Tilberg: University Press.

Briassoulis, H. (2002) 'Sustainable tourism and the question of the commons'. *Annals of Tourism Research* 29(4): 1065–1085.

Bruce, D. and Cantallops, A.S. (1996) 'The walled town of Alcudia as a focus for an alternative tourism in Mallorca'. In Briguglio, L., Butler, R., Harrision, D., and Filho, W.L. (eds.) *Sustainable Tourism in Island and Small States: Case Studies*, 241–261. London: Pinter.

Butler, R.W. (1999) 'Sustainable tourism: A state-of-the-art review'. *Tourism Geographies* 1(1): 7–25.

Caffyn, A. and Jobbins, G. (2003) 'Governance capacity and stakeholder interactions in the development and management of coastal tourism: Examples from Morocco and Tunisia'. *Journal of Sustainable Tourism* 11(2/3): 224–245.

Choy, D.J.L. (1991) 'Tourism planning: The case for "market failure"'. *Tourism Management* 12(4): 313–330.

Coccossis, H. (1996) 'Tourism and sustainability: Perspectives and implementation'. In Priestley, G.K., Edwards, J.A., and Coccossis, H. (eds.) *Sustainable Tourism: European Perspectives*. London: CAB International.

Coles, T.E. and Church, A. (2007) *Tourism, Power and Space*. London: Routledge.

Cooper, C., Fletcher, J., Gilbert, D., and Wanhill, S. (1998) *Tourism: Principals and Practice*, 2nd ed. London: Pearson Education Ltd.

Crosby, B. (1996) 'Policy implementation: The organizational challenge'. *World Development* 24(9): 1403–1415.

De Kadt, E.B. (1992) 'Making the alternative sustainable: Lessons from development for tourism'. In Smith, Y. and Eadington, R. (eds.) *Tourism Alternatives*, 47–75. Chichester: John Wiley and Sons.

Dodds, R. (2007a) 'Malta's tourism policy—Standing still or advancing towards sustainability?' *Island Studies Journal* 2(1): 44–66.

Dodds, R. (2007b) 'Sustainable tourism and policy implementation: Lessons from the case of Calviá, Spain'. *Current Issues in Tourism* 10(1): 296–322.

Dodds, R. (2008) 'Sustainable tourism policy—Rejuvenation or critical strategic initiative', *Anatolia: An International Journal of Tourism and Hospitality Research* 18(2): 277–298.

Dye, T. (1992) *Understanding Public Policy*, 7th ed. Englewood Cliffs N.J.: Prentice Hall.

Eber, S. (ed.) (1992) *Beyond the Green Horizon: A Discussion Paper on Principals for Sustainable Tourism*. Goldalming, Surrey: Tourism Concern/WWF.

Edgell, D.L. (1999) *Tourism Policy: The Next Millennium*, Vol. 3. USA: Sagamore Publishing.

Elliot, J. (1997) *Tourism: Politics and Public Sector Management*. NY: Routledge.

Godfrey, K. (1998) 'Attitudes towards sustainable tourism in the UK: A view from local government'. *Tourism Management* 19(3): 213–224.

Goeldner, R.C., Ritchie, J.R.B., and McIntosh, R.W. (2000) *Tourism, Principles and Practices*, 8th ed. New York: John Wiley and Sons.

Gunn, C.A. (1994) *Tourism Planning: Basic Concept Cases*, 3rd ed. NY: Taylor and Francis.

Hall, C.M. (1994) *Tourism and Politics: Policy, Power and Place*. Chichester: John Wiley and Sons.

Hall, C.M. (1999) 'Rethinking collaboration and partnership: A public policy perspective'. *Journal of Sustainable Tourism* 7(3/4): 274–289.

Hall, C.M. (2000) *Tourism Planning: Policies, Processes and Relationships*. Singapore: Pearson Education Ltd.

Hall, C.M. (2001) 'Trends in coastal and marine tourism: The end of the last frontier?' *Ocean and Coastal Management* 44(9–10): 601–618.

Hall, C.M. (2008) *Tourism Planning: Policies, Processes and Relationships*, 2nd ed. London: Pearson Prentice Hall.

Hall, C.M. and Jenkins, J. (1995) *Tourism and Public Policy*. NY: Routledge.

Hardin, G. (1968) 'The tragedy of the commons'. *Science* 162: 1243–1248.

Hashimoto, A. (1999) 'Comparative evolutionary trends in environmental policy: Reflections on tourism development'. *International Journal of Tourism Research* 1: 195–216.

Healy, R.G. (1994) 'The "common pool" problem in tourism landscapes'. *Annals of Tourism Research* 21(3): 596–611.

Hunter-Jones, P.A., Hughes, H.L., Eastwood, I.W., and Morrison, A.A. (1997) 'Practical approaches to sustainability: A Spanish perspective'. In Stabler, M.J. (ed.) *Tourism and Sustainability: Principles to Practice*, 263–273. London, CAB International.

ICLEI (1995) *European Local Agenda 21 Planning Guide—How to Engage in Long-Term Environmental Action Planning Towards Sustainability?* Brussels: European Sustainable Cities and Towns Campaign.

Inskeep, E. (1991) *Tourism Planning: An Integrated and Sustainable Development Approach*. New York: Van Nostrand Reinhold.

Ioannides, D. (1996) 'A flawed implementation of sustainable tourism: The experience of Akamas, Cyprus'. *Tourism Management* 16(8): 583–592.

Ioannides, D. and Holcomb, B. (2003) 'Misguided policy initiatives in small-island destinations: Why do up-market tourism policies fail?' *Tourism Geographies* 5(1): 38–48.

Jackson, G. and Morpeth, N. (2000) 'Local agenda 21: Reclaiming community ownership in tourism or stalled process?' In Richards, G. and Hall, D. (eds.) *Tourism and Sustainable Community Development*, 119–134. London: Routledge.

Jenkins, C.L. (1980) 'Tourism policies in developing countries: a critique'. *International Journal of Tourism Management*, 22–29.

Johnson, P. and Thomas, B. (1999) *Perspectives on Tourism Policy*. London: Biddles Ltd.

Manning,T. (1999) 'Indicators of tourism sustainability'. *Tourism Management* 20: 179–181.

McNutt, P. and Oreja-Rodriguez, J.R. (1996) 'Economic strategies for sustainable tourism in islands: The case of Tenerife'. In Bruguglio et al. (eds.), *Sustainable Tourism in Island and Small States*, 261–280. London: Pinter.

Meethan, K. (1998) 'New tourism for old? Policy developments in Cornwall and Devon'. *Tourism Management* 19(6): 583–593.

Middleton, V. and Hawkins, R. (1998) *Sustainable Tourism: A Marketing Perspective*. Oxford: Butterworth-Heinemann.

Parlato, M. (2004) *Ecotourism and the Maltese Islands: A Case Study*. Rural Recreation and Tourism of the Royal Agricultural College, Cirencester, unpublished master's dissertation.

Pigram, J.J. (1980) 'Sustainable tourism—Policy considerations'. *Journal of Tourism Studies* 1(2): 2–9.

Poierer, R. (1995) 'Tourism and development in Tunisia'. *Annals of Tourism Research* 22(1): 157–171.

Pollacco, J. (2003) *In the National Interest: Towards a Sustainable Tourism Industry in Malta*. Malta: Progress Press.

Portney, K.E. (2003) *Taking Sustainable Cities Seriously: Economic Development, the Environment, and Quality of Life in American Cities.* Cambridge, MA: MIT Press.

Reed, M.G. (1999) 'Collaborative tourism planning as adaptive experiments in emergent tourism settings'. *Journal of Sustainable Tourism* 7(3 and 4): 331–355.

Richter, L.K. (1989) *The Politics of Tourism in Asia.* Honolulu: University of Hawaii Press.

Ryan, C. (2002) 'Equity, management, power sharing and sustainability—Issues of the "new tourism"'. *Tourism Management* 23: 17–26.

Sharpley, R. (2000) 'Tourism and sustainable development: Exploring the theoretical divide'. *Journal of Sustainable Tourism* 8(1): 1–19.

Sharpley, R. (2003) 'Tourism, modernization and development on the island of Cyprus: Challenges and policy responses'. *Journal of Sustainable Tourism* 11(2 and 3): 246–265.

Singh, T.U. and Singh, S. (1999) 'Coastal tourism, conservation and the community: case of Goa'. In Singh et al. (eds.) *Tourism Developments in Critical Environments*, 65–76. Elmsford: Cognizant Communications Corporation.

Smith, T. (1973) 'The policy implementation process'. *Policy Sciences* 4: 197–209.

Tosun, C. (2001) 'Challenges of sustainable tourism development in the developing World: The case of Turkey'. *Tourism Management* 22: 289–303.

Twinning-Ward, L. (2002) *Monitoring Sustainable Tourism Development: A Comprehensive Stakeholder-Driven Adaptive Approach*, unpublished Thesis. Surrey, UK: University of Surrey.

UNEP (1996) *Guidelines for Integrated Planning and Management of Coastal and Marine Areas in the Wider Caribbean Region.* Kingston, Jamaica, Caribbean Environmental Programme.

UNEP and ICLEI (2003) *Tourism and Local Agenda 21: The Role of Local Authorities in Sustainable Tourism.* France, UNEP.

UNEP, WTO and Plan Blue (2000) *Final Report: International Seminar on Sustainable Tourism and Competitiveness in the Islands of the Mediterranean.* Italy, May 17–20.

Vera, F. and Rippin, R. (1996) 'Decline of a Mediterranean tourist area and restructuring strategies'. In Priestley, G.K., Edwards, J.A., and Coccossis, H. (eds.) *Sustainable Tourism: European Perspectives.* Wallingford, UK: CAB International.

Weaver, D.B. (1998) *Ecotourism in the Less Developed World.* Wallingford, UK: CAB International.

Wight, P. (1993) 'Ecotourism: Ethics or eco–sell?' *Journal of Travel Research* 31(3): 3–9.

Wilkinson, P.F. (1997) *Tourism Policy and Planning: Case Studies from the Commonwealth Caribbean.* Elmsford, NY: Cognizant Communications Corporation.

Williams, A.M. and Shaw, G. (1998) 'Tourism policies in a changing economic environment'. In Williams A.M. and Shaw, G. (eds.) *Tourism and Economic Development*, 3rd ed., 375–391. Chichester: Wiley.

Wong, P.P. (1998) 'Coastal tourism development in South East Asia: Relevance and lessons for coastal zone management'. *Ocean and Coastal Management* 38: 89–109.

Younis, T. (1990) *Implementation in Public Policy.* Hampshire, UK: Billing and Sons Ltd.

5 Transport and Tourism in Scotland
A Case Study of Scenario Planning at VisitScotland

*Stephen Page, Ian Yeoman, and
Chris Greenwood*

INTRODUCTION

VisitScotland (VS) is the lead organisation for tourism in Scotland, responsible to the Minister for Tourism in the Scottish Executive (Government) for destination marketing, tourism policy issues, and economic advice. Tourism in Scotland is a £4.5bn industry, representing 6 per cent of gross domestic product (GDP) and 9 per cent of all employment. VS's main domestic markets are Scotland, the north of England, and the south of England. In 2002, domestic UK tourists spent £3.7bn and made 18.5 million trips to Scotland. Overseas tourists spent £811m and took 1.6 million trips, with the main overseas markets being North America, Germany, France, and the Benelux countries. Scotland is predominantly a UK weekend leisure destination, with UK tourism representing 92 per cent of all trips and 83 per cent of revenue.

The vision for VS is, with the support of the Scottish Executive, to increase the value of Scottish tourism by 50 per cent, by 2015. This poses challenges for sustainable tourism growth and development and one of the cornerstones of any expansion of tourism must be the provision of transport infrastructure and accessibility to connect visitor markets with the destination. This chapter examines the process of scenario planning at VS, highlighting some of the processes used to try and understand future issues likely to affect the sustainability of tourism, such as how transport and tourism will interact to shape future destination management and marketing issues. The chapter initially examines the process of scenario planning and how it has been developed over the last 5 years at VS and then outlines some of the results from a scenario planning project on transport and tourism to 2025.[1] The example of Scotland is fairly unique since it is one of the few National Tourism Organisations (NTOs) to engage in scenario planning on a regular basis.

SCENARIO PLANNING AND VISITSCOTLAND

The Bali and Madrid bombings, 9/11, war in Iraq, SARS, and foot and mouth disease (FMD) highlighted the uncertainty that has come to dominate the discourse of contemporary media and international affairs (Bierman 2003). These incidents have had an impact on tourism and led to the importance of communications and crisis management in tourism destination management organisations (Baral, Baral, and Morgan 2003), particularly at VS given the impact of FMD on the Scottish tourism economy. VisitScotland's history in scenario planning and futures thinking is rooted in the 'Futures' department of the organisation. This department undertook futures thinking and scenario planning, but lacked a full-time professional futurologist or scenario planner to carry this out. An example of this was observed by Seaton and Hay (1998) when VS engaged the Henley Centre in an environmental scanning exercise. The centre invited sixty key informants drawn from both the public and private sector to identify 'drivers of change' and create a 'vision' for Scottish tourism and methods for achieving the vision. The exercise, while motivated by a genuine attempt to get to grips with the future, tended to produce rather predictable platitudes and orthodoxies and therefore was not taken any further.

One way to deal with this problem in a holistic and systematic manner is to use the process of scenario planning (Hiejden et al. 2002), where terrorism is put into the context of tourism, in which it is explored as one variable that influences the tourism demand and supply. If this is not done, a language of crisis and negativity is created that blurs policy-makers' decision-making ability (Sparrow 1998). Rather, policy-makers should be using a scenario planning process that can look at the future and make sense of it, then be able to test shocks such as terrorism incidents against a range of scenarios to understand consumer behaviour, market segmentation, and economic behaviour. By adopting a holistic, creative, and future thinking process an organisation can create a culture of learning and change, where forward thinking becomes the norm rather than a one-off project syndrome or crisis management ethos; hence the reason for designing a scenario planning process.

The use of scenario planning in tourism has received little attention (aside from recent studies by VS), although early work by the Singapore Tourist Board (Yong, Keng, and Leng 1989) examined a methodological process and marketing implications of a series of events using a Delphi forecast. Tress (2003) used a scenario planning methodology for a participatory landscape study of Denmark. Eden and Ackermann (1998) used scenario planning techniques in strategy building for Scottish Natural Heritage. More recently, the UN-World Tourism Organization has used scenario planning techniques when dealing with contingency planning (Glaesser 2003). The main problem has been that, although scenario planning has been used in tourism, much of the work is confidential. Therefore, there is a lack of research publications in this field.

DESIGNING A SCENARIO PLANNING PROCESS AT VISITSCOTLAND

Scenario planning is the capability of VS to perceive what is going on in the business environment, thinking of the consequences of what this means, and taking action. The objective is to give Scotland a competitive edge by seeking to develop products and a tourism experience which is ahead of its competitors. The importance of scenario planning is to allow VS to make sense of the future in a structured manner, from understanding trends to developing policy. The development of a scenario planning process started with the appointment of a dedicated scenario planner. The appointment came about due to external comments from the tourism industry that VS was not paying enough attention to the external environment and assessing how this affected tourism. These comments were more insistent after a number of shocks such as 9/11 and FMD, and demonstrated that tourism was the first industry to be affected by these traumatic events. Two distinct elements of the various scenario planning tools used at VS include environmental scanning and the use of the Moffat model.

Environmental scanning is a qualitative process (not just one technique) of capturing shocks, surprises, trends, and drivers that will influence and shape tourism in a systematic and sensible manner. In contrast, the Moffat Model is an economic and forecasting model that paints a picture of the present, produces forecasts of the future, and measures the economic impact of shocks and surprises on the tourism sectors. The purpose of the Moffat Model (Blake et al. 2004) is to quantify scenarios. It was developed partially in response to a demand for 'real figures', as most of the policy-makers felt more comfortable with quantitative data, rather than qualitative measures. This demand for 'figures' was also partly driven by the need for a comfort zone, where policy-makers could more easily get to grips with some of the uncertainties associated with future tourism growth and development.

The Moffat Model (Blake et al. 2004) allows VS to quantify the future by explaining the present state of tourism through economic indicators. Forecasting the future using econometrics and measuring the impact of shocks or changes in policy, it was computable general equilibrium modelling, which is an integrated economics tool. An integrated approach is necessary as traditional forecasting techniques are of limited use in predicting shocks or changes in variables (Blake et al. 2004; Prideaux, Laws, and Faulkner 2003). The scenario analysis element of the model uses the principles of computable general equilibrium (CGE) modelling of the Scottish economy. CGE models are a well-established methodology for measuring changes or shocks in the economy and the effect they have on tourism (Blake and Sinclair 2003). For example, the model allows for events such as SARS, changes in value added (sales) tax or air passenger duty/tax, as well as a range of optimistic and pessimistic scenarios relating to the future of the Scottish economy. The model provides macroeconomic effects of alternative scenarios on income, employment, and welfare, but the most important feature of the Moffat Model is the integration of the indicators, forecasting and scenario analysis as user-friendly software. The model does not give a final answer about the future, just a means to explain and interpret it. It must be remembered that interpretation is an art, not a science, where subjectivity reigns, not objectivity.

Many of the elements of VS use of scenario planning tools and methodologies are essential to ensure a holistic and systematic approach to scenario planning, as seen in Figure 5.1. This approach to scenario planning has come about after extensive consultation and observations of other organisations' approaches. Advice was sought from Scottish Enterprise and Shell International about scenario construction. The Scottish Executive gave advice on economic modelling and Strathclyde University on environmental scanning. The three clusters are a representation of the organisation's needs to deliver a process that can create change both within the organisation and the industry. For example scenario planning at Royal Dutch/Shell allows executives to think about the future and events that may impinge upon strategy; it provides them with the skills needed to manoeuvre the organisation and, over time, to create change. de Geus (1988) called this 'Adaptive Organisational Learning', where the organisation creates a capability to understand, create, and manage change that leads to competitive advantage.

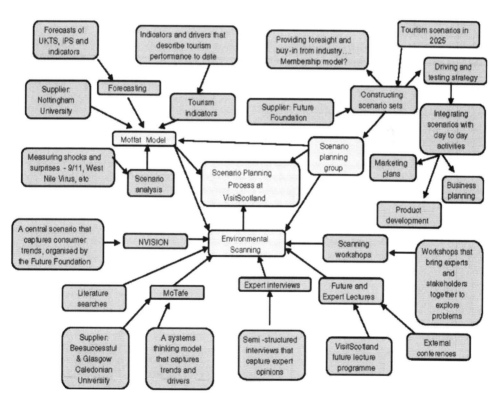

Figure 5.1 Scenario planning at VisitScotland.

As VS is the lead advisory body for tourism in Scotland, it has a responsibility to create a strategic conversation about tourism with industry and policy-makers from a wide range of public sector organisations. VS is a public sector agency whose prime responsibility is the marketing of Scotland as a destination; therefore it can bring players and stakeholders (Eden and Ackermann 1998) together to focus their minds on increasing the value and importance of tourism. The aim of scenario planning projects in broad terms at VS is to find answers to a number of questions, including: 'What actions does VisitScotland and its stakeholders need to take to ensure tourism is the first and everlasting industry of Scotland to 2025?'

This question is based on VS's vision of tourism as Scotland's 'First and everlasting industry' (Yeoman and Lederer 2004). This comes from the desire to be Scotland's premier industry in terms of GDP, employment and status; the main reason to live, work, and play in Scotland. Those who have responsibility for tourism now must take the right decisions to ensure a lasting future.

The choice of the year 2025 ensures that the scenarios show real change and are not just an extension of the present. One of the trends that will have an impact on Scotland and tourism by 2025 is the change to Scotland's demographics. Research highlights (Hay and Yeoman 2004) that Scotland will have more people over sixty-five than under twenty-one by 2025 and a critical dependency ratio for those in work supporting those out of work. This means less economic activity in Scotland by 2025, which will affect business tourism, labour supply, and fiscal policy.

In addressing this core question scenario planning examines a number of key questions :

- Which markets will thrive and which will decline to 2025?
- Which product offers will thrive and which will decline to 2025?
- How will consumer needs and wants change to 2025?
- How will supply-side and structural issues assist or hinder the development of Scottish tourism to 2025?

One useful tool used in this context is the systems thinking workshops.

SYSTEMS THINKING WORKSHOPS

The purpose of the systems thinking workshops is to bring together stakeholders and experts in order to explore and share thinking on a specific topic that will have an impact on tourism. Many of the topics examine the degree of disruption of tourism caused by shocks; these include 'Scenarios for the G8 Summit' (Yeoman and Lennon 2004), 'West Nile disease' (Yeoman, Lennon, and McMahon-Beattie 2004), 'Foot and mouth' (Yeoman and Lennon 2004), and avian and pandemic flu (Page et al. 2006). The workshops follow a format of presentations, discussion of issues, visual modelling of issues, policy implications, and action points. Two interactive visual modelling techniques are used: cognitive mapping or hexagons.

Cognitive mapping enables a modeller to represent a participant's individual thoughts on a problem situation (see Eden and Ackermann 1998). The map itself shows a series of linked ideas, with arrows indicating how one idea might lead to another (i.e., it is a sign-directed map expressing chains of cause and effect among the issues comprising and relating to the problem area). The approach is supported by a computer package called 'Decision Explorer' (Jones 1993).

Hexagons, or their commercial name of IDONS (Hodgson 1992) assist in the process of introducing and structuring ideas. Hodgson (1992, 227) states that:

> Concept mapping with IDONS is the process of rendering tacit models sharable by use of representation mapping. This mapping is done by means of a variety of techniques which are like moving diagrams. The fundamentals of the process involve, in a group session, individuals noting down ideas on magnetic, coloured hexagons, which are placed on white-board. These hexagons are then clustered to show related concepts and connections to ideas. The flexibility of the method means that it can be used in various contexts and provides a means to stimulate creative thinking and eliciting a collective view of ideas.

Through a series of workshops this allows the exploration of issues and the surfacing and challenging of assumptions, through using visual models, which act as holding devices or cryptic labels of knowledge where participants explore their own, and others' mental models of the problem situation. The workshops bring structure to messy problems in which these mental models can be negotiated, leading to a range of clear and agreed actionable outcomes (Yeoman 2004). With these issues in mind, attention now turns to an example of scenario planning undertaken in 2007 by VS in relation to transport and tourism.

TRANSPORT AND TOURISM IN SCOTLAND

Transport is acknowledged as one of the most significant factors to have contributed to the development of tourism as noted by the Tourism Framework for Change (TFFC; Scottish Executive 2006). Historically it has been responsible for facilitating travel and movement, which are critical to tourist mobility in Scotland (Page 2005). Scotland has a dispersed tourism product, based outside of the main cities and accessibility to rural and island product bases are entirely dependent upon transport infrastructure. As infrastructure investment is one of the drivers of growth for Scottish tourism's ambition to grow the industry by 50 per cent by the 2015, transport is a key element of that growth agenda. Considering the importance and interconnectivity of transport and tourism, a scenario planning exercise took place to untangle the key uncertainties surrounding this subject given pending projects such as Edinburgh Trams, Forth Road Bridge (the main bridge connecting the south and east of Scotland by crossing the Forth Estuary), energy supply, and environmental determinants. In terms of a time horizon, 2025 was adopted considering completion of major infrastructure and investment projects as well as recognising that transport was not a short-term issue which was confined to the planning horizon of the 2015 50 per cent growth target of the *Tourism Framework for Change*. Although in this short chapter it is only possible to give a flavour of the study and more detail can be found elsewhere,[2] the study set out to:

- review the importance of travel and tourism in Scotland by 2025.
- identify the key drivers that will shape travel to and around Scotland by 2025.
- provide four case studies of how transport has or may hinder/facilitate the economic development of tourism.
- exemplar the drivers in two scenarios, including their economic impact
- assess supply (constraints) and demand for each sector, where feasible. Analyse the implications for different transport sectors and tourism.
- undertake a risk assessment.
- illustrate how transport impacts on the different tourism market.
- make recommendations for the future in a considered manner i.e., importance and timescale.

HOW TRANSPORT IS A FACILITATOR OF TOURISM: UNDERSTANDING A TOURISM GROWTH AGENDA

Across the world transport is both a product and necessary means of achieving tourism growth. In Switzerland, the World Economic Forum (2007) found Switzerland to have the most competitive travel and tourism sector. A key contributing factor for this recognition was transport. The public transport network is one of the most efficient in the world, stretching 25,612 km with 27,300 stopping points. The network is integrated with timetables coordinated to the extent that postal-bus services arrive prior to the departure of the next train. Transport can also be a key form of innovation for tourism development as the example of British Columbia's seaplanes show, which is both a tourism product and necessary passenger transport. In Victoria, on the Southern tip of Vancouver Island, sixty scheduled daily flights connect the state capital with Vancouver, Seattle, and western Canada. In addition, a number of scenic tourist flights take place over the island. Morocco has the ambition to increase its visitor numbers by 10 million by 2010, which means the tripling of room capacity. To achieve that fast track investment, the government has an open skies policy, which saw twenty-three new routes established from the UK, France, and Spain in 2006, along with a modern rail network, which is described as the best in Africa, connecting the country's major cities, in addition to 60,000 km of all-weather highways. This illustrates one example of how a destination is investing in its transport infrastructure to develop a competitive destination.

Within Scotland, like the rest of the world, the low cost carrier has facilitated growth both domestically and internationally. Since 2000, Scotland has seen considerable growth from visitors arriving from London and South East of England based on the connectivity of London's regional airports or European visitors flying directly to Scotland for weekend breaks. What these examples also suggest is that a destination, such as Scotland, needs to understand the future transport investment and planning requirements for tourism given its constant neglect by policy-makers and planners who tend to assume that domestic transport infrastructure can easily be utilised to facilitate the tourism sector's growth ambitions. As the case of both Canada and Morocco show, the investment in the infrastructure, route development, and transport-related tourism products are vital in seeking to address issues of accessibility to tourism markets and long-term growth ambitions.

Transport is not the panacea for the future development of a country's tourism industry by itself (as a holistic view of the entire sector would suggest), but transport does emerge as a key enabling factor when seeking to attract visitors in a competitive environment. For this reason, VS undertook a wide ranging review of its key infrastructure requirements to achieve the 50 per cent growth agenda by 2015 and identified two areas for further analysis—the accommodation sector and the transport sector. The transport sector was examined using a scenario planning exercise as a means to help understand what impact transportation would have on the sector's growth potential beyond 2015 and the wider implications for the country as a tourist destination. The exercise was designed to generate not only industry-led thinking on this critical issue but also a range of policy responses and actions to help raise the transport issue higher up the tourism agenda within Scotland. It closely followed the methodology and approach used to try and understand future change in transport in the UK to 2050 (Office for Science and Technology [OfST] 2006). This was seen as a logical corollary to the recent in-house work undertaken by VS on the country as a sustainable tourism destination in 2025 and the consequences for destination planning and development in view of current concerns over the challenges of balancing tourism growth with sustainability goals. In other words, the scenario planning exercise fits neatly with a wider debate on how destinations achieve growth management objectives in the medium to long-term without compromising the resource base. This is a major challenge for any organisation given the problems of understanding changes which are occurring in transport as a mode of tourist travel, particularly the impact of the climate change and sustainability agenda and potential constraints on policies which may be developed globally (though in the case of Europe, by the European Union and Member States such as their individual sustainability plans). In this respect, transport and tourism cannot be seen in isolation from the wider sustainability agenda that is gathering considerable momentum among policymakers outside of the tourism sector which will certainly impact upon tourist transport in the medium to long-term. Attention now turns to the scenario construction exercise.

TRANSPORT AND TOURISM IN 2025

Two scenarios were constructed that represented plausible but different futures or outcomes[3]. *Urban Metropolis* is based on urban growth and the decline of rural and island tourism due to changes in transport and *Pragmatism Prevails* is based on continued development of tourism within more limited constraints imposed by transport infrastructure. Each scenario was given an economic value based on changes to Gross Value Added (GVA), employment and tourism markets using the VS's Moffat Model.

These scenarios have been shaped by the following drivers after detailed interviews with twenty industry stakeholders, based on a series of in-depth discussions on key issues shaping tourism and transport in Scotland now and to 2025. The scenarios were also shaped by the analysis of other secondary data such as nVision data. nVision is a subscription online service comprising of an online resource, telephone consultancy, seminars, and workshops. It provides a comprehensive understanding of social, cultural, and economic trends as well as focused analysis and raw data. The service is available at www.nvisiononline.co.uk and is provided by the Future Foundation. The advantage of the service enables a one-stop shop for latest consumer and related trends that are available instantly. The importance of the service from a scenario planning perspective is that it allows employees at VS to understand the underlying assumptions behind the main trends that are shaping tourism.

By inviting staff at VisitScotland to the training seminars run by nVision, and encouraging them to use this information resource directly without going through the scenario planner, it has also provided for a 'quick win', thus helping to change people's understanding of scenario planning and their understanding of the drivers that shape the process. The interviews were analysed alongside the nVision drivers of change to try and collate a series of broad themes and issues which were then input into a proprietary systems thinking software system to try to identify the principal parameters which may be key to change in the transport sector and its impact on tourism to 2025. Whilst this is not an exact science, other studies such as the Office for Science and Technology (2006) scenario planning study of transport in the UK to 2050 helped provide a benchmark of the types of issues and parameters which may shape transport in a very long-term horizon. On the basis of the previous literature published on transport and scenario planning in the UK, a number of key drivers emerged from both the interviews and background literature. In terms of supply-led drivers, the following were identified:

- uncertainty in future environmental policy and its application to tourism
- infrastructure provision
- pricing models (see Table 5.1)
- the future of energy and oil and changes in EU climate policy
- air transport
- urban gateways and tourism concentration in Glasgow-Edinburgh and rural areas
- monopoly of provision in public transport

This shows that a range of factors will shape the future evolution of transport supply and accessibility in Scotland. These drivers also highlight a key concern in Scotland that whilst transport and accessibility are critical to destination development, in the case of energy and climate policy (which will impact upon air travel), this is within the remit of the UK and not the Scottish government. There are also much wider concerns for long-haul travel as Gössling, Peeters, and Scott (2008) suggest, which may involve innovations in carbon-smart tourist travel to overcome the potential impact upon outbound markets in the next ten to twenty years of changes in energy and climate policy at a destination level. As Table 5.1 shows, future changes in policies for environmental taxation on air travel are not fully understood even though they are likely to be a key area of action by 2010 for the European aviation sector.

This was a key concern raised by a number of industry respondents, where EU and UK policy as well as the impact of UK policy that would shape the future development of accessibility given the dependence upon UK hubs for air access over and above the point-to-point European and low-cost transatlantic flights. A number of demand-side drivers were also identified which highlight general concerns with the factors shaping tourist travel to 2025:

- wealth and costs
- consumer spending priorities
- declining cost of transport
- changing consumer aspirations and expectations for travel

On the basis of these drivers and looking at the scenario storylines constructed by the Office for Science and Technology (OFST 2006), two contrasting scenarios were written to incorporate these drivers and themes in two wide ranging futures. Whilst it is acknowledged that these two storylines may be extreme and it is likely that only part of the suggested changes that may be depicted in the scenarios may in fact occur, leading to moderate changes in tourist arrivals as Scotland has never promoted itself as a mass market. In fact

Table 5.1   Overview of Policies Evaluated by the EC on Environmental Taxation of Air Travel

Options rejected at this stage	Options concerning existing actions that need to be continued/ strengthened	Options considered in detail for implementation
Restrictions on air traffic volumes	Raising awareness of air transport users	En-route charges or taxes on aircraft emissions and impacts
Regulatory standards	Improving air traffic management (ATM)	Emissions trading for aviation through inclusion in the EU Emissions trading scheme (EU ETS)
Restrictions on access to EU airports for the least efficient aircraft	Research and development in air transport technology and operations	
Voluntary agreements with airlines to reduce emissions*	Applying energy taxes to commercial aviation	
Departure/arrival taxes, VAT on air transport, removal of public subsidies	Improving the competitiveness of rail transport	

Source: European Commission (2005)
*It is to be noted, however, that such voluntary agreements are already in place between airlines and aircraft manufacturers.

the current marketing strategy of VS is one based on a range of niche products as opposed to mass products which may have a moderating influence on the scenarios if this focus is retained. The scenarios are reproduced in Table 5.2. Following on from the storylines, the Moffat model was used to attempt to extrapolate the likely impact of specific assumptions about how these scenarios would lead to changes in specific parameters in the model. Much of the choice of parameters was based on the experience of changes induced by specific events in Scotland's tourism sector as well as experience in other destinations. Whilst a degree of subjectivity is built into this modelling process, it should be stressed that it is not about giving a definitive answer: more to show how changes in certain aspects of tourism (e.g., transport and accessibility) resulting from a current series of assumptions and policy changes may impact upon the tourism economy. This is designed to illustrate the degree or change or magnitude of impact which could occur rather than would occur, as the caveat already discussed, is that time may have a mediating and moderating impact on the degree of change which occurs, removing the extreme impacts. Yet for policy-makers and planners to understand the interconnections in the tourism and transport system and the relationship to the tourism economy, the Moffat model helps to demonstrate these relationships.

Table 5.2 Scenarios 1 and 2

Scenario 1: Urban Metropolis

In this world there is a realisation that the exponential growth of tourism as predicted by the UN World Tourism Organization in 2005 was unsustainable. No longer could it be justified both on moral grounds and resources for consumers to over-indulge themselves with holidays. Although tourism is the world's largest industry and China is the world's leading destination, things have changed, as climate change policies are embedded in international trade agreements. It all started in 2010 when the government realised that emission reduction targets couldn't be met and the Middle East countries decided to stop selling oil to the West. So it was necessary to encourage people to travel less and this in turn required a public policy intervention to help households and individuals change their lifestyle. By 2025, we have seen a major attitudinal change towards travel and tourism.

In fact transport is only allowed if it is green, clean and generates a low carbon footprint. For example, new cleaner technologies have made road-based car transport viable for a relatively short distance of up to 150 miles. Car distances beyond that point are heavily taxed. Public transport—electric and low energy is efficient and widely used. This efficiency is typified by the Metropolitan Glasgow—Edinburgh transport network. Two cities joined together by a Japanese style bullet train. The rail network links to the UK's high speed network. However the system doesn't extend further than Glasgow or Edinburgh. This metropolitan hub has joined the two cities together blurring the boundaries of the old divide. Glasgow and Edinburgh are suburbs of each other. Visiting this metropolitan hub means hotels are sustainable, which includes self contained power generation from solar panel cells. The hub has an excellent IT infrastructure needed for a striving financial services industry. Corporate meetings are now confined to virtual worlds ever since the introduction of telepresencing in 2014. Telepresencing combines video conferencing and virtual reality to create three dimensional, high speed, fluid interaction across different geographical locations. Business people even have their own hologram to give that physical presence. The metropolitan hub has invested heavily in conference and exhibition space, combined with an efficient land-based transport system making this metropolitan hub an ideal association meetings destination. In 2025 association meetings are still important for face to face situations. Not everything can be done by telepresencing.

Technology in this scenario lets the world deal with change and overcome many of the environmental and economic challenges of climate change and energy policy. In this scenario it is about transporting people from A to B in mass numbers therefore reducing the need for individual journeys. Urban hubs have organised themselves to minimise travel and carbon footprints. For example, Edinburgh is now a UNESCO tourism colony with award winning features such as the skyscraper Botanical Gardens. The city's green credentials stretch from the connectivity of its ULRS (Urban Light Rail System), connecting the airport with the city's business and leisure districts, making the city centre a car free zone, to the novel use of *Segways* for the elderly and infirm tourists. Globally, competition is increasingly between cities and not countries, and the winners in the competitive environment are those able to link high value-knowledge assets with a desirable workforce, good quality of life, and appropriate public assets such as cultural and educational resources. That's one of the reasons why cities are changing and Scotland's metropolitan hub is at the forefront.

(continued)

Table 5.2 *(continued)*

However, Scotland's cities have grown up at the expense of rural communities. The rail network is practically non existent outside cities. Many rural communities have become isolated and unsustainable due to the lack of tourists combined with immigration towards urban hubs. Many of the rural areas that surround the cities have become gardens for the urban populations, providing them with food for farmers' markets and adventure playgrounds for day trippers. The story for Scotland's islands has been mixed. Arran, once 'Scotland in miniature' is now a five star resort with golf courses and second homes for celebrities and the mega rich. Rhum, Muck and Eigg were all abandoned in 2018 due to the high cost of transport to the islands. Air travel is relatively expensive as it still dependent upon carbon fuels, meaning it is heavily taxed at 40 per cent VAT and vulnerable to oil shocks. Government plans to expand both Edinburgh and Glasgow airport have been curtailed, with investment being switched to rail networks and urban hubs. Scottish tourism is predominantly a city based product with rural destinations offering an exclusive experience for those who can afford to travel to the islands. In this scenario resource use is now a fundamental part of the tax system and people are more careful in their use of resources. So is this what it could look like?

Scenario 2: Pragmatism Prevails

In this world the exponential growth of tourism as predicted by the UN World Tourism Organization in 2005 has happened. International arrivals have grown from 800m in 2005 to 1.75bn by 2025. However China's exponential growth isn't what it used to be; this is typified by Beijing's smog and dust storms. In 2025 snow no longer falls on the top of Kilimanjaro, the Caribbean coral reef is no more, Polar bears no longer wander the Arctic, the Maldives doesn't exist as a country, lack of water on the African plains has meant the end of the Wildebeest migration and the Cairngorms plateau, once an arctic tundra for ptarmigans and snow buntings are no more. Global warning has changed the areas' ecology. The great, green rolling carpet of hardy moss that once covered the plateau is obliterated. But still, people still want to travel as on average, disposal income has doubled in real terms per person in the UK over the last twenty years. People are aware of climate change and the issues of sustainability but it is recognised that climates change and everyone has to get on with their own daily life. In this scenario the key drivers are market economics, mitigation and adaptation strategies.

There is no such thing as a perfect world. Individuals, business and governments have had to comprise as politics is dominated by coalition governments. Edinburgh Airport has expanded rapidly with a dedicated rail terminal direct to the city. Glasgow Airport has rapidly expanded because of the budget carriers. Campbeltown and Lossiemouth are Scotland's exclusive space ports. The new Forth Road Bridge crossing was opened in 2012. No new roads have been built and expansionist plans for a railway infrastructure were blown away by the Eddington report. It was once suggested that expanding the railway network would act as substitute for air travel, but capacity constraints meant that this was a non starter. Green and clean buses dominate public transport systems in Scotland's cities, with a few exceptions. Edinburgh has a skeleton tram system which hasn't been able to expand due to public expenditure constraints—which is shaped by demographics. Glasgow's clockwork orange underground system closed in 2014 for health and safety reasons, meaning the city is the only major European city without a light railway system. Successive governments in Scotland and the UK haven't been able to tackle the growing threat imposed by climate change fearing the electorate still wants the right to travel, where, when and how they want.

(continued)

Table 5.2 *(continued)*

However certain measures have been introduced. Everyone in Scotland has their own voluntary carbon account. The purpose of the account sets out to improve sustainability and make consumers think before they travel. Even the introduction of a 17.5 per cent VAT rate on air travel has had little impact on the desire to travel to far-away places. The consumer in 2025 knows about climate change, thinks of themselves as an eco tourist but stills takes at least four short breaks and one long distant holiday per annum. However, aircraft manufacturers have been able to offset carbon emissions by 60 per cent due to technological improvements in aircraft design. Similar improvements by car manufacturers, with the introduction of intelligent cars, which adhere to speed limits using the latest hybrid / hydrogen technologies, combined with the government's tough national sustainable surveillance system ensures that everyone knows the cost of travel as road pricing is now the norm. The car is still the number one and most important form of travel to and around Scotland. Hybrid and/or hydrogen cars are the preferred choice of most households. Arran was the first community in Scotland to ban the petrol car in 2016 as the island's electric car scheme allows tourists to get around the island because of relatively short distances.

Businesses are energy efficient. They can't afford not to be, as the penalties for using old technologies are huge. For example, all hotels of over twenty-five bedrooms have to provide a sustainable hot water system that doesn't draw upon the national grid. Solar panel cells are the norm on the skyline today and many communities have their own windmill. In today's society, consumerism still drives our desire for travel, even with climate change. But government incentives and legislation has led to innovation in business. For Scotland, climate change has been positive. Consumers to a certain extent have been put off by China's smog and population. Although the Far East may be relatively cheap at least Scotland is clean and green. Pragmatism wins in this case.

Source: VisitScotland

As 2025 was used as a baseline, it was assumed that tourism has grown by 3.6 per cent per annum (in line with the Scottish Framework for Change 50 per cent growth ambitions), therefore the economic value of Scottish tourism in 2025 is £10.7 bn as illustrated by Table 5.3. The two scenarios were then measured against this economic analysis for variance and disruption.

Scenario One: Urban Metropolis

The main economic assumptions in this scenario were as follows:

- Domestic tourism demand by car and air declines due to inhibitive taxation on these areas.
- Domestic train, coach, and bus travel increases with growth of public transport and concentration of travel within central belt of Scotland.
- The rest of UK transport for tourism sees significant drop in car and air travel to come to Scotland, again for inhibitive green taxation.
- Train, coach, and bus travel are principle hub-to-hub transport options to reach Scotland.

Table 5.3   Moffat Model for Scottish Tourism in 2002 and 2025

	2002 Demand £ million value	2025 Demand £ million value[4]
GDP	69,360	153,016
GVA	51,109	112,752
Welfare	56,000	123,542
Employment (FTE jobs)	1,831,846	4,041,262
Government revenue	22,648	49,964
Daytrips expenditure	2,277	5,023
Domestic Tourism expenditure	1,376	3,036
Rest of UK Tourism expenditure	2,307	5,090
International Tourism expenditure	1,209	2,667
Domestic plus Rest of UK Tourism expenditure	3,683	8,125
Overnight Tourism expenditure	4,892	10,792
Tourism plus Daytrips expenditure	7,169	15,816

Source: VisitScotland

- Boat travel from rest of UK to Scotland grows with resurgence of nautical touring and cruising as an environmentally friendly tourist option.
- International travel sees air travel decline with European Visitors increasing use of most cost effective tunnel and ferry routes to reach the UK and Scotland.
- Domestic Tourism Demand falls in total by -24 per cent, the rest of UK tourism demand increases by 0.9 per cent, and international Tourism Demand drops by 77 per cent.
- Productivity increases by 5 per cent across the Tourism industry reflecting improvements in efficiency.
- Public transport increases capital stock by 10 per cent showing growth in the number of buses, coaches, and trams in line with increased incentive towards public transport use.
- Air travel shows a decline in the overall number of aircraft by -25 per cent compared to 2025 figures with the price of air travel increasing 300 per cent over the same period. A 40 per cent tax increase on air travel has also been applied.

- A 25 per cent increase in transport services has also occurred as entrepreneurs provide guided tour services as personal transport is costly.
- Other Industry Sectors show a 5 per cent increase in productivity reflecting improvement in efficiency.
- Coal production shows a -15 per cent drop in coal reserves and subsequent increase in world price of 50 per cent
- World Price of oil has increased by 150 per cent through demand limitation and political sales to preferred countries by suppliers.
- Cars and other land transport (trains, trams, buses, etc.) see efficiency increases of 10 per cent a technology response to increasing hydrocarbon prices whilst the number of cars on the roads declines by -15 per cent as an increase of 7 per cent tax is applied to new car purchases.
- Electricity generation improves efficiency by 10 per cent as alternate sources are harnessed for its production whilst global gas supplies decline by -25 per cent.
- Both gas and electricity see 25 per cent increases in world prices directed to consumers.

The economic impact of *Urban Metropolis* is observed in Table 5.4 and summarised as the following:

- The value of tourism in Scotland declines by 31.17 per cent with employment dropping by a similar proportion.
- Under these assumptions, travel for tourism contracts considerably from all markets in Scotland and the economic consequences of an Urban Metropolis model with only urban based tourism and a limited amount of travel to other destinations highlights the constraints imposed upon tourism as an activity based on the freedom to move and travel without any restraints being applied.
- A major drop in most forms of tourist travel is observed, with a significant impact on inbound tourism.
- Overall, overnight tourism declines as a major contraction in tourism spending filters through the tourism economy.

The main reasons for these changes are that transport is reconfigured from its current pattern as the spatial patterns of activity are focused on urban development and transport becomes the key determinant of tourism activity.

What were the main reasons for this outcome? Are all aspects equally important? The key impacts of *Urban Metropolis* would be:

- The key change to tourism is the end of rural tourism in the absence of infrastructure to access rural areas, with the exception of exclusive resorts.
- The converse effect is the strengthening of urban tourism and a dependence upon domestic markets and a decline in inbound markets from overseas.

Table 5.4   Macroeconomic Results for *Urban Metropolis*

Macroeconomic Results	£ million change	% change
GDP	5732.96	3.75
GVA	-158.22	-0.14
Welfare	2311.91	1.87
Employment (FTE jobs)	-1446.20	-0.08
Government revenue	2298.92	4.60
Daytrips expenditure	-163.72	-3.26
Domestic Tourism expenditure	-670.11	-22.08
Rest of UK Tourism expenditure	-219.53	-4.31
International Tourism expenditure	-2474.70	-92.75
Domestic plus Rest of UK Tourism expenditure	-889.64	-10.95
Overnight Tourism expenditure	-3364.40	-31.17
Tourism plus Daytrips expenditure	-3528.10	-22.31
Summary TSA Results	change	% change
GVA attributed to tourism (£ million)	-648.04	-31.86
Employment attributed to tourism (FTE.)	-36107.64	-31.86
GVA in tourism related sectors (£ million)	-257.02	-3.53
Employment in tourism related sectors (FTE.)	-10543.95	-3.25

Source: VisitScotland

- Public policy interventions have shifted the balance on environmental sustainability to limit tourism options for Scotland, by requiring a lifestyle and attitude change with the focus on clean and green travel.
- The geopolitical environment and availability of oil would lead to a sea change in the current dependence upon car-based travel for tourism. It is the end of the car for journeys over 150 miles.
- Air travel is heavily taxed to create a change in behaviour through the imposition of a 40 per cent on VAT, meaning public transport becomes the most viable option
- The shift in focus and city-based meetings has implications for business and conference tourism and its potential location in the major cities.
- Technology has allowed behavioural change to occur.
- The geographical focus of Scottish tourism has come down in scale from a national to a city perspective: Competition is between cities and places rather than countries and so the brand has to change to reflect the more micro place-based marketing requirements. Similarly, economic development for tourism is focused on urban

hubs and gateways with the local hinterland the focus of short breaks whilst the Scottish Highlands and Islands remain isolated and inaccessible.

- The overall effect on the economic significance of tourism is massive due to the loss of tourism in rural and island economies, highlighting whether Scotland can really afford to allow such a scenario to occur.
- In political terms, this scenario has a high degree of state regulation, limiting personal freedom to travel to destinations.
- The greatest impact of this scenario is on the high yield markets, notably international tourism, especially business tourism as Scotland is not a global player in its leading services such as the financial service sector and biomedical research.

Yet in terms of the key determinant of inbound travel from outside of the UK (i.e., air travel), it is not clear how travel flows would be influenced in the future which may question some of the assumptions and explanations associated with this scenario. Low-fare airlines are likely to be affected even by moderate price increases, as they work largely on the concept of "free travel" perceptions, but in the case of scheduled flights, the situation may be different. For example, as part of a sensitivity analysis study of the consequences of EU climate policy for a sample of tourism-dependent island states (Gössling et al. 2008) it was evident that there will be a decline in demand the year aviation is integrated in the EU ETS, but where demand for island holidays is growing, the post-2011/2012 scenario is that visitor arrival numbers will continue to increase. Therefore, assuming decline in relation to some markets associated with air travel to 2025 may inevitably change as policy initiatives evolve and shape travel in the future. Nevertheless there are a number of key themes which we can derive from scenario 1.

Key learning points of this Urban Metropolis scenario are:

- The relationship between transport and tourism now is complex, and the government needs to make key decisions to set the framework for future investment and direction for Scottish tourism—they need to invest and not just plan. From a policy perspective, talking down air travel could lead to this scenario becoming a long-term reality given the debates occurring on tourism and sustainability, if environmentalism dominates the agenda.
- The decision to focus on an urban gateway style of tourism experience in terms of continuing the existing allocation of investment and development would lead to the urban concentration scenario for Scottish tourism as the remaining products are not viable due to access and transport problems.
- Much of the focus in the Eddington report (HM Treasury 2006) would simply lead Scotland down this route since it is about air and

road and not rail. The Eddington report is a pro-air travel and air-port development statement and examines the supposed relationship between economic growth and air travel.

- Opportunities exist for new transport-tourism experiences in this scenario, with the development of space tourism, offering Scotland a new luxury transport product that could be developed within the next five years to position Scotland as a key destination.
- Lack of investment in transport infrastructure and inadequate main-tenance and re-investment in the public sector could lead to long-term Health and Safety issues and EU Directives that could make non-urban trips more restrictive if the infrastructure is no longer fit for purpose. Conversely too much investment and making Scotland too accessible could lead to a rise in outbound travel or the rise of mass tourism with a Lake District style experience transposed to the High-lands of Scotland.

Scenario 2: Pragmatism Prevails

In this scenario, the following assumptions were made to model the impact of the storyline on the tourism economy:

- Domestic air travel increases 5 per cent, although inhibitive in price, public preference for journeys to outer isles is still by plane.
- Rest of UK tourism transport demand for travel to Scotland sees increases in car use for personal transport.
- Public transport for the rest of the UK for travel to Scotland shows air travel increase by 10 per cent and represents the most popular form of public transport. Train travel declines as the rail infrastructure decays, with budget public transport fulfilled by increases in demand from bus and coach travel.
- International travel to Scotland is dominated by a demand increase in air and sea travel (ferry travel).
- Overall, Scottish tourism experiences domestic tourism demand increases 11 per cent, rest of UK tourism demand increases 9 per cent, and international tourism demand increases 19 per cent.
- Due to improvements in efficiency driven partly by policy and tax incentives, productivity in tourism sectors increases by 10 per cent.
- Transport sectors show a mix in productivity with a declining railway infrastructure showing a -10 per cent demand, Hybrid engines power cars and buses seeing a 30 per cent increase, and aircraft improving productivity by 15 per cent.
- Due to the underinvestment in railways and decline in the service the volume of trains declines by 10 per cent, cars, buses and trams increase in number by 25 per cent, however car taxation (in terms of road pricing) increases by 2.5 per cent.

- World price of air travel increases by 40 per cent and for UK passengers VAT has been applied at 17.5 per cent.
- Productivity (efficiency) in other industry sectors has shown gains of 10 per cent over 2002 levels.
- The voluntary carbon taxation has seen the forestry industry see productivity gains of 15 per cent, and resultant increase in woodland of 150 per cent. The carbon taxation system has also incorporated a tax break for the forestry industry of 10 per cent as an incentive.
- Oil industry experiences world price increases of 300 per cent.
- There is a 10 per cent increase in the number of cars although an increase in purchase tax of 5 per cent is incurred.
- Gas and Electricity world prices have increased by 25 per cent leading to the effects shown in Table 5.5.

Table 5.5 Macroeconomic Results for *Pragmatism Prevails*

	£ million change	% change
GDP	17941.30	11.47
GVA	8939.80	7.75
Welfare	26413.30	20.91
Employment (FTE jobs)	37497.40	2.05
Government revenue	7479.79	14.64
Daytrips expenditure	213.09	4.15
Domestic Tourism expenditure	-212.65000	-6.85400
Rest of UK Tourism expenditure	162.53200	3.12319
International Tourism expenditure	-670.47000	-24.57700
Domestic plus Rest of UK Tourism expenditure	-50.12100	-0.60338
Overnight Tourism expenditure	-720.60000	-6.53000
Tourism plus Daytrips expenditure	-507.51	-3.14
Summary TSA Results	*£ million change*	*% change*
GVA attributed to tourism (£ million)	-123.85	-6.09
Employment attributed to tourism (FTE)	-6900.46	-6.09
GVA in tourism related sectors (£ million)	496.63	6.81
Employment in tourism related sectors (FTE)	25275.94	7.80

Source: VisitScotland

The main economic changes which Table 5.5 shows are that:

- A more optimistic economic scenario exists for tourism compared to the Urban Metropolis scenario, but still 6.53 per cent less than the baseline 2025 forecast, which has assumed 3.6 per cent.
- Underinvestment in transport infrastructure dampens overall growth from its real potential by £720m.
- Tourism is more overtly affected by government policy than any other industry, highlighting the impact of transport policy on Scottish tourism. In fact, Scottish GDP increases by £1.8bn or a 11.47 per cent increase in GDP. This increase is attributed to productivity gains due to technology innovation.
- International tourism is the most affected market in this scenario due to modelling assumptions of 17.5 per cent VAT and a 40 per cent increase in world prices in real terms. The 25 per cent lost in revenue would have being higher if assumptions haven't being made about a 10 per cent productivity gain in the aviation sector.

The key impacts of *Pragmatism Prevails* would be:

- A tourism product change associated with climate change would create a Costa Caledonia, with great dependence upon a domestic market and an ageing population (i.e., a declining market base). This would also lead to some displacement of tourism from other destinations to Scotland.
- There is less change to tourism in this scenario than the *Urban Metropolis*.
- Scottish tourism continues to grow in this scenario, but environmental change caused by climate change leads Scotland to remain a competitive destination due to its sustainable image.
- The three key features of tourism in this scenario are the result of mitigation, adaptation, and pragmatism so that sustainability remains a key feature of Scottish tourism in areas where the transport infrastructure permits.
- Fiscal constraints on investment mean limited transport investment so tourism will only be able to develop in ways that the infrastructure allows. The case of business tourism and Glasgow with inadequate underground rail travel is a case in point.
- The political environment has not constrained the right to travel as pragmatism has retained a balance with the environment and economic development. As a result a voluntary carbon tax and attitude change via education rather than tax is a key factor shaping tourism and travel propensity.
- The type of people travelling, their behaviour and thinking would indicate a shifting tourism proposition as carbon rationing became

the norm and a new set of products challenge the development of Scotland's brand with space travel, rural, and urban tourism growing. In contrast to the *Urban Metropolis* scenario, rural tourism and rural areas would grow.

- The example of Arran in the scenario offers an important example of best practice, should Scotland wish to trial a car-free island destination.
- Much of the tourism activity associated with the scenario is possible due to technological solutions allowing tourism and transport to continue as business as usual, with greater squeezing of the transport capacity. In the case of technology, energy efficiency and enabling travellers to see the real cost of travel is a cultural shift. Carbon tax becomes the metaphor for currency restrictions on travel, but tourism in this scenario cannot grow to its full potential due to the constraints imposed by inadequate investment in infrastructure. It reflects a current philosophy in political circles that transport infrastructure is largely a matter for the resident population and that specific infrastructure for tourist use is not the focus.

Key learning points of the Pragmatism Scenario are:

- This scenario was seen as realistic mainly because it was an evolution of what was expected as opposed to a radical revolution in the nature of travel and tourism in the timeframe.
- The Scotland tourism brand was unlikely to change radically.
- Technology alone is unlikely to solve the problems of the transport sector and by default, the nature of tourist travel.

KEY POLICY IMPLICATIONS OF THE
SCENARIOS FOR SCOTTISH TOURISM

Transport is the one key factor which will make or break the growth ambitions target for Scottish tourism and has significant implications for sustainability. There is evidence in both scenarios of current debates about tourism and pleasure travel as conspicuous consumption; consequently tourism could begin to bear the brunt of public opposition to the growing green lobby, causing sustainability to become weighted towards restriction rather than growth. This is further highlighted in the *Urban Metropolis* and the draw backs of pure environmental sustainability policy. The *Pragmatism Prevails* scenario highlights the failure to invest in infrastructure and consequences, for example, Glasgow being unable to bid for major business and sporting events due to the lack of a light transit system. The benchmark is moving upwards—if Morocco or Estonia has appropriate structures and Scotland does not, business tourism markets will simply go elsewhere.

Transport needs to be integrated into tourism thinking, development, and planning, and vice versa across the entire public and private sector to realise that it is not a passive element which is taken for granted as someone else's responsibility. At destination level, greater engagements via Area Tourism Partnerships and local authorities to ensure local infrastructures are adequate to achieve the growth ambitions of the TFFC. Tourists are not and never will be the general travelling public; therefore at local authority level tourism development should be an explicit element of road planning, service development, and capacity building.

The urban redevelopment of Scottish cities combined with new sustainable transport solutions (e.g., trams) offers a new tourism product opportunity for Scotland. This means that Scotland needs to engage with innovative ideas about the future, such as a range of environmentally friendly transport solutions as highlighted by Arran as a no CO_2 car destination or Scotland as a space tourism port like Sweden and New Mexico.

CONCLUSIONS: THE GROWTH AGENDA, SUSTAINABILITY, AND TRANSPORT

Transport is a vital element underpinning the ambition for growth for Scotland to 2015 (Scottish Executive 2006) and also to 2025. To lever a 20 per cent increase in visitors by 2015, transport infrastructure will need to be able to accommodate growth. There is certainly evidence from the transport infrastructure in the major cities (i.e., Glasgow and Edinburgh) that it is reaching its optimum peak capacity (though tourist use out of peak times is a positive feature). There is also evidence that urban growth elsewhere has put additional resident pressure on transport infrastructure, squeezing the tourist capacity further. Transport is the one immediate area where incremental improvements in product offering can yield additional spending and value adding through greater integration of supply with demand. Arguably, without a greater focus on transport the entire agenda will be constrained by inadequate capacity to:

1. Increase access to Scotland from UK and Overseas markets via air and road.
2. Assist in geographically dispersing visitors from the Glasgow-Edinburgh-Central Belt triangle to explore further afield.
3. Allow greater access and travel within the destination to spread the local economic benefits of tourism to new areas and assist in local regeneration (e.g., waterfront and waterway areas).

Combined with economic development, transport is the principal driver of tourism development where it is either supply-led to open up access such as the Route Development Fund or at a destination level, by linking visitor and destination as exemplified by the current difficulties in the ferry services exploiting the Highlands and Island's real tourism potential.

No growth agenda, even where it is about adding value rather than sheer numbers, can ignore the evidence that transport is of global as well as national importance and that local activity is changing rapidly. This is illustrated by the debates on global warming, emissions trading, and the growing awareness of the sustainability debates. The relationship between these issues is far from clear and whilst the tourism sector will not want to be constrained by climate policy and developments related to energy and pollution that could impact upon the volume of visitors, it raises an important area for policy analysis: the trade-offs which will be made between tourism, transport, climate policy, and perceived commitments to sustainability at UK government level and nationally within Scotland. These relationships are far from well developed and understood at this stage and they are likely to unfold in the coming years as these relationships begin to impact upon tourism activity.

The main concern for the TFFC is that transport is taken for granted as a publicly provided form of infrastructure and will continue to receive adequate investment. This is far from assured in the current policy environment. While other forms of supply (e.g., accommodation) are vital to fulfilling demand, transport must be viewed as the first priority and not delegated to transport planners who have a history of ignoring or misunderstanding tourism and its needs. The competition is investing in transport infrastructure as a matter of course and so it is a key priority to achieve further growth in tourism.

With transport vital to regional and national economic growth, infrastructure investment should be easier to leverage where the tourist utility can be demonstrated in public spending reviews, even though this is not necessarily commonplace. The return on investment in transport is well established in the private sector and it needs to be used more fully in public sector lobbying for fundamental and complementary transport investment (e.g., for roads and destination access).

To achieve the TFFC ambitions for a vibrant tourism economy, transport needs to be more fully understood in terms of its facilitating, constraining, and generative effects on tourism activity in Scotland and deserves to take its rightful place in policy debates. The chapter also shows that some of the key lessons of using a scenario planning process at VS are that if you are going to carry out scenario planning, do it properly, not as a one-off project but as a process of thinking. Both industry and policy-makers need a mechanism to make sense of the future in a holistic and systematic manner. Scenario planning is that mechanism. Ultimately, you need a scenario planner who will champion the cause and connect to powerful people with a futuristic and visionary outlook. As most stakeholders are talking and acting in the present they also want to prepare for the future. As scenario planning is resource intensive, it requires buy-in from industry and the organisation. This is a constant battle as policy-makers, people, and politicians will always change

Furthermore, scenario planning is a process, not an answer, and so it must use both quantitative and qualitative methods, as means to interpret the future.

NOTES

1. In the limited space available, only a cursory overview is available, but more detail is available from the authors.
2. More detail is available from the authors.
3. There is not space within this chapter to provide the detailed storylines that were used in guiding the development of the two scenarios—these are available from the authors.
4. The value of tourism in 2025 assumes a growth of 3.6% growth per annum, on the baseline figures for 2002 Tourism Satellite Account (TSA).

REFERENCES

Baral, A., Baral, S., and Morgan, N. (2003) 'Marketing Nepal in an uncertain climate: Confronting perceptions of risk and insecurity'. *Journal of Vacation Marketing* 10(2): 186–192.

Bierman, D. (2003) *Restoring Tourism Destinations in Crisis: A Strategic Marketing Approach*. Oxford: CABI Publishing.

Blake, A. and Sinclair, M.T. (2003) 'Tourism crisis management: US responses to September 11'. *Annals of Tourism Research* 30(4): 813–832.

Blake, A., Eugenio-Martin, J.L., Gooroochurn, N., Hay, B., Lennon, J., Sugiyarto, G., Sinclair, M.T., and Yeoman, I. (2004) 'Tourism in Scotland: The Moffat Model for tourism forecasting and policy in complex situations'. *Tourism: State of the Art Conference Proceedings*. Strathclyde University, Glasgow, June.

De Geus, A. (1988) 'Planning as learning'. *Harvard Business Review* 66(2): 70–74.

Eden, C. and Ackermann, F. (1998) *Journey Making*. London: Sage.

European Commission (2005) *Reducing Climate Change Impact of Aviation*. COM (2005) 495 Final.

Glaesser, D. (2003) *Crisis Management in the Tourism Industry*. Oxford: Elsevier.

Gössling, S., Peeters, P., and Scott, D. (2008) 'Consequences of climate policy for international tourist arrivals in developing countries'. *Third World Quarterly* 29(5): 869–897.

Hay, B. and Yeoman, I. (2004) 'Our ambitions for Scottish tourism'. *Tourism: State of the Art Conference Proceedings*. Strathclyde University, Glasgow, June.

Hiejden, K., Van Der Bradfield, R., Burt, G., Cairns, G., and Wright, G. (2002) *The Sixth Sense: Accelerating Organisational Learning with Scenarios*. Chichester: Wiley.

HM Treasury (2006) *The Eddington Transport Study*. HMSO: London.

Hodgson, A. (1992) 'Hexagons for systems thinking'. *European Journal of Operational Research* 59: 220–230.

Jones, M. (1993) *Decision Explorer: Reference Manual Version 3.1*. Kendal: Banxia Software Ltd.

Office for Science and Technology (OfST) (2006) *Transport—Intelligent Futures Project*. London: OfST.

Page, S.J. (2005) *Transport and Tourism: Global Perspectives*, 2nd ed. Harlow: Prenctice Hall.

Page, S.J., Yeoman, I., Munro, C., Connell, J., and Walker, L. (2006) 'A case study of best practice—Visit Scotland's prepared response to an influenza pandemic'. *Tourism Management* 27(3): 361–393.

Prideaux, B., Laws, E., and Faulkner, B. (2003) 'Events in Indonesia: Exploring the limits to formal tourism trends forecasting in complex crisis management'. *Tourism Management* 24: 475–487.

Scottish Executive (2002) *Tourism Satellite Account*. Edinburgh: Scottish Executive.

Scottish Executive (2006) *Tourism Framework for Change*. Edinburgh: Scottish Executive.

Seaton, A.V. and Hay, B. (1998) 'The marketing of Scotland as a tourist destination, 1985–96'. In MacLellan, R. and Smith, R. (eds.) *Tourism in Scotland*. London: Thomson Business Press.

Sparrow, J. (1998) *Knowledge in Organisations*. London: Sage.

World Economic Forum (2007) *Travel and Tourism Competitiveness Report*. WEF, Geneva.

Yeoman, I. (2004) *Developing a Conceptual Map of Soft Operational Research Practice*, PhD Thesis. Napier University, Edinburgh.

Yeoman, I. and Lederer, P. (2004) 'What do you want Scottish tourism to look like in 2015?'. *Quarterly Economic Commentary* 29(2): 31–47.

Yeoman, I. and Lennon, J. (2004) *Terrorism and Disruption: Scenarios for the G8 Summit*. Internal VisitScotland Working Paper.

Yeoman, I., Lennon, J., and McMahon-Beattie, U. (2004) *West Nile Disease: Scenarios or not for Scottish Tourism*. Internal Visit-Scotland Working Paper.

Yong Y.W., Keng, K.A., and Leng, T.L. (1989) 'A Delphi forecast for the Singapore tourism industry: Future scenario and marketing implications'. *European Journal of Marketing* 23(11): 35–46.

6 Tourism and Climate Change Mitigation
Which Data is Needed for What Use?

Jean-Paul Ceron and Ghislain Dubois

INTRODUCTION

The goals of emissions reductions, so as to maintain climate change within manageable boundaries, at national, supranational (e.g., the EU) or international scales have become more stringent over time. For developed countries the necessary reductions greenhouse gas (GHG) emissions to 2050 were assumed to be of 60 per cent a few years ago (Tyndall Centre 2005), now they can be as high as 85 per cent (IPCC 2007). Indeed, the refining of scientific knowledge has lead to the substitution of a 450 ppmv concentration goal for GHG in the atmosphere for the 550 ppmv objective which was hitherto deemed to be compatible with a stabilisation of temperatures at 2°C over preindustrial levels (Bows et al. 2006; Meinhausen 2006; Stern Review 2006). The methodological approaches regarding the ways to reach the goals are also evolving. Until recently the focus was put on the final outcome; e.g., a fourfold reduction of emissions in 2050 (Radanne 2004). However, there is currently a shift from this approach to reasoning in terms of a carbon budget of cumulated emissions until 2050, which is more consistent on scientific grounds. The objective then is just as much the final goal as following a diminishing track of emissions through time.

Though such work on emissions targets ultimately only makes sense at a global level, it calls for research blocks for activities and territories at different scales. The importance of GHG emissions from tourism has now been recognised for some years (Agnew 1995; Ceron 1998; Viner and Agnew 1999; Rechatin and Dubois 2000; Agnew and Viner 2001; Perry 2001; Gössling 2002). More recently a global worldwide assessment has been achieved for the second world conference on tourism and climate change (UNWTO-UNEP-WMO 2008). All these studies show that the contribution of tourism to GHG emissions is both significant (around or over its contribution to GDP) and growing fast, owing both to the development of the activity and to its increasing use of transport (and more particularly of the most polluting mode, i.e., aviation). Until recently, however, at a political level, tourism was not identified as one of the key activities to focus on so as to meet these objectives in terms of reduction of GHG emissions. This is partly due to a lack of data, knowledge, and thus awareness on tourism emissions, all the more that some key points of this issue, e.g., the actual contribution of air transport to radiative forcing (Peeters et al. 2007) are still debated.

What is the data necessary to contribute to a better understanding, and a better involvement of tourism within climate change policies? To answer this, a preliminary step is to assess critically the data used and produced within previous exercises on GHG emissions from tourism, whether they have been used to produce scenarios or not. Following this, the paper presents the main features of a project based on the processing of a French national tourism and transport survey, the SDT survey. This constitutes one of the best samples in the world for national tourism demand monitoring, and thus allows determining what kind of results can be derived from such sources. In particular, relying on individual household data could help understanding travel consumption patterns, which the UNFCCC transport emissions inventory (Peters 2008)—based on fuel sales—does not allow.

ASSESSING TOURISM GHG EMISSIONS: LESSONS LEARNT

As previously noted, past studies have shown the importance of the contribution of tourism to current emissions and also its expected growth in the future. The potential role of research in building a global awareness on the issue of tourism and climate change is obviously important and can be expected to shape the perception of the sector contribution to climate change, and thus future mitigation policies affecting it.

Evolving Objectives

Before recommending some type of surveys or data processing, it is necessary to come back to the users' needs: To what extent can databases help to elaborate climate change mitigation policies? An overview of the literature shows that data needs evolve with objectives.

Understanding Present Emissions

In the field of tourism GHG emissions, the first assessments (EPA 2000; Høyer 2000; Rechatin and Dubois 2000; Becken, Frampton, Simmons 2001; Hoyer and Noess 2001; Becken 2002; Becken and Simmons 2002; Gössling 2002) corresponded to the need to *estimate* the share of tourism transportation, accommodation, and activities, in national or global emissions. The ultimate purpose was to raise awareness on the contribution from tourism. This does not require detailed precision, and can be achieved through proxies, ratios applied for example to passenger km travelled, or through specific surveys. While Rechatin and Dubois (2000) insisted on transport for France, especially by car, deriving from existing transport surveys a share of tourism, Becken (2002) reached a first estimate of the total contribution of tourism (transportation, accommodation, activities), for New Zealand, through a specific survey. At this step, however, the data needs remain limited.

Modelling the Tourism and Transport System

Modelling is a second step, preparing an insight into future emissions, which is required for long term policies such as climate change policies. Models are all the more present in the tourism and transport literature as assessing transport requires a strong component of quantification. Models can be more or less precise, but require in general a segmentation of tourism demand in sub markets. The data is obviously helpful, to provide a starting point and a baseline, but also to confront model results with the reality (on this step of calibration, see Dubois and Ceron 2005; Ceron and Dubois 2006).

Analysing Future Trends and Elaborating Scenarios

The general framework for analysing the contribution of tourism to climate change is set by the major international exercises. The IPCC scenarios (IPCC 2000), the periodical Global Environmental Outlook (UNEP 2002; UNEP 2007) or the Millennium Ecosystems Assessment by UNEP (Millennium Ecosystems Assessment 2005) offer benchmarks to which new exercises refer, so as to be conveniently read and understood, even if some of them just pick a particular item or result (e.g., a level of GHG emissions), and ignore other aspects of the construction which can be contradictory with their own design. For example the frequent reference to the IPCC A2 scenario just because it yields high emissions, but not the highest of the four families and is felt as supporting a concerned but not extremist outlook on the future, illustrates this process which pays little if no attention to the demographic, economic, etc., hypotheses on which A2 is built. In a first step the work on scenarios focused on forecasting.

Forecasting tends to predict difficulties or even doomsday, for example through business as usual scenarios, whereas backcasting aims to show what could be done to avoid such outcomes. Backcasting exercises have developed more recently as a logical follow up to the shortcomings of forecasting. In climate research, forecasting scenarios (IPCC 2000; Pepper et al., 2005; also see Schafer and Victor 1999) in the field of personal transport and backcasting (stabilisation) scenarios coexist (Swart et al., 2002). Work on the latter tends to develop at a fast pace, following the setting of goals at the beginning of this decade (e.g., the European Union's 2°C above preindustrial level [COM 2007]) which are translated (although not without uncertainties and debates [e.g., Meinhausen 2006] into GHG concentrations at the end of the period, a carbon budget for the whole period, and consequently one or several curves of yearly emissions (Bows et al., 2006).

In backcasting scenarios transport is granted importance owing to its contribution to emissions, to the difficulty to decouple it from carbon fuels (especially for aviation) and to its expected growth in volume (both for passengers and freight). In addition to its contribution to national back-casting exercises (Radanne 2004; Tyndall Centre 2005), some research focuses exclusively on transport. Most of these deal with both passengers and freight (WBCSD 2004; Timms, Kelly, Hodgson 2005; Laboratoire d'économie des transports and ENERDATA 2008). In national exercises, domestic transport is often privileged and better dealt with than international transport; this is often due to an institutional bias: Institutions aiming to comply with international obligations have felt less concerned with what is not included in the Kyoto Protocol. For example, international air transport is explicitly excluded or very crudely included (Conseil Général des Ponts et Chaussées 2006) and is not given the full consideration it deserves (Laboratoire d'économie des transports and ENERDATA 2008).

Regarding GHG emissions, the future of tourism is mainly linked to mobility and transport. Tourism and leisure are usually paid attention to in outlooks on mobility, primarily through qualitative discourse (ENER-DATA 2004; Futuribles 2005; Laboratoire d'économie des transports and ENERDATA 2008). However, this is not surprising owing to the lack of quantitative data (Ceron and Becken 2006) and to the difficulty to bridge them and to establish a coherence with data on overall mobility, both at an international and at a national level (Peeters, van Egmond, Visser 2004; Ceron and Dubois 2006).

As a result of these circumstances very few long term scenarios referring to climate change for tourism, including forecasting and backcasting, exist. If the MusTT exercise is excluded (it projects trends to 2020; Peeters et al. 2004), we are left with:

- the scenarios for the Davos conference, including a 'business as usual' scenario to 2035, whose sensitivity to technological progress and to behavioural change is crudely assessed (UNWTO-UNEP-WMO 2008).
- a work on the domestic and international tourism of the French including a BAU (forecasting) scenario to 2050 and a construct of a sustainable image of French tourism in 2050 (Ceron and Dubois 2006). The latter is built by analysing the impact relative to the BAU of a rather large variety of factors (fourteen factors) which subsumes some construction of economic and societal pathways trajectories to 2050 but does not constitute a proper backcasting in the sense that clear intermediate steps are not defined.

The Problem of Data Availability and Estimates

When looking at the detail of sources and calculations, the data produced and used by past studies—despite their utility for decision-making or raising of awareness—appear to have some considerable shortcomings (Ceron and Becken 2006; Gössling and Hall 2006). These assessments have been confronted with:

- the absence of certain types of data: for example the lack of generalised data on domestic tourism worldwide, which is roughly five times (WTO 1997; UNWTO-UNEP-WMO 2008) the volume of international tourism in terms of number of trips at a global level but demonstrates substantial variation at national levels (Hall 2005);
- inadequate statistical categories regarding the issue tackled; i.e., splitting tourism between national and international trips is not relevant when the focus is distance (Ceron and Dubois 2006; Gössling and Hall 2006). For example, a trip from Paris to Brussels is considered as 'international', whereas one from New York to Los Angeles is domestic;
- calling on data coming from different bases when the categories, the collecting, and treatment methodologies do not coincide from one base to another (Peeters et al. 2004; Ceron and Dubois 2006; Dubois and Ceron 2007).

The authors of the various studies have dealt with such difficulties by implementing various solutions:

- asking official institutions to produce the missing data. This has been the case for the study for UNWTO (UNWTO-UNEP-WMO 2008) whose statistical office handed in no time complete statistics of domestic tourism for all countries. One can of course wonder about the reliability of such data but they have an official source and allow the authors to avoid criticism;
- filling the gaps themselves, for example, by:
  - interpolating/extrapolating data from one year to another,
  - analogy: if tourism in country A is rather similar to that of country B they use data from the first corrected according to population or another parameter,
  - iterating when a figure from an expert source does not seem to fit, it is modified and the model it feeds is run as many times as necessary to be calibrated with observed data (Ceron and Dubois 2006).

After such processes have been conducted it is particularly appreciated to see the authors describe their effects on the reliability of their results as it is done in the UNWTO study (UNWTO-UNEP-WMO 2008).

Table 6.1   Emissions of Global Tourism Transport, Accommodation and Activities and their Uncertainty

	CO_2		Contribution to RF (W/m^2)	
	Mt	*Share in tourism (%)*	*Excluding cirrus*	*Including maximum cirrus impact*
Air transport	515	39	0.0398	0.0982
Car	396	30	0.0168	0.0199
Other transport	74	6	0.0031	0.0199
Accommodation	274	21	0.0116	0.0116
Other activities	48	4	0.0019	0.0019
Total Tourism	1,307	100	0.0734	0.1318
Total world	26,400	-	1.6	1.7
Share of tourism in total world (%)	5.0	-	4.6	7.9

Estimates include international and domestic tourist trips, as well as same-day visitors (base year 2005). Colors represent the degree of certainty with respect to the data and underlying assumptions. Light grey represents a degree of uncertainty of +/-10 per cent, medium grey +/-25 per cent and dark grey +100 per cent/-50 per cent.

The Need to Go One Step Beyond

Such a critical outlook as highlighted earlier could foster scepticism regarding the results and their usefulness for strategic planning. It should probably be noted that the situation is not much better in many other fields. For example, the uncertainties on radiative forcing figures for aviation do not only concern tourism transport but also airborne freight and military aviation. Similarly, it also applies to GHG emissions from residential buildings.

Previous studies give a fairly general outlook and do not deliver a detailed assessment of the contribution of various forms of tourism to emissions. They outline the impacts of the various transport modes, of some trends such as the shortening of stays and the increase in distances, but they do not go as far as assessing in detail various patterns of tourism which differ widely regarding emissions. Who generates what? What are the psycho-sociological characteristics of those people who are hypermobile? Indeed, in order to elaborate adequate GHG reduction strategies, tourism needs to be 'unpacked': researchers must reach a more thorough understanding of production and consumption patterns behind emissions profiles, identifying types of travellers, and causal links (Dubois and Ceron 2006; Gössling and Hall 2006).

This chapter describes a research project that aims at improving the knowledge base on which mitigation strategies can be based. It focuses on the emissions of transport from the origin to the destination which has been demonstrated to generate at least three quarters of the emissions and quite probably more if only net impacts of accommodation (during a tourist trip the energy consumption of the home diminishes) are taken into account (Gössling et al. 2005; UNWTO-UNEP-WMO 2008).

METHODS

Description of the Surveys

Our study is based on the panel 'Suivi des déplacements touristiques' (SDT), a tourism travel survey concerning the domestic and international tourism of residents in France, i.e., the French and Foreign residents in metropolitan France, thus excluding the tourism of Foreign visitors to France, and that of the French living in Foreign countries or in French territories overseas. This survey is derived from the 'Metascope' panel of SOFRES, a major poll institute in France. Metascope includes 30,000 households and 53,000 individuals. Around 25 per cent of panellists change each year so as to reduce professionalisation and weariness effects among the respondents. The rate of answer to questionnaires is in the region of 75 per cent. Representativeness is ensured through a stratification employing five criteria: the place of residence, the age of the family leader, his or her socio-professional category, the size of the household, and the size of the township. Biases of the panel such as the under representation of the less than 30 year olds or the over representation of people over sixty-five and of single persons are corrected. Among the 30,000 households, 20,000 are sampled and one individual per household is questioned each month. Table 6.2 indicates some of the components of the survey.

Around the end of the century, the results of the SDT showed a decrease in the number of trips which contrasted with the increase of mobility assessed by transport surveys and concerning all modes of transport. These results were not necessarily incompatible since the boundaries of the statistics differed and thus the figures could not be strictly compared. The diminishing number of tourists was not contradictory with the increase of the number of trips since the latter resulted from an increase of day trips which are not accounted for by the SDT. This is why a survey of day trips was initiated in 2004. This survey deals with 8,000 individuals drawn from the SDT with the same criteria for representativity and periodicity, for their trips over 100 kms from home (Table 6.3).

Table 6.2   Components of the Suivi des Déplacements Touristiques (SDT) Survey

Description of individuals	
Demographics	Age, gender, size of household, Number of children, family status (e.g., married . . .)
Geographical	Region of residence, size of township
Sociological and economic	Profession, educational level, income, housing conditions (type of home, number of rooms), number of cars, second home

Description of trips and stays	
Trip characteristics	Number of trips which ended within the last month, and for three of them; departure and return date, main means of transport used, motives (19 choices)
Stay characteristics	Number of stays for each trip and for the two longer stays: length of stay, main reason, place (commune or departement for France, country if abroad), main accommodation used, if relevant, share of professional and non professional bed nights within the stay, number of persons accompanying, type of accommodation (17 choices), type of geographical space (sea-shore, mountain, countryside, urban), main activities (19 choices)

Limitations

The two surveys, in spite of their size, their monthly periodicity and of the period of time they cover (part of the data are followed since 1990) suffer from a certain number of limitations.

An Under Estimation of the Number of Tourists and Bed-Nights

Apart from the fact that not all trips (only three) and stays (only two) per month are described, it is suspected that the proportion of persons that travel a lot (a characteristic that is not and cannot be directly checked) is lower in the panel than among the French population. Busy people might be reluctant to fill in a large questionnaire every month and if they do, might skip some trips to gain time. This is all the more problematic as these people are those who travel the most. Additionally, the surveys only include the trips of individuals

Table 6.3   Components of the 'Enquête des Déplacements à la Journée'

Description of same-day trips
Date
Motive
Destination
Main means of transport
Number of persons travelling with . . .

under fifteen when they travel with their parents, thus disregarding the autonomous travel of children. Also, before 2004, foreign residents were excluded from the panel which implies a slight discontinuity in volume.

Reliability and Size of the Samples

Dealing with subcategories inside the panel—which is indispensable to 'unpack' tourism—implies diminishing the reliability of the results. This means that analyses on an annual basis are more reliable than on a monthly basis; yet regarding the number of trips, of stays and bed-nights, monthly data are considered as reliable. This type of difficulty also arises on a geographical scale when dealing with minor destinations. This concerns destinations abroad (the panel does not reveal a significant number of people going to New Zealand for example), which can be dealt with by aggregating destinations. It is also the case for some regions of France. For example, the SDT captures better the volume of travel to the Riviera and its characteristics, than to less appealing regions.

Time Related Limitations

The survey on same day trips only started in 2004 as did the inclusion of foreign residents in the SDT. Even though the SDT started in 1990, there have been some important methodological changes in 1999 which would render comparisons with data prior to that date very complicated even if the ancient data were accessible . . . which is not the case. Even within a shorter period, the possibility to track the changes in the behaviour of individuals is limited by the renewal of the sample; e.g., some people change residence or give up filling the questionnaire. To these factors must be added the fact that the SDT is not a 'true' panel as within the household the individual answering might change over time.

Adding a GHG Calculation Module

Associating Distances to Trips

Since 1999, SOFRES associates a distance matrix (from commune to commune, and there are 36,000 communes in France) to the trips (both direct distance and road distance). Yet this leaves aside distances to foreign countries and to French territories overseas. For neighbouring countries, the data have been completed using as a starting point the regional capital cities. For distant countries and French overseas territories, Paris has been considered as the starting point. For neighbouring destinations the capital city of the country has been retained insofar as it seems reasonable. For very long distance destinations the web site used considers the barycentre of the country. In addition, a certain number of trips for which the destination cannot be straightforwardly identified in the answer to the questionnaire as it is a commune (e.g., a ski resort), although the small number of cases has allowed to deal with each one of them. Finally, another problem arises with trips that include multiple stays; here the data base has been corrected to avoid counting them as two trips to separate destinations which would have over evaluated emissions.

Emissions Coefficients

The almost unique source for French emissions coefficients is Ademe, a French governmental source. The coefficients concern emissions from well to wheel, i.e., including indirect emissions but excluding a life cycle approach. For air transport emissions Ademe uses an uplift factor of twice the amount of CO_2 emissions. A few coefficients reflect specific French situations (a highly electrified rail network fed by near to carbon neutral electricity). Some of the coefficients have been modified to take into account specificities linked to tourism travel such as the size of cars (larger than the average) or the occupancy coefficients. This leads to the figures presented in Table 6.4.

Table 6.4 Emission Factors

Transport modes	Emission per passenger.km (g CO_2-e)	RFI coefficient (non carbon effects)	Detour coefficient	Final emissions factors (g CO_2-e)
Car	223.7	1.05	223.7	
Coach	38.5	1.05	38.5	
Minibus	74.6	1.05	74.6	
Motorbike	113.7	1.05	113.7	
Train	9.5	1.05	9.5	
TGV 1st class	2.9	1.05	2.9	
TGV 2nd class	2.2	1.05	2.2	
Regular train	12.8	1.05	12.8	
Regional train	37.4	1.05	37.4	
Plane	210,8	2	1.05	442
Plane short haul– economy class	146.7	2	1.05	308
Plane short haul– business class	330	2	1.05	693
Plane long haul– economy class	110	2	1.05	231
Plane long haul– business class	256.7	2	1.05	539
Boat	60	1.05	63	

FIRST RESULTS

Some of the results for 2006 are summarised below, so as to exemplify the questions that can be answered through this type of process.

An Assessment of National Tourism Transport GHG Emissions

The assessment provides a reliable picture of the contribution of leisure and 'visiting friends and relatives' tourism to national emissions. Domestic tourism transport (excluding professional trips) accounts for 3 per cent of national GHG emissions in 2006, and 6 per cent when international tourism is added. Further adding professional trips leads to 8 per cent. Conversely, it does not provide business tourism figures that allow comparison of its contribution to that above, owing to the underestimation of business trips in the answers to the questionnaire.

A Better Understanding of the Drivers of GHG Emissions

Owing to the dimension of the database and to the highly detailed questionnaire, a very precise light is shed on the influence of transport modes and of distance on emissions. In 2006, 7 per cent of tourist trips have been based on the plane while 75 per cent used the car; the former are responsible for 62 per cent of the impact on climate change related to origin/destination transport whereas the latter emit in the region of 10 million tonnes of CO_2-e, i.e., 36 per cent of emissions.

A tourist travelling by plane in business class or first class (706 000 trips in 2006, 0.5 per cent of the total of trips for personnal motives) emits 3.7t CO_2-e for his transport, whereas for the same distance a tourist in economy class emits three times less. A person using the high speed train (TGV) will emit on average for his trip 3kg CO_2-e. This can also be considered in terms of distance per kg of CO_2-e emitted: in 2006 a tourist travelled 400km if taking the TGV and 1km in business class by plane (see Figure 6.1).

Air transport is not only linked to long distance but also to a gain in time (Schafer and Victor 1999; Hall 2005). The distance over which the plane tends to substitute for other modes of transport is rather short: over 2000km for a return trip (e.g., from Lille to Marseille) air transport almost has no competitor. For trips shorter than 2000km the car remains the most used means of transport: about 85 per cent of stays up to 500km, 75 per cent between 500 and 1000km and still 50 per cent from 1500 to 2000km. The share of rail increases up to 2000km, without ever exceeding that of cars. Over 1000km the train is used for approximately one quarter of stays.

The contribution of long haul tourism to GHG emissions appears to be considerable. Trips within metropolitan France represent 36 per cent

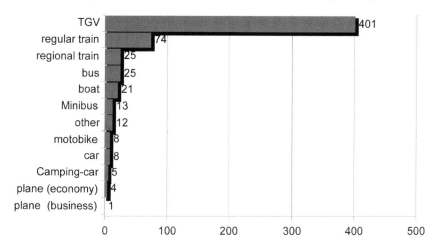

Figure 6.1 Distance travelled per stay for 1 kg of CO_2-e in 2006.

of the emissions of French residents, and trips to Europe and the south Mediterranean 21 per cent. Trips to the rest of the world (2 per cent of the total and 5 per cent of bed-nights) reach 43 per cent of the emissions. Among the destinations that welcome more than 50,000 tourists from France those that are associated with the highest emissions per stay are: Mexico (more than 4t CO_2-e), la Reunion, China, the United States (around 3.5t), the Dominican Republic, Brazil and Martinique (between 3 and 3.5 t).

Furthermore, the processing of the database allows to establish a link between types of activities (see Figure 6.2) or accommodation and the emissions from transport that are associated to them. The tourists that stay in the most comfortable accommodation (three stars and more) are those whose emit the most: 36 per cent of the total emissions. This is a consequence of a high share of trips abroad in that category. Conversely family tourism is associated with low emissions from transport: Tourists that visit friends and relatives or stay in a second home emit less than 100kg CO_2-e per stay (the latter travel though frequently). Marine tourism (surfing, sailing, diving) is generally associated with high emissions (owing to the share of long distance destinations) as is the visit to natural sites (three times the average). In contrast winter sports activities are generally associated with low emissions: as the French stay in France and frequently use the high speed train.

Finally the database provides for the mapping of emissions according to the different regions tourists originate from (see Figure 6.3). The emissions are high in the north, the east, and the Paris basin, which can be related to various factors such as climate, urban conditions, the proximity of a major airport, or high income (which probably also explains the high emissions for the South East).

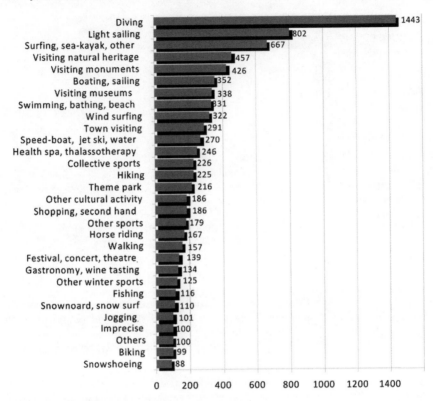

Figure 6.2 Ranking of activities according to the average GHG emission per trip of associated transport.

Social Stratification of Emissions

Probably the main output of the research was that it provided for the assessment of the contributions to emissions of different categories of individuals. Figure 6.4 indicates emissions by social strata and, from this data, a typology of travellers has been developed (Figure 6.5). For example, a small number of tourists are responsible for a large share of emissions: 5 per cent of tourists generate 50 per cent of emissions and 10 per cent generate almost two-thirds of the impact. In other terms slightly more than three million French residents emit 15Mt CO_2-e, which is as much as the remaining 60 million inhabitants (including those who do not travel).

DISCUSSION AND CONCLUSION

The processing of a database such as the SDT can yield a huge amount of information. It should not be forgotten that though this information deals with an essential aspect of emissions from tourism, it does not take into

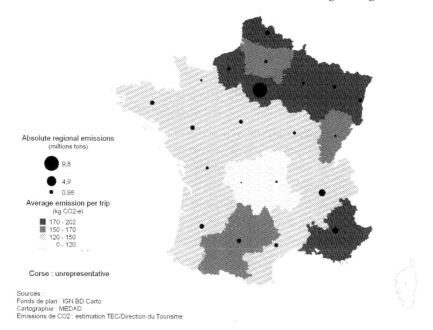

Figure 6.3   GHG emissions of tourism transport and average emissions per trip, according to the tourist's region of main residence.

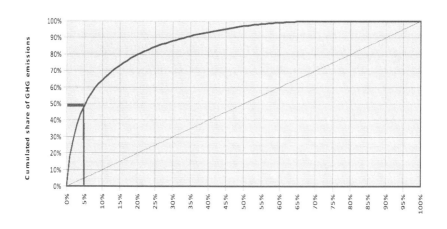

Figure 6.4   Cumulated share of tourism GHG emissions.

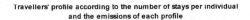

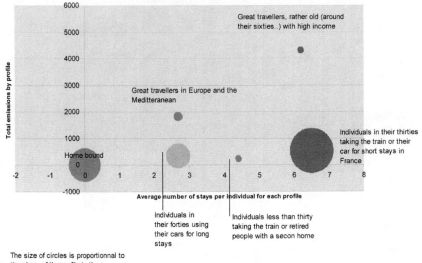

Figure 6.5 Typology of French travellers, regarding their emission profiles and behaviour.

account all the dimensions. For example, it excludes the emissions from accommodation and activities and also those from visitors to France.

Further work in these directions calls for other research projects. Studies of the emissions of international tourism to France (77 millions arrivals) could use the national database survey on arrivals (EVE). The EVE survey is recent and for the moment gives results for a single year. This survey follows the old 'Enquête aux Frontières' which yielded reliable results for the last time in 1996; figures since then had been estimated in a more or less disputable way. EVE, however, is far from being as detailed as the SDT. With respect to accommodation and activities we currently have no precise idea on the methods that could be used to approach their emissions for France. Integrating these three main dimensions (transport emissions of the French, transport emissions from international inbound tourism, emissions from accommodation and activities) into a comprehensive and reliable picture will be a serious challenge. Additionally, any attempt to draw an evolution of the picture through time (e.g., for the last ten years) seems out of reach.

Nevertheless, this research sheds some light on the current situation, of the climatic implications of the tourism of the French, on some driving forces such as potential effect of an ageing population and extensively on the trends when the analysis of longitudinal data is completed. It should be stressed that producing figures is not a goal in itself and the extent to which they are both useful and reliable needs to be assessed.

Regarding reliability, the limitations linked to the size of the samples of subcategories have already been pointed out. The reliability of the conclusions that can be drawn is particularly significant for trend analyses or comparisons between subcategories. This means that working with a general database on tourism, even if it is rather a large one, has real limits when dealing with specific issues, and often cannot substitute for ad hoc studies. If, for example, the emissions of the French heading for eco-tourism in the Caribbean or for diving in the southern hemisphere are to be assessed, specific studies are clearly needed, not so much because the sample would be larger (though it might be) but because the questionnaire would need to go much further into relevant details than a general study can do. Nevertheless, hopefully, the size of the samples should be sufficient to assess some specific issues such as, for example:

- to characterise the travel habits and the personal features of a some particularly interesting groups of people such as those who have no car, those who have a car but do not use it much, those who almost only use the train, those who have taken the plane in 2006, the owners of a second home (average number of trips to the home, average distance between the homes, average length of stays, transport modes) etc.
- the impact of the recent TGV link on the modal distribution of tourist flows from Paris to the Riviera, etc.

Regarding usefulness, such detailed results might help policy-makers elaborating more precise and less generic climate policies, for example with subtle trade-offs between long haul and short haul mobility, or with the introduction of social equity. For scenario development, the increasing amount of modelling to inform climate policies increases the demand for data, whose supply does not seem to follow (on transport surveys see Stopher and Greaves 2007). It is hard to figure out to what point this work will help, but there cannot be a total divide between data producers on the one hand and modellers on the other, the latter complaining of the poor quality or quantity of data and yet often using them without the necessary caveats, since this might shed some discredit on their findings (see also Gössling and Hall 2006). It is our opinion that much progress in the field will call for some kind of modelling (not necessarily sophisticated) and some building of adequate data from primary sources.

REFERENCES

Agnew, M. and Viner, D. (2001) 'Potential impact of climate change on international tourism'. *Tourism and Hospitality Research* 1(3): 37–60.

Agnew, M.D. (1995) 'Tourism'. In Palutikof, J., Subak, S., and Agnew, M.D. (eds.) *Economic Impacts of the Hot Summer and Unusually Warm Year of 1995*, Norwich, 139–147. United Kingdom: Department of the Environment.

Becken, S. (2002) 'Analysing international tourist flows to estimate energy use associated with air travel'. *Journal of Sustainable Tourism* 10(2): 114–131.

Becken, S., Frampton, C., and Simmons, D. (2001) 'Energy consumption patterns in the accommodation sector. The New Zealand case'. *Ecological Economics* 39: 371–386.

Becken, S. and Simmons, D.G. (2002) 'Understanding energy consumption patterns of tourist attractions and activities in New Zealand'. *Tourism Management* 23: 343–354.

Bows, A., Mander, S., Starkley, R., Bleda, M., and Anderson, K. (2006) *Living with A Carbon Budget*. Manchester, Tyndall Centre.

Ceron, J. and Becken, J. (2006) 'Tourism statistics and databases for climate change research'. E-CLAT *Climate change and tourism conference: Tourism and Climate Change Mitigation*. Tilburg, NHTV: University of Breda.

Ceron, J.P. (1998) 'Tourisme et changement climatique'. *Impacts Potentiels du Changement Climatique en France au Xxième siècle*, 104–111. Paris, Premier ministre: Ministère de l'aménagement du territoire et de l'environnement.

Ceron, J.P. and Dubois, G. (2006) *Demain le Voyage. La Mobilité de Tourisme et de Loisirs des Français Face au Développement Durable. Scénarios à 2050*. Paris: Ministère des transports, de l'équipement, du tourisme et de la mer.

COM (2007) *Limiting Global Change to 2°Celsius: The Way Ahead for 2020 and Beyond*. Final edition. Commission of the European Communities.

Conseil Général des Ponts et Chaussées (2006) *Démarche Prospective Transports 2050*. Paris: Ministère des Transports de l'Equipement, du tourisme et de la Mer.

Dubois, G. and Ceron, J.P. (2005) 'Tourism/leisure greenhouse gas emissions forecasts for 2050: Factors for change in France'. *Journal of Sustainable Tourism* 14(2): 172–191.

Dubois, G. and Ceron, J.P. (2006) 'Tourism and climate change: proposals for a research agenda'. *Journal of Sustainable Tourism* 14(4): 399–415.

Dubois, G. and Ceron, J.P. (2007) 'How heavy will the burden be? Using scenario analysis to assess future tourism greenhouse gas emissions'. In Peeters, P. (ed.) *Tourism and Climate Change Mitigation. Methods, Greenhouse Gas Reductions and Policies* 6, 189–207. Breda, The Netherlands, NHTV.

ENERDATA (2004) *Un Scenario de Transports Ecologiquement Viables en France en 2030*. Grenoble.

EPA (2000) *A Method for Quantifying Environmental Indicators of Selected Leisure Activities in the United States*. Washington: US Environmental protection agency.

Futuribles (2005) *Rapport d'Etude Prospective Pour l'Elaboration de Scénarios Exploratoires sur les Transports en 2050*. Paris: Futuribles.

Gössling, S. (2002) 'Global environmental consequences of tourism'. *Global Environmental Change* 4(12): 283–302.

Gössling, S. and Hall, C.M. (eds.) (2006) *Tourism and Global Environmental Change*. London: Routledge.

Gössling, S., Peeters, P., Ceron, J.P., Dubois, G., Patterson, T., and Richardson, R. (2005) 'The Eco-efficiency of Tourism'. *Ecological Economics* 54: 417–434.

Høyer, K. and Næss, P. (2001) 'Conference tourism: a problem for the environment, as well as for research?'. *Journal of Sustainable Tourism* 9(6): 451–470.

Høyer, K.G. (2000) 'Sustainable tourism or sustainable mobility'. *Journal of Sustainable Tourism* 8(2): 147–160.

IPCC (2000) *Special Report. Emissions scenarios*, Summary for policy makers. Geneva, Switzerland: IPCC.

IPCC (2007) 'Summary for policymakers'. In Solomon, S., Qin, D., Manning, M., Chen, Z., Marquis, M., Averyt, K.M., Tignor, M., and Miller, H.L. (eds.) *Climate Change 2007: The Physical Science Basis, Contribution of Working Group I to the Fourth Assessment Report of the Intergovernmental Panel on Climate Change*, 1–18. Cambridge, UK and New York, NY, US: Cambridge University Press.

Laboratoire d'économie des transports and ENERDATA (2008) *Comment Satisfaire les Objectifs de la France en Terme d'Emissions de Gaz à Effet de Serre et de Pollution Transfrontières?* Paris, PREDIT: Ministère chargé des transports.

Meinhausen, M. (2006) 'What does a 2°C target mean for greenhouse gas concentrations. A brief analysis based on multi gas emission pathways and several climate sensitivity uncertainty estimates'. In Schellnhuber, H.J., Cramer, W., Nakicenovic, N., Wigley, T., and Yohe, G. (eds.) *Avoiding Dangerous Climate Change*, 253–279. Cambridge, UK: Cambridge University Press.

Millenium Ecosystems Assessment (2005) *Ecosystems and Human Well-Being: Synthesis*. Washington, D.C.: Island Press.

Peeters, P., van Egmond, T., and Visser, N. (2004) *European Tourism, Transport and Environment*. Final version. Breda, NHTV CSTT.

Peeters, P., Williams, V., and Gössling, S. (2007) 'Air transport greenhouse gas emissions'. In Peeters, P. (ed.) *Tourism and Climate Change Mitigation. Methods, Greenhouse Gas Reductions and Policies*, 29–50. Breda, NHTV.

Pepper, W.J., Xing, X., Chen, R.S., and Moss, R.H. (2005) *Scenarios 92, A to F*. IPCC: Columbia University.

Perry, A. (2001) *More Heat and Drought—Can Mediterranean Tourism Survive and Prosper*. Proceedings of the 1st International Workshop on Climate, Tourism and Recreation. International Society of Biometeorology, Commission on Climate Tourism and Recreation.

Peters, G. (2008) 'From production-based to consumption-based national emission inventories'. *Ecological Economics* 65(1): 13–23.

Radanne, P. (2004) *La Division par 4 des Emissions de Carbone en France d'Ici 2050*. Paris: Mission interministérielle de l'effet de serre.

Rechatin, C. and Dubois, G. (2000) *Tourisme Environnement Territoires. Les Indicateurs*. Orléans: IFEN.

Schafer, A. and Victor, D.G. (1999) 'Global passenger travel: implications for carbone dioxyde emissions'. *Energy* 24: 657–679.

Stern Review (2006) *The Economics of Climate Change*. Cambridge, UK: Her Majesty's Treasury, Cambridge University Press.

Stopher, P.R. and Greaves, S.P. (2007) 'Household travel surveys: where are we going?' *Transportation Research Part A* 41: 367–381.

Swart, R.J., Mitchell, J., Morabito, T., and Raper, S. (2002) 'Stabilisation scenarios for climate impact assessment'. *Global Environmental Change* 12(3): 155–165.

Timms, P., Kelly, C., and Hodgson, F. (2005) *World Transport Scenarios Project*. Norwich, UK: Tyndall Centre.

Tyndall Centre (2005) *Decarbonising the UK. Energy for a Climate Conscious Future*. Available at: http://www.tyndall.ac.uk/media/news/tyndall_decarbonising_the_uk.pdf (12 November 2008).

UNEP (2002) *Outlook 2002–2032*: 321–400.

UNEP (2007) *Global Environment Outlook: GEO4*. Valetta, Malta: UNEP.

UNWTO-UNEP-WMO (*United Nations World Tourism Organization, United Nations Environment Programme, World Meteorological Organization*) (2008a) 'Climate change and tourism: Responding to global challenges'. [Scott, D., Amelung, B., Becken, S., Ceron, J.-P., Dubois, G., Gössling, S., Peeters, P., Simpson, M.] United Nations World Tourism Organization (UNWTO), *United Nations Environmental Programme (UNDP) and World Meteorological Organization (WMO). UNWTO, Madrid, Spain*.

UNWTO-UNEP-WMO (2008b) *Climate Change and Tourism: Impacts, Adaptation and Mitigation*. Final draft before publication. Madrid and Paris, World Tourism Organization and UNEP.

Viner, D. and Agnew, M. (1999) *Climate Change and Its Impact on Tourism*, report prepared for WWF-UK. Norwich, United Kingdom: University of East Anglia, Climatic Research Unit.

WBCSD (2004) *Mobility 2030: Meeting the Challenge to Sustainability*. Conches, Genève: World Business Council for Sustainable Development.

WTO (1997) *International Tourism. A Global Perspective*. Madrid: WTO.

7 Theory and Practice of Environmental Management and Monitoring in Hotel Chains

Paulina Bohdanowicz

INTRODUCTION TO ENVIRONMENTAL MONITORING, PERFORMANCE INDICATORS, AND BENCHMARKING

Hotels constitute one of the main, and still expanding, pillars of the tourism sector. Resource intensive and frequently inefficient systems and operational routines result in a considerable environmental impact. It is estimated that hotel facilities worldwide consume about 100 TWh of energy (Gössling 2002) and 450–700 million m³ of water per annum (estimates based on figures in Davies and Cahill 2000) and generate millions of tons of waste (IHEI 2002). A certain level of cost-driven activities to reduce resource use can be seen in the sector. The currently observed price increases of basic commodities, such as energy resources and water, further encourage the implementation of resource use efficiency and conservation measures in hotel facilities (Stipanuk 2001). In addition, hoteliers now realise that they can market their environment-driven attitudes and initiatives to a growing cadre of green consumers, as increasingly noted in professional publications (Anavo and STI 2005; Sustainable Travel Report 2005; Hauvner, Hill and Millburn 2008; Putting out the green message 2008; Scoviak 2008). The so-called ecological lifestyles market (LOHAS—Lifestyles of Health and Sustainability) in the United States alone is estimated at over 80 billion USD, with over 60 million consumers willing to use services and products of environmentally responsible companies (Anavo and STI 2005). Furthermore, according to a Green Hotel Association survey almost 43 million American travelers are interested in the environmental issues in the hospitality sector, while 80 per cent of those interviewed by Travelocity claim to be willing to pay extra to visit an eco-friendly destination or business (Hauvner et al. 2008).

In consequence, many of the hotel companies now take on environmental responsibilities, develop relevant policies and programmes, and implement environmental monitoring and benchmarking of facility performance. Hotel-oriented NGOs, such as International Tourism Partnership, Green Hotel Association and others, frequently assist hoteliers by creating framework programmes and guidelines. Nowadays, examples exist worldwide showing that it is viable to incorporate environmental practices into contemporary business models and still provide high quality hospitality services. Scandic can be mentioned as one of the companies capable of putting a price tag on sustainability. Resource efficiency and conservation implemented over the past twelve years saved over 20 million dollars in energy, water, and waste bills, while the brand became well established in the Nordic market and associated with responsible performance and management. Two years into the environmental programme allowed Hilton to avoid spending over 9 million dollars on energy (Breaking Travel News 2008), while Marriott claims to have reduced its carbon dioxide emissions by 70,000 tonnes in 2006 (Marriott 2007). Fairmont Hotels & Resorts and Xanterra Parks and Resorts are also among the corporations with strong environmental inclinations. While most of the hotel companies have their internal construction/refurbishment standards increasingly following the directives of sustainability, Starwood is an example of an operator launching a 'green' brand. The luxury hotels '1' will be built according to the Leadership in Energy and Environmental Design Green Building Rating System (USGBC LEED), providing a high level of comfort and luxury at reduced resource consumption. However, despite these large scale initiatives, still most of the leading and state-of-the-art environmental practices are seen among individual hotels (i.e., Sånga Säby Course and Conference Centre in Sweden, Couran Cove Island Resort in Australia, The Orchid in India, and many other hotels). These independent establishments are very vivid examples that being "green" is actually good for business.

To achieve environmental management and reporting success, reliable performance assessment tools are needed and relevant quantitative and qualitative information from individual facilities must be collected, often on a monthly basis. This includes energy, water and chemicals consumption, waste generation, turnover, number of customers, outdoor and indoor temperature/humidity conditions, investment within local communities, and so on. Data also needs to be collected on attitudes, pro-environmental measures and social initiatives implemented on-site, which can be obtained from questionnaires/interviews with staff and managers, and recommendations made by various stakeholders. The most commonly used environmental indicators include energy and water use per guest-night, as well as waste generation and carbon dioxide emissions per guest-night (Despretz 2001; Stipanuk 2003a, 2003b; WTO 2004). In addition many economic and social indicators can also be collected, such as information on demand seasonality, investments in local economy, employment, poverty alleviation, competitiveness, leakages, as well as well-being of host communities, gender equity among the workforce, health and safety, and conserving cultural and natural heritage (Leslie 2001; GRI 2002; Kozak 2004; WTO 2004).

Although there is currently a relatively extensive literature available on performance indicators and benchmarks in the international hotel industry (reviewed in Bohdanowicz 2003; Green Globe 21 2004; CI and IBLF 2005;

Bohdanowicz and Martinac 2007; Nordic Ecolabelling 2007 v.3.0), the accuracy and validity of the published figures is widely debated due to large variations in the data reported (de Burgos-Jiménez, Cano-Guillén, and Céspedes-Lorente, 2002; Becken and Cavanagh 2003; Warnken-Bradley, and Guilding 2005). Among the reasons for such a situation, the most commonly mentioned are the differences in methodology used to collect the data as well as characteristics of the establishments used as a reference (weather conditions, facilities, type of customers served, occupancy, building size). Although some of the benchmarks suggest the corrective factors to be used to account for these, the overall reliability of such tools is questionable. In addition, Warnken et al. (2005) argued that for global benchmarks to be reliable, too many hotel sub-categories would be required, or extensive databases would need to be created. Most of the reported indicators (for example in environmental reports, and academic studies), do not provide a perspective on indicators as related to the total use of resources (input), the company's activity (output), or goals and objectives of the management (what should be the level of consumption according to the management plan). Many guidelines are being published on how to collect the information and construct indicators (GRI 2002; Kozak 2004; WTO 2004), but the need for good and reliable metrics and comparison schemes in pursuing sustainability still remains unresolved.

EXISTING TOOLS FOR ENVIRONMENTAL MONITORING AND BENCHMARKING IN THE HOSPITALITY SECTOR AND THEIR MARKET PENETRATION

Recently, a number of environmental reporting/benchmarking tools for hotels have been developed by international organisations, branch associations and even hotel corporations (APEC 1999; Green Hotelier 16 1999; Despretz 2001; Bohdanowicz, Simanic and Martinac 2005a; WTO 2004; Matson and Piette 2005; Bohdanowicz 2007). Some of these tools rely on conventional methods of data collection and result communication (Green Globe 21, Hong Kong Hotel Building Assessment Scheme, Nordic Swan eco-label initiated by Nordic Council of Ministers and coordinated by Nordic Ecolabelling board—a web-based application is currently being tested and will be publicly available in 2008; Fahlin 2007), while others are web-based interactive tools, such as International Tourism Partnership BenchmarkHotel (www.benchmarkhotel.com, formerly known as International Hotel Environmental Initiative BenchmarkHotel, updated in 2007, renamed into Environmental Bench and to be launched in mid-2008), Hospitable Climates Hospitality Energy Analysis Tool—HEAT Online (www.hospitableclimates.org.uk), FHRAI Energy and Environment E^2 Benchmark for Hotels (www.fhrai.com/BenchMark/), Canadian Green Globes (www.greenglobes.com), as well as US EPA/DOE EnergyStar Portfolio Manager (www.energystar.gov/index.cfm?c=evaluate_performance.bus_portfolio-manager). Among the major hotel corporations Accor International, Fairmont Hotels & Resorts, InterContinental Hotels Group, Hilton Hotels Corporation (Hilton International section), Hyatt International, Rezidor SAS, Shangri-La Hotels & Resorts, Xanterra Parks & Resorts have developed their own reporting schemes, and include information about those on the company websites, and in internal and/or environmental reports (Green Hotelier 1999, 2000). Some reference to the existence of such tools can also be found in best practice compilations, journal and conference papers (Wilson 1999; WTO 2004; Bohdanowicz et al. 2005a, 2005b; Bohdanowicz 2006, 2007).

All these systems aim to assist the hotel manager in evaluating the environmental performance of the facility and frequently offer an indication of possible improvements. Most of these tools are for-profit instruments or internal applications and, as a result, there is limited information publicly available about them. The EnergyStar Portfolio Manager is the exception that provides a document explaining the methodology used. Generally the reliability of such instruments is questionable and there is no guidance available on how to develop and successfully incorporate similar systems elsewhere. All tools require the input of information pertinent to the hotel facility, and also key figures on resource consumption and occupancy. In return, these systems offer a tabular or graphical presentation of the facility performance in time, compared to industry benchmarks, and, in the case of corporations, compared to other establishments within the portfolio.

The comparison of some of the schemes is presented in Table 7.1, while the history of the tool developed by Addsystems for Hilton (Hilton Environmental Reporting—HER) and Scandic (Scandic Sustainability Indicator Reporting—ScandicSIR) is presented in Box 1.

Each of the tools has its unique way of presenting the results, although the final Key Performance Indicators (KPIs) are typically displayed in a normalised form, e.g., consumption per guest-night or per square meter. It is also a frequent practice to correct the final KPIs or the benchmarks to account for the weather conditions/anomalies, as well as special services offered at the hotel.

Green Globe uses an inverted Y-axis in a graph, and presents hotel performance in time in comparison to the baseline and best practice performance for a particular country and type of establishment. Both baseline and best practice are established based on national statistics for a given region. Total energy consumption in mega Joules (MJ) is presented per guest-night, likewise, water consumption is also calculated per guest-night.

The ITP BenchmarkHotel presents final results in a table comparing the hotel individual score to benchmarks adjusted to hotel characteristics. It also offers the information on the saving potential should the facility perform in a more efficient manner. Energy is normalised per square meter of the area and presented separately for electricity use and for the use of other

Box 7.1   Case Study of Environmental Reporting Software Developed by Addsystems for Scandic (ScandicSIR) and for Hilton (HER)

Computerised environmental reporting systems, as the one developed by Addsystems in 1996, can become powerful tools in supporting performance improvements at an individual as well as a corporate level. The first version of the software—Scandic Utility System (SUS) was developed for the Scandic Hotels AB 'Resource Hunt' program. Following the acquisition of Scandic by Hilton International in 2001, a more sophisticated version of SUS, applicable to all Hilton International and Scandic facilities was developed and launched in 2004—Hilton Environmental Reporting (HER). In 2007, as a result of the acquisition of Scandic by Nordic private equity group EQT, the original database was separated and now each company has its own platform: Scandic Sustainability Indicator Reporting (ScandicSIR) and Hilton Environmental Reporting (HER). ScandicSIR is used by team members of all 130 hotels, while HER serves 160 properties mainly located in Europe, United Kingdom and Ireland, and Africa.

The HER/SIR tool is a web-based application that provides a flexible and robust way of collating, recording, and monitoring environmental and resource use and cost information. The system relies on accepted methodology for collection of information and key performance indicators (KPIs) calculations and undergoes periodical audits by a third party company. The three objectives of the tool are to provide clear feedback to members in individual hotels and across the business, to provide a global mechanism for reporting operational, resource consumption, and cost data, and to collate data for the purpose of environmental management and Corporate Social Responsibility reporting. Both applications are being continuously developed according to the needs of individual users at various levels of the company. Reports for the following KPIs are available in the system: total and per guest-night (per area) energy and water consumption and carbon dioxide emissions, waste generation (sorted and unsorted)—at individual hotel level as well as country and regional levels, energy and water league tables for regions and areas, as well as utility cost reports. In the 'Monitoring and Targeting' report hotel resource consumption is predicted based on the past performance and chosen indicators and then compared to actual performance, helping to indicate problems such as e.g., not optimised heating/cooling systems. A section with customised reports where the user selects the hotels, the time span, as well as types of KPIs to be shown in reports has also been developed. Another development is a Balance Score Card report measuring each hotel's performance against its individual target.

At Scandic, this tool has been used for a number of years to nominate the annual best environmental performer in the categories of water and energy use within the 'Resource Hunt' initiative. The database was of

(continued)

Box 7.1 (continued)

great assistance during the Nordic Swan labeling process of the hotels, has been utilized in the process of collecting the data for the 'Energy Passport for buildings' project, and provides environment related information requested by the corporate customers. The information collected in the platform served to prepare estimates of the overall savings/avoided impacts of Scandic's ten-plus years of sustainability work for the sake of a 'Better World' campaign launched in 2007. Numerous other activities were undertaken and reported in the SIR system.

At Hilton, HER has been the key measuring tool of the *we care!* environmental program launched in January 2006 for Europe & Africa. The goal of the *we care!* program is to reduce the energy consumption over the period of three years by 15 per cent as compared to 2005, through the education and empowerment of all hotel team members to implement environmental initiatives in their workplace. To encourage the team member participation a Mountain Bike competition was announced, with all the team members of the best performing hotel in each operational region receiving a mountain bike. HER was the official monitoring channel used to collect the information, and display the energy performance league tables for each competing region (weather and occupancy corrected KPIs). The initiative is seen as a great success and the combined efforts of all the team members in Hilton Europe and Africa resulted in the reduction of the unitary (per guest-night and per area) energy consumption by 6.5 per cent and thus avoidance of USD 3.2 million in additional energy costs. In addition, more than 1800 team members can switch to a more environment friendly and healthier vehicle: bike. The challenge continued in 2007 and the delivered reduction in energy consumption was over 10 per cent across the European hotels (based on a cumulative total achieved between 2006 and 2007 versus 2005), while water consumption fell by 5 per cent. The team members from the five winning hotels will again be awarded mountain bikes.

Last, but not least, the implementation of these tools, in combination with other elements of the 'sustainability smörgåsbord' (the Swedish buffet—with many options available) is continuously helping to raise awareness of both, the team members and the customers. These instruments are especially useful in visualising performance of the facility in the form understandable by most of the team members—understanding that in the past, this was a task frequently limited to the technical services department. This also allows to quantify the results of various initiatives—further convincing individuals that it is worth a try.

More information: Bohdanowicz et al. 2005b; Bohdanowicz and Martinac 2007; Bohdanowicz 2007; Scandic Better World (www.scandichotels.com/betterworld); Hilton *we care!* (www.hiltonwecare.com); Addsystems (www.addsystems.com).

Table 7.1 Comparison of Some of the Benchmarking Schemes

	Green Globe 21	ITP BenchmarkHotel (version 2007)	HEAT Online version 2.0	Hilton's HER / ScandicSIR (status as of March 2008)	US EPA/DOE Energy Star Portfolio Manager
Type of tool	Traditional reports	On-line, password protected (www.benchmarkhotel.com)	On-line, password protected (www.hospitableclimates.org.uk)	On-line, password protected	On-line, password protected (www.energystar.gov)
Geographical coverage/Climatic zones	Energy consumption weather adjusted with temperature calculations	Temperate, Mediterranean, tropical, individual input of degree days (used for benchmark adjustments)	UK only, based on postal codes	Energy consumption weather adjusted with the degree day calculations	USA, energy consumption weather corrected with the degree day calculations (based on postal codes)
Different types of hotels	Yes, 5 types: business hotels, vacation hotels, motels, bed & breakfast, hostels	Intended for luxury hotels mainly, various facilities and services accounted for by internal correction factors	Yes, 3 types: large luxury, medium, small	All brands within company portfolio, each brand forms a separate class	Yes, 5 types: upper upscale, upscale, mid-scale with food & beverage, mid-scale w/w F&B, economy & budget
Building characteristics, engineering systems	No	Included in the hotel general information form—separate form for each year	Included in hotel type and detailed additional questions	Included in the annual hotel profile form and in types of resources reported monthly	Mostly included in hotel types, additional information on on-site services and types of spaces required

Resource consumption	Building only	Building only	Building only	Building and transport vehicles where fuel is paid by the hotel	Building only
Periodical reporting/ benchmarking	Annual	Annual	Annual (suggested), number of months can also be chosen	Monthly, month-on-month and year-to-date comparisons	Monthly for 12 or 24 months
Type and level of input data	Seven core earth-check™ indicators were identified and are assessed: presence of sustainability policy (yes/no), social commitment, consumption of water, recycled materials, chemicals and energy (all types of fuels), solid waste generation.	Characteristics of hotel building and services offered, systems included and operational statistics, water and energy consumption and costs, waste generation	Building information, energy systems and efficiency improvements, services provided, energy consumption and cost data	Occupancy characteristics, environmental initiatives, consumption of water, energy, chemicals and refrigerants, generation of wastes, costs	Occupancy characteristics, energy and water consumption and cost data
Data quality issues	Data verification by an external organization: Earth Check Ply Ltd	Manual data verification possibility by corporate user for each company (planned)	No data verification for the user (there is only a note on a possibly limited accuracy provided)	Manual data verification by the system coordinator; automatic data verification system being developed	Internal verification system used in weather normalization of data (based on E-Tracker tool developed for US EPA/DOE), thus floor area & energy consumption data for individual observations may be estimated or rounded
Level of detail	Low	Moderate	Moderate	Moderate	Moderate/high

(continued)

Table 7.1 (continued)

	Green Globe 21	ITP BenchmarkHotel (version 2007)	HEAT Online version 2.0	Hilton's HER / ScandicSIR (status as of March 2008)	US EPA/DOE Energy Star Portfolio Manager
Level of expertise from the person performing the reporting	Low	Moderate (good knowledge of the hotel required)	Moderate	Moderate (good knowledge of the hotel required)	Moderate
Report type	Graph (Figure 7.1)	Table (Figure 7.2)	Graph (Figure 7.3) and description	Numerous graphs and league tables (Figure 7.4)	Table (Figure 7.5)
Benchmarking	Against country baseline and best practice (developed from published statistics and normalized according to local/national conditions)	Against benchmarks based on empirical data from large hotel companies, new representative benchmark to be available by the end of 2007	Against other hotels with similar characteristics in Energy Measures database	Against other buildings in the database (divided into brands, countries, regions and areas)	Against American buildings data collected in Commercial Buildings Energy Consumption Survey database
Indication of possibilities of saving and suggestions for improvements	No	Yes, both resource and cost savings provided that hotel performed better, links to Green Hotelier Know-How documents	Yes, comprehensive	Not in this program, information on possible initiatives available on company intranet and via Technical Services department	Not in this program, information available under EnergyStar framework, possibility of setting reduction targets
Cost	US$ 20 (guidelines only)	Not known yet	Free of charge	Classified information	Free of charge

Source: adapted from Bohdanowicz et al. 2005a, Matson and Piette 2005, Bohdanowicz 2006 & 2007, International Tourism Partnership BenchmarkHotel software 2005–2008, HEAT Online 2006, EnergyStar Portfolio Manager software 2006–2008, Hilton's HER software 2005–2008, and ScandicSIR software 2005–2008.

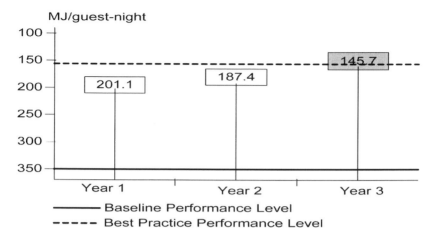

Figure 7.1 Example of Green Globe 21 earthcheck™ energy indicator and benchmark for accommodation (NOTE: the Y axis is inverted).

fuels, while water and waste are calculated per guest-night. Benchmarks are based on the literature data and empirical information from a number of large and well established hotel corporations. If historical data is available for a particular facility, a special report showing performance over time can be prepared.

HEAT Online uses a guestroom as a normalising factor for presenting hotel performance, and compares it to a typical hotel from the Energy Measures database, as well as best practice facility. Provided that historical data is included in the platform, a report showing hotel performance over time is created and displayed.

Hilton Environmental Reporting offers a large variety of output reports available at hotel- as well as regional level (Bohdanowicz 2006, 2007). The energy consumption is normalised on guest-nights and square meters of hotel area, while water, waste and CO_2 emissions are presented per guest-night. Hotel performance is evaluated over time and also compared to the country and corporate average.

EnergyStar Portfolio Manager allows for the selection of an individual baseline period to which performance is compared. It is also compared to the benchmarks established within the EnergyStar framework and to the industry average performance as available within the Commercial Buildings Energy Consumption Survey database. Both point scale and energy normalised per square foot of area are used. In addition, energy costs and CO_2 emissions are used as benchmark criteria.

A survey performed among hotel corporate managers in 2006 provided some insight into the existence of reporting instruments and their types (Bohdanowicz 2006). Nine of the thirteen hotel corporations surveyed confirmed

BenchmarkHotel   The Benchmarking Tool

Your Benchmarking Area | Logout

Account details

Step 1: Update Hotel General Information

Step 2: Update Hotel Operational Data

Step 3: Benchmark your performance

Step 4: Your Benchmark Results

Step 5: Improve your performance

Log off

- Full Benchmark Report 2006

▶ Go back   ▶ Print report   ▶ Generate PDF

Electricity	Excellent	Satisfactory	High	Excessive
Target Benchmark (KWH/sqm)	< 135	< 145	< 170	> 170
Corrected Benchmark (KWH/sqm)	< 103.5	< 113.5	< 138.5	
Your score (KWH/sqm)				138.9
Saving Potential (KWH/sqm)	35.4	25.4	0.4	
SEK	SEK251.014	SEK180.187	SEK3.051	
US $	$32.223	$23.128	$392	

Other Energy	Excellent	Satisfactory	High	Excessive
Target Benchmark (KWH/sqm)	< 150	< 200	< 240	> 240
Corrected Benchmark (KWH/sqm)	< 70.5	< 120.5	< 160.5	
Your score (KWH/sqm)		88.0		
Saving Potential (KWH/sqm)	17.5	0.0	0.0	
SEK	SEK134.827	SEK0	SEK0	
US $	$17.262	$0	$0	

Water	Excellent	Satisfactory	High	Excessive
Target Benchmark (m3/guest night)	< 0.5	< 0.65	< 0.9	> 0.9
Corrected Benchmark (m3/guest night)	< 0.315	< 0.375	< 0.715	
Your score (m3/guest night)	0.150			
Saving Potential (m3/guest night)	0.000	0.000	0.000	
SEK	SEK0	SEK0	SEK0	
US $	$0	$0	$0	
Energy for Hot Water SEK	SEK0	SEK0	SEK0	
Energy for Hot Water $	$0	$0	$0	

Waste by volume	Excellent	Satisfactory	High	Excessive
Target Benchmark (litres/guest night)	< 3	< 5	< 7	> 7
Your score (litres/guest night)				
Saving Potential (litres/guest night)	0.0	0.0	0.0	

Waste by weight	Excellent	Satisfactory	High	Excessive
Target Benchmark (kg/guest night)	< 0.6	< 1.2	< 2	> 2
Your score (kg/guest night)	0.5			
Saving Potential (kg/guest night)	15168.0	196128.1	437408.2	

Disclaimer | Terms & Conditions

Figure 7.2 Facility performance report generated by IBLF/ITP BenchmarkHotel (BenchmarkHotel 2007).

HEAT Online example based on own data from a hotel in London

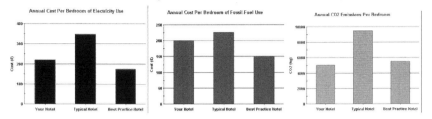

Example provided by HEAT Online

Figure 7.3 Energy performance report generated by HEAT Online for a hotel facility in London (own analysis, and examples from HEAT Online 2006).

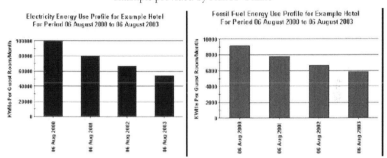

Figure 7.4 Regional league table showing normalised performance of individual hotels and % changes in consumption as compared to the base year.

NOT FOR USE WHEN APPLYING FOR THE ENERGY STAR

Statement of Energy Performance
FACILITY SUMMARY REPORT
Las Vegas

For 12-month Period Ending: December 31, 2006
Date Generated: March 20, 2008

This document was generated using EPA's Portfolio Manager system. All information shown is based on data provided by the Portfolio Manager account holder. Depending on the use of the SEP Facility Summary, building owners or managers may want to have a professional engineer (PE) verify that the underlying data is accurate. Blank space has been left intentionally on the SEP Facility Summary for a PE stamp.

Dr.
Las Vegas, NV 89119

Year Built: 1997
Gross Floor Area: (ft²) 105,617

Facility Space Use Summary

Hotel (Upscale)

Space Name	Gross Floor Area (ft2)	Number of Rooms	Food Preparation Facility
Single Building	105,617	190	Y

Energy Performance Comparison

Results	Current (12/31/2006)	Baseline (12/31/2006)	Delta	Target	Industry Average	ENERGY STAR
Energy Performance Rating	83	83	0	91	50	75
Energy Intensity (kBtu/ft2)						
Site	109	109	0	98	144	119
Source	242	242	0	218	320	265
Energy Cost						
$/year			0		301079	248940
$/ft2/year			0.00		2.86	2.36
CO₂ Emissions (tons/year)	1411	1411	0	1268	1866	1543

More than 50% of your building is defined as Hotel (Upscale). Please note that your rating accounts for all of the spaces listed. If you cannot see a rating, you will be compared to the national average of Hotel (Upscale).

♨EPA
United States
Environmental Protection
Agency

Figure 7.5 EnergyStar Portfolio Manager output table for a hotel facility in Las Vegas (own analysis), energy intensities in kBtu/feet² (EnergyStar 2008).

using a performance reporting tool. Five of those used their intranet or Internet to collect the information, while the remaining companies relied on e-mail submissions to the headquarters of Excel files from individual hotels. Half of the corporations questioned required monthly reports on resources used, their costs and operational data, with two requiring only quarterly figures. One of the companies required additional and more comprehensive annual reports. Some reported to only collect financial information, while others required figures on carbon dioxide emissions, recycled materials, as well as reductions in utility consumption. The two oldest systems dated back to the early 1990s, with most other tools introduced post-2000 (at times these were upgraded versions of the previously existing systems). All companies communicate the results obtained to their personnel and seven out of nine also share the information with their customers and suppliers.

It can be concluded that a considerable number of hotel corporations collect information on the individual performance of hotels. This information is however typically limited to cost figures, necessary for the preparation of annual financial reports. In addition, the information collected is rarely used to continuously monitor performance of individual facilities. Nevertheless, in recent years an increasing number of companies have decided to develop systems that would not only allow for the periodical collection of financial information, but also for continuous monitoring of the actual performance of individual facilities as well as the company as a whole.

THE PROBLEM

The problems currently faced by hotel managers and corporate managements can be summarised as a lack of solid and proven strategies and user-friendly tools to monitor and benchmark hotel performance. Although some monitoring tools are commercially available and successful case studies can be found in the sector, their market penetration is still limited and there is no collective statistics in the field. Various problems were faced and studied by the companies that already have reporting tools, but the resulting reports are internal documents not available to the public or competing establishments. The market lacks experience-based manuals on how to implement environmental monitoring, while the reliability, accuracy and universal applicability of monitoring and benchmarking instruments is still in question. It is thus of utmost importance to provide the hoteliers and other tourism managers with clear and well-proven guidelines and tools on how to successfully incorporate performance monitoring in their facilities.

STRATEGIES FOR THE IMPLEMENTATION OF MORE RESPONSIBLE BUSINESS PRACTICES

Types of strategies for the implementation of environmental management in the hotel industry differ depending on the type of stakeholder, as well as the individual's position within the company. The primary options are the same as for any other initiative and include legislation that can be utilised to encourage/impose improved and more nature-friendly business practices, and limit resource wastage, economic incentives or disincentives (such as fees for wastage), as well as the provision of knowledge through the awareness raising campaigns. Although increasing numbers of companies begin the 'green talk' and slowly embark on the 'green walk', it is too early to say that initiatives are being incorporated purely for the sake of saving the environment. Following the experience from hospitality enterprises the following rules of thumb for the development of the environmental management programme can be listed:

- serious commitment
- partnerships
- clear goals and implementation plans
- communication and knowledge transfer
- continuous improvement

The first condition is a sincere desire to improve the environmental performance of the company supported by written environmental policy. The preparation of the policy should be preceded by a general understanding of what the environment means to the company's operations. In addition, the costs incurred and expected benefits (economic, marketing, social, etc.) of the environmental profiling of the company and the implementation of a relevant programme need to be evaluated against each other to reach a consensus.

All types of tourism organisations may find it beneficial to establish partnerships with environmental organisations. There are a multitude of NGOs and commercial institutions that offer such services (i.e., The Natural Step, International Standardisation Organisation, the Eco-Management and Audit Scheme [EMAS], etc.), with some of them solely focused on the tourism industry (International Business Leaders' Forum/International Tourism Partnership [ITP]). Alternatively, published standards, guidelines, and best practice case study compilations can be used in the development of environmental programmes (publications by ITP, Conservation International, United Nations Environmental Programme, and others). Information published on the websites of other tourism companies, and in their environmental reports may also be used to this end, as it is believed that tourism operators frequently initiate their environmental programmes as a response to the actions of competitors in the region (Karagiorgas et al. 2006). Involvement of all employees in the early stages of the process would possibly positively affect their willingness to participate in the programme once launched, and can bring invaluable input as regards what needs to be done and how.

The next step should involve the development of goals and an action plan. The goals ought to be long-term, clear and realistic time- and technology-wise (i.e., any reduction goals should consider local conditions, economic, and technical feasibility). It is thus advised to first establish an understanding of the current status of operations via an audit, and create a benchmarking baseline in terms of resource consumption. Initially it may be more appropriate to set optimisation goals relative to one's own performance, rather than published standards. Furthermore, in setting goals, it is frequently better to conservatively assess possible resource reductions and achieve better results than anticipated, rather than vice versa. Companies are advised to start with easy actions requiring relatively little effort, low capital input, but still providing visible results. In this way, employees will become acquainted with the concept in a relatively stress-free manner, and will be more willing to take on more loads when the time comes. Some argue that because the resource consumption improvements are typically based on relative scales (i.e., energy use reductions per guest-night), the impact is not as significant since absolute numbers continue to increase due to the sector growth. However, the business growth would take place in response to the demand anyway, and without these relative savings the overall resource consumption, and thus environmental impacts of the sector, would have been significantly higher. It has to be underlined that these small steps are important, and with time and growing awareness may actually translate into absolute resource savings. The appointment of an environmental champion, responsible for reporting and coordinating all actions, may enhance the success of the initiative, however, the responsibilities of improving the performance should involve all employees.

The proper transfer of knowledge and know-how is crucial for the success of any environmental programme. The launch of any such initiative ought to be preceded by an intensive and wide ranging information campaign among all staff members. Classroom training may be supplemented by interactive computer programmes and more conventional materials. The entire process should be presented in a transparent manner, allowing all employees to understand it and identify with it. The participants should then be given tools allowing them to proceed with the implementation of the action plan. These may include financial resources, technical assistance, or managerial support.

All staff members need to be aware of the importance and convinced about the usefulness of the initiative, as well as the reason for doing it. Otherwise, less committed individuals or individuals not really believing in the concept will, after some time, probably start to refrain from applying the rules in their daily operations. Organising competitions between departments or facilities within the portfolio may be a good means to keep the initiatives alive, under the condition that the competing teams are comparable and goals fair. Environmental commitment and actions undertaken should also be communicated externally to customers and supporting businesses, as this may have important implications on marketing. Once the programme is operational, the environmental initiatives and resource management solutions, as well as all the assisting tools, ought to continuously receive the attention and support of the corporate management at all levels of the company. In addition, the programmes should regularly be reviewed and updated to ensure the continuous improvement of the facility's performance.

ENGAGING IN ENVIRONMENTAL MONITORING

Developing and implementing an environmental reporting programme is not an easy task, which partially explains the limited use of such tools. It is thus important to present the procedure and important stages in the creation of such an initiative. The procedure outlined here is based on hotel-specific

Table 7.2 Timeline for the Implementation of an Environmental Reporting and Resource Management Program

Stage	Time, months	Actions
1 The development of the reporting tool	6–8	Defining the scope and form of the environmental management program and reporting tool
		Consultation with company representatives, external consultants
		Development of the program and supporting materials
		Testing the software
		Providing infrastructure at all units (i.e., computers)
2 Introduction of the program and reporting tool	1–2	Introduction of the concept of environmental reporting and monitoring (its benefits and limitations) by the top management
		Presentation of the environmental program and reporting system (performed during a company meeting, management conference or via post/mail)
		Nomination of environmental champions at all business units
		Installation of the reporting system at all units (or providing the necessary support materials)
3 Training	2–4	Two training sessions for the users (personnel):
		1. detailed presentation of the system (data sources, data acquisition methodology, reporting procedure, output reports, possible limitations of the tool), either classroom-type workshop or interactive training
		2. time for the users to get acquainted with the tool, verify if they have access to all the required information within the time limit specified in the system, etc.
		3. classroom-type Question & Answer training session 2–3 weeks after the first meeting (virtual conference meeting also an option) lead by the tool developer
		If there are different levels of users in the system—separate training sessions should be organized for each group, as different aspects of the tool may be of interest
4 Creation of the reporting culture	3–6	Careful monitoring of the data quality provided by individual hotels and their reporting status
		Regular reminders of the need to report on time
		Crucial to offer the individuals support and encouragement in order to raise the awareness and develop a routine of regular and correct reporting
		Management support for the initiative

(continued)

Table 7.2 *(continued)*

Stage	Time, months	Actions
5 Data collection	Up to 12	Constant collection of data in the system
		Management support for the initiative
		Environmental program and training can be initiated
		If centralised data available—this can be uploaded to the system to speed up the process
6 Initiation of resource saving program	When 12 months of data available	Setting of environmental (resource reduction) goals (with the past 12 months as a base year)
		The tool used to monitor progress and performance of individual units

systems but the same guidelines can be adapted for a system developed for any other tourism-related enterprise. Table 7.2 presents possible time spans of particular stages, however these are just estimates, each phase can take less or more time depending on a number of factors.

The available literature on sustainability indicators provides guidance on the choice of relevant performance parameters and the information required. There are however difficulties arising when system users at individual facilities are not able to collect the requested information due to its limited availability on-site. One of the possible solutions is to involve future users in the design and development of the system. Local staff members have knowledge and experience of the situation, access to real-life data and information, and can thus help in assessing what outputs from the system are most useful at department- and facility levels. Finally, when participating in the creation of the system they will identify with the concept and will probably be more willing to use it. Of course, in the case of larger companies it is virtually impossible to involve all personnel in such a project. Therefore, a representative sample of future users and data providers ought to be chosen, preferably having different backgrounds, level of technical and environmental knowledge and skills, as well as functions within the company structure. An external consultant experienced in environmental management and reporting in hotel businesses may be contracted at this stage.

It is important to define the physical boundaries of the system considered and the frequency of data collection. Depending on the type of business and the range of environmental commitment, reporting may be limited to the accommodation establishment's consumption only, or cover impacts related to the goods production and transport to and from the facility. The frequency of data collection can be decided, based on weighting the cost of staff-time and the installation of meters on major utility end-users (i.e., kitchen, swimming pool, conference centre) against possible benefits. While for the purpose of general benchmarking it may be enough to collect annual data from invoices, performance monitoring requires monthly figures. In the case of individual end-users responsible for considerable energy and water use, such as kitchen or laundry for example, it may be beneficial to adopt daily reporting routines.

To ensure efficient utilisation of the system, methodologies for data collection and reporting procedures should be standardised and users provided with definitions of various terms, especially energy units, as these tend to be confusing. When developing a performance monitoring system for a multi-user corporation there may be a conflict between the expectations of user-friendliness, universal applicability, and relative simplicity of the tool with the general flexibility and reliability requirement. Currently most of the existing instruments collect detailed information about the hotel characteristics to allow for a more accurate benchmarking of performance, through the use of correction factors (see Figure 7.6).

If the possibility exists, in the case of corporate systems, data could be introduced into the database centrally from one source, i.e., an energy broker, reservation system, or outsourced laundry provider (centrally populated data). Frequently, centralised updating of such a database may be a more efficient solution, which would ensure the prompt reporting of high quality information (especially if combined with automated data logging). However, such an arrangement could have a reduced educational purpose, and would not be likely to encourage individuals to be concerned about the performance of their facility. A combination of centrally populated data with information reported by individual users may prove to be the optimal solution. Regardless of automatic or manual data upload procedure, there must be constant and prompt technical support as well as a data verification system (a check that the information in the system is correct as emphasised by Wöber 2002). A lack of support in solving technical problems of the users or questionable data quality may reduce the willingness to use the system and thus lead to the failure of the initiative. During the development of the corporation-wide system the issue of data security needs to be addressed, especially with regard to sensitive proprietary data on cost and/or occupancy rates. A procedure needs to be designed and enforced to ensure that only company employees have access to the database, and that the level of access is related to their role in the enterprise. Care needs to be taken to have up-to-date information on the business units' status in the database, such as renovations, sales, and acquisitions, in order to have all operational units from the portfolio actually reporting to the system.

The transparency of the system ought to be ensured by detailed information on the computer system's status, as well as the conversion factors and mathematical models used in the creation of output reports. Preferably conversion factors and computational procedures developed and accepted by internationally recognised organisations should be used. Climate differences among various locations also need to be considered for the purpose of comparisons, and properly accounted for (either by the sub-classification

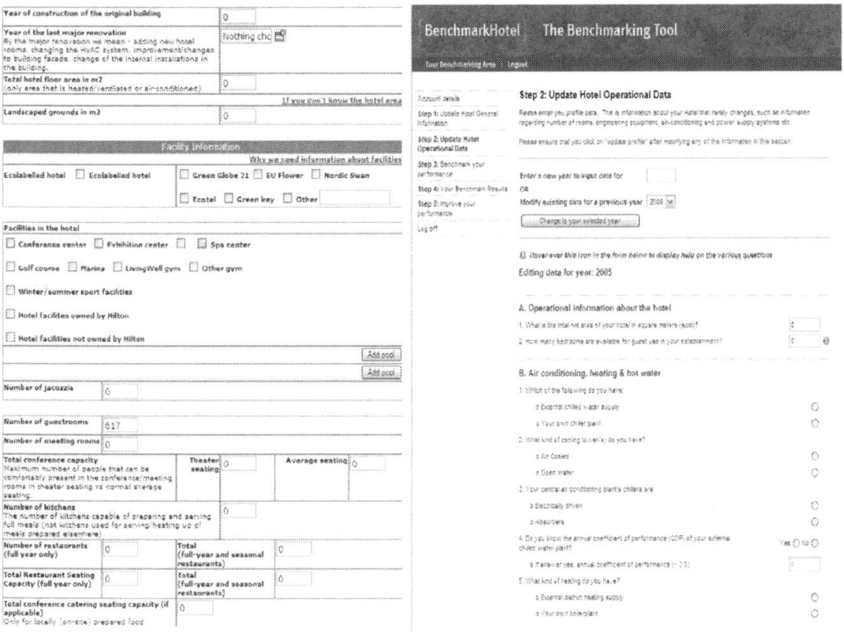

Figure 7.6   Section of a hotel profile developed in HER, and hotel operational data from IBLF/ITP BenchmarkHotel (BenchmarkHotel 2007).

of hotels into regional groups, or by incorporation of degree day data or average monthly outdoor and indoor temperatures—EnergyStar Portfolio Manager, HEAT online, HER, ScandicSIR, BenchmarkHotel).

The software interface ought to be user-friendly and attractive, both in terms of navigation as well as general layout. It should also be in line with company image and policy. In addition, the system ought to be adapted to the needs and possibilities of users at all levels of the organisation. In the case of companies with a differing portfolio (the number of hotel standards/ brands represented) a hotel sub-classification within the database may be a necessary development, in order to ensure realistic comparisons and bench- marks within the brand. Finally, the tool should be interactive and provide instantaneous feedback to the user.

THE PROCEDURE AND BENEFITS OF ENVIRONMENTAL REPORTING

Preferably the monitoring system should be designed to serve as a frame- work for a comprehensive environmental programme with clear goals. The type of output provided by the system ought to be well defined and relevant

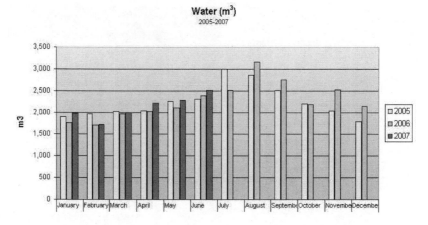

Figure 7.7   Standard report in ScandicSIR showing total water per month consumption for an individual hotel (2005–2007 figures) (ScandicSIR 2007).

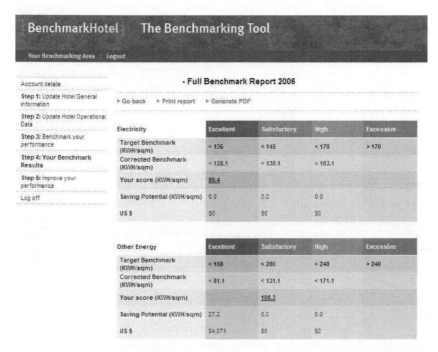

Figure 7.8   Full benchmark report in IBLF/ITP BenchmarkHotel (Note: corrections are based on hotel location and facilities offered; BenchmarkHotel 2007).

12 Months Ending	Current Rating (1-100)	Target Rating (1-100)	Current Site Energy Intensity (kBtu/Sq. Ft.)	Energy Reduction per Sq. Ft. (kBtu/Sq. Ft.)	Current Weather Normalized Source Energy Intensity (kBtu/Sq. Ft.)	Total Energy Cost per Sq. Ft. (US Dollars ($))	Indoor Water Use per Sq. Ft. (kGal (thousand gallons))
November 2006	87	N/A	89.5	1.4	167.1	$1.30	0.06
November 2005 (Baseline)	86	N/A	90.9	No Savings	169.7	$1.21	0.06
Change	-1		1.4	-1.4	2.6	($0.09)	0.00

Figure 7.9   Facility performance report in EnergyStar Portfolio Manager, comparing current consumption with a baseline year (EnergyStar 2007).

to the needs of the users, as different information may be of interest at different levels within the company. Typically recognised feedback levels in hotel companies include individual hotel, brand, country/region, and entire portfolio. These are also typical benchmarking groups.

For a user at an individual hotel the major interest is to monitor one's own performance over the years (Figure 7.7), and compare it with company or industry benchmarks (Figure 7.8). Typically, these reports are available for energy (divided into types), water, unsorted waste, chemicals and emissions of CO_2, and ozone depleting substances at the hotel level, either as total quantities or normalised per guest-night, square meter of the hotel area, or revenue. Most instruments allow the reports to be downloaded for future reference and off-line use. Various tools may have some additional options, i.e., BenchmarkHotel offers estimates of possible individual savings, both monetary and in resources (Figure 7.8), while EnergyStar Portfolio Manager allows the choice of the baseline year and shows if the hotel is eligible for the EnergyStar (Figure 7.9).

Users with corporate responsibilities typically require more aggregated feedback reports that allow them to (a) see the overall performance of the portfolio, and (b) indicate facilities that may require assistance or are eligible for awards. Most of the tools offer, or are in the process of development of such reports. These may take a form of graphs (Figure 7.10), lists, league tables, or collated reports (Figure 7.11). In the case of manual monthly reporting it is also important to have an overview of the data reporting status of all hotels in the portfolio, to be able to take appropriate actions. In corporate feedback reports colour coding may act as a visual enhancer of the information provided. Once the performance goals are set, it becomes beneficial to have a report showing individual facilities' performance in relation to the targets set, as is the case of Balanced Score Card developed in ScandicSIR (Figure 7.12), as well as in the Portfolio Manager.

The provision of feedback is a crucial aspect of the overall success of the reporting system's implementation. Users need to see that their efforts in collecting and reporting information are appreciated and used to produce valuable feedback. Further to this, it is of utmost importance that once the

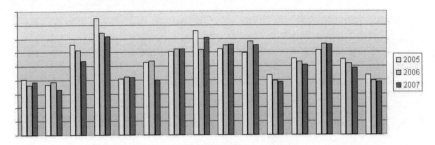

Figure 7.10 Regional report showing total energy per guest-night for individual hotels (X axis) in the region for the past 3 years (year-to-date information, January-June).

system is operational it constantly receives strong corporate support. The continuity of the system's utilisation may be achieved by frequent reference being made to it by the top management, while hotel managers or environmental coordinators may be encouraged to report and discuss the hotel environmental status with all staff members on a monthly basis. It can also be used to evaluate the commitment level of area and hotel managers. For example, within Hilton Europe, HER is used to monitor the progress of the *we care!* environmental programme and identify the winners of the Mountain Bike Competition. ScandicSIR has served similar purpose at Scandic for almost a decade, and now the platform is being prepared for a live export of hotel performance information online to provide full transparency.

To further encourage the use of the system and promote the sustainability and efficiency concept through its framework, this tool can contain a forum where users can exchange ideas and experiences, or present their best practices and environmental initiatives. Furthermore, this instrument may be used to provide support and advice to individual managers on how to improve the performance of their facilities, change utility contracts to more beneficial ones, or indicate which awards and eco-labels to apply for.

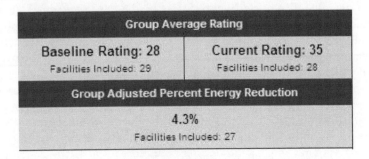

Figure 7.11 Group average EnergyStar Portfolio Manager rating report for a group of hotels (EnergyStar 2007).

Hotel	Actual	Target	Last Year YTD
Scandic	32.22	31.06	35.00
Scandic	54.75	58.98	64.68
Scandic	48.81	49.65	52.87
Scandic	42.35	40.24	47.37
Scandic	32.11	29.39	32.56
Scandic	52.33	49.93	58.22
Scandic	51.84	57.27	62.45
Scandic	42.41	52.63	57.53
Scandic	50.86	56.71	64.00
Scandic	38.62	48.65	52.26
Scandic	52.82	59.12	65.45
Scandic	70.01	82.19	93.61
Scandic	45.47	43.34	47.27
Scandic	50.84	56.93	67.66
Scandic	30.63	26.86	28.50
Scandic	46.75	47.54	51.94
Scandic	35.34	35.70	38.18

Figure 7.12   Balance Score Card report showing hotel actual performance compared to the target (ScandicSIR 2007).

Such services can be provided via an interactive discussion forum, a FAQ list, or an external consultant.

A well designed and implemented environmental and resource management programme may certainly bring considerable benefits at an individual as well as corporate level. These benefits can be referred to as both business and non-business, and include increased profitability due to reduced operational costs, the potential for an improved market-share, and preservation of limited natural resources to promote sustainable development (see Box 2).

FINDINGS

It is concluded that despite the relatively wide range of available tools and guidelines for the benchmarking of a hotel's environmental performance, their reliability and accuracy for individual units continues to be questionable (Warnken et al. 2005). A high heterogeneity among the facilities in the sector limits the applicability of uniform benchmarks. Nevertheless, reporting/benchmarking tools that allow for a hotel's benchmarking against its own performance and also provide guidance/advice on possible improvements are of high value for hotel managers and engineers. Further to this, the increasing popularity of reporting and benchmarking tools within large hotel corporations implies their usefulness in monitoring performance. Unfortunately, detailed information on the nature of such systems, procedures

Box 7.2 Sustainability in Practice—A Case Study of 10+ Years of Sustainability Initiatives at Scandic

Scandic is one of the better researched and documented success stories of sustainability commitment on a corporate scale. The company has been at the forefront of sustainability work in the hotel industry with its environmental program dating back to 1994. After more than a decade of continuous efforts in improving the performance of its facilities and working with and for local communities Scandic is in a unique position of being a Scandinavian pioneer of many successful actions, and being able to put a dollar sign on various initiatives.

The activities undertaken by Scandic over the years represent a very comprehensive approach to the problem of sustainability. These range from increasing the use efficiency and conserving natural resources within the 'Resource Hunt' program, eliminating the use of single packaged items, reducing and sorting waste, responsible construction and management of hotels, through educating team members in various sustainability related issues (including the disability problems), taking part in local community events and providing for charity organisations, to banning jumbo prawns from all its kitchens because of the unsustainable farming practices. As a result of all the above actions an average per guest-night energy and water consumption (only hotel related consumption) at Scandic Nordic was reduced by 21 and 16 per cent, respectively, between 1996 and 2007, carbon dioxide emissions dropped by 34 per cent and unsorted waste by 66 per cent, as reported in a Scandic Sustainability Indicator Reporting system—Scandic-SIR. These reductions resulted in significant 'avoided costs' which since the implementation of the 'Resource Hunt' program added up to millions of dollars. In addition, Swedish Scandic was the first in Scandinavia to serve the fair trade coffee in all its hotels, to have a Disability Coordinator, as well as to eco-certify all hotels in a country with the Nordic Swan label. In 2007 Scandic set a goal of becoming carbon neutral by 2025 (direct activities at the hotels), while in 2008 they announced to phase out foreign and local bottled water from its restaurants and conferences in Scandinavia (to reduce the impacts from transport), instead filtered tap water will be served.

A decade of comprehensive work has allowed Scandic team members and management to learn a number of valuable lessons which—when shared with others in the sector—could greatly contribute to the greening of this industry. All these actions had no negative affect on the economic bottom line, to the contrary, the company continued to grow steadily and is now among those typically mentioned when the subject of sustainability is raised.

More information: Bohdanowicz et al. 2005b, Scandic Better World (www.scandichotels.com/betterworld)

and group member participation is still limited, thus limiting the transparency of results published. Of the commercial platforms available, the EnergyStar Portfolio Manager addresses some of the heterogeneity issues, and ITP

BenchmarkHotel certainly aims to do so, with new updated benchmarks being developed and validated in the course of 2008. Furthermore, all the commercial tools are designed to be used by individual facilities, with a provision of a corporate report being an additional option.

The experience with hotel-specific reporting systems provides a number of valuable lessons, which can be considered by others planning to create their own systems. If a company decides to create own reporting tools, care needs to be taken during the development process to ensure that the information required from individual departments or units is relevant, and relatively easily available. The development process and the system itself ought to be transparent and easily understood by the users. Otherwise there is a risk that it will not be used as designed. Training and information related to the system need to be provided to all potential users, with additional self-study materials. The quality of input data ought to be verified constantly or trustworthy external sources should be used for collecting the information for central scripting. Continuous technical support is crucial to the success of the initiative. All users need to be provided feedback on their actions and business unit performances.

Users at various levels in the company may require differentiated feedback and output reports from the system and those need to be readily available. Individual users and independent hoteliers are interested in the performance of their own facilities, while national coordinators or area managers want to have a quick overview of the situation in all their units. Graphical representation of results, as well as colour-coded tabulated feedback reports can be seen as user-friendly outputs from such systems. Both normalised (per service output, revenue or service area) and absolute consumption figures ought to be monitored to ensure a comprehensive picture of the situation. Intra-company benchmarks for groups of similar facilities in the portfolio may be created or best-in-the-class standards established. A 'best practice' section or a discussion forum where users can exchange ideas and experiences could be created to facilitate improvements at an individual level. In addition, once the system is operational it requires constant support from the central management. To ensure the continuity of the reporting system's utilisation it can be used as a framework and tool for all environmental actions within the company. Such an instrument can also be used to identify winners of any environmental competitions arranged within the company, or even among the departments (if the system allows sub-categorisation of reporting to individual departments). The companies that have implemented computerised reporting systems typically claim to have high participation and positive attitudes among their portfolio members and personnel, and treat these instruments as important management tools within the corporate social responsibility framework.

Implementation of environmental and resource management programmes is not very difficult but needs to be well-organised and fully supported by the company management.

On a sectoral level legislation and economic incentives/disincentives typically serve as driving factors for change. On a company level it is typically the will to change and the commitment expressed by management that drives the implementation. Basic features of a well-developed and successful environmental programme can be summarised as commitment, partnership and co-operation, achievable goals, communication, and finally, continuous improvement (with increasingly demanding goals). On a global scale there exist many examples showing that conducting responsible business brings economic benefits, and is increasingly appreciated by the customers. This message should be widely disseminated among hotel businesses to achieve the goal of both 'greening the Goliaths and multiplying the Davids'.

REFERENCES

Anavo Group and STI (2005) *Travelers Want a Greener Way to Travel by Air.* Available at: www.my-climate.com/anavosti.pdf (31 March 2006).

APEC (1999) *Institutionalisation of a Benchmarking System for Data on the Energy Use in Commercial and Industrial Buildings.* Hawaii: Asia-Pacific Economic Co-operation.

Becken, S. and Cavanagh, J.-A. (2003) *Energy Efficiency Trend Analysis of the Tourism Sector.* Lincoln: Energy Efficiency and Conservation Authority, Landcare Research.

Becken, S., Frampton, C., and Simmons, D. (2001) 'Energy consumption patterns in the accommodation sector—the New Zealand case'. *Ecological Economics* 39: 371–386.

BenchmarkHotel (2005, 2007) *Hotel Benchmarking Software Developed by the Prince of Wales International Business Leaders Forum International Tourism Partnership* (formerly known as International Hotel Environmental Initiative). Available at: www.benchmarkhotel.com (July 2007).

Bohdanowicz, P. (2003) *A Study of Environmental Impacts, Environmental Awareness, and Pro-Ecological Initiatives in the Hotel Industry.* Stockholm: Royal Institute of Technology.

Bohdanowicz, P. (2006) *Responsible Resource Management in Hotels—Attitudes, Indicators, Tools and Strategies.* Stockholm: Royal Institute of Technology.

Bohdanowicz, P. (2007) 'A case study of Hilton Environmental Reporting as a tool of corporate social responsibility'. *Tourism Review International* 1(2): 115–131.

Bohdanowicz, P. and Martinac, I. (2007) 'Determinants and benchmarking of resource consumption in hotels—case study of Hilton International and Scandic in Europe'. *Energy and Buildings* 39: 82–95.

Bohdanowicz, P., Simanic, B., and Martinac, I. (2005a) 'Sustainable hotels-environmental reporting according to Green Globe 21, Green Globes Canada/GEM UK, IHEI benchmarkhotel and Hilton Environmental Reporting'. In *Proceedings of Sustainable Building (SB05) Conference,* 27–29 September 2005, 1642–1649. Tokyo, Japan.

Bohdanowicz, P., Simanic, B., and Martinac, I. (2005b) 'Environmental training and measures at Scandic Hotels, Sweden'. *Tourism Review International* 9(1): 7–19.

Breaking Travel News (2008) 'Hilton claims ten percent energy cut'. *Breaking Travel News,* April 1, 2008. Available at: www.breakingtravelnews.com (3 April 2008).

CI and IBLF (2005) *Sustainable Hotel Siting, Design and Construction.* London: Conservation International & the Prince of Wales International Business Leaders Forum, Nuffield Press.

Davies, T. and Cahill, S. (2000) *Environmental Implications of the Tourism Industry.* Washington DC: Resources for the Future.

de Burgos-Jiménez, J., Cano-Guillén, C.J., and Céspedes-Lorente, J.J. (2002) 'Planning and control of environmental performance in hotels'. *Journal of Sustainable Tourism* 10(3): 207–221.

Despretz, H. (2001) *Green Flag for Greener Hotels.* Valbonne: European Community, ADEME, ARCS, CRES, ICAEN, IER, SOFTECH.

EnergyStar (2006–2008) *Energy Star Portfolio Manager Software,* developed by the US EPA/DOE. Available at: www.energystar.gov/index.cfm?c=evaluate_performance.bus_portfoliomanager (March 2008).

Fahlin, M. (2007) personal communication with Marie Fahlin, SIS Miljömärkning, 12 June 2007.

Gössling, S. (2002) 'Global environmental consequences of tourism'. *Global Environmental Change* 12: 283–302.

Green Globe 21 (2004) *Travel and Tourism Industry Earthcheck™ Baseline and Best Practice Indicators.* Available at: www.greenglobe21.com (11 November 2004).

Green Hotelier Magazine, (1999 & 2000) London: International Hotels Environmental Initiative, now International Tourism Partnership.

GRI (2002) *Sustainability Reporting Guidelines.* Boston: Global Reporting Initiative.

Hauvner, E., Hill, C., and Milburn, R. (2008) *Here to Stay: Sustainability in the Travel and Leisure Sector.* Hospitality Directions Europe 17. London: PricewaterhouseCoopers.

HEAT Online (2006) *Hospitable Climates Hospitality Energy Analysis Tool— HEAT Online.* Available at: www.hospitableclimates.co.uk (April 2006).

HER (2007) *Hilton Environmental Reporting.* Internal web site (June 2008).

IHEI (2002) *Hotels Care: Community Action and Responsibility for the Environment.* London: International Hotel Environmental Initiative.

Karagiorgas, M., Tsoutsos, T., Drosou, V., Pouffary, S., Pagano, T., Lopez Lara, G., and Melim Mendes, J.M. (2006) 'HOTRES: renewable energies in the hotels. An extensive technical tool for the hotel industry'. *Renewable and Sustainable Energy Reviews* 10: 198–224.

Kozak, M. (2004) *Destination Benchmarking: Concepts, Practices and Operations.* Cambridge: CABI Publishing.

Leslie, D. (2001) 'Serviced accommodation, environmental performance and benchmarks'. In Pyo, S. (ed.) *Benchmarks in Hospitality and Tourism,* 127–148. Binghampton: Haworth Press.

Marriott (2007) 'Marriott energy conservation reduces greenhouse gases by 70,000 tons–equivalent to removing 10000 cars from U.S. streets', *Marriott News,* January 11, 2007. Available at: www.marriott.com (8 April 2008).

Matson, N.E. and Piette, M.A. (2005) *High Performance Commercial Building Systems–Review of California and National Benchmarking Methods, Working draft.* Berkeley: Ernest Orlando Lawrence Berkeley National Laboratory.

Nordic Ecolabelling (2007) *Svanenmärkning av Hotell loch Vandrarhem,* Förslag till Version 3.0, Juni 2006–Juni 2012 (Nordic ecolabelling of hotels and youth hostels. Criteria document, June 2007—June 2012, v.3.0). Stockholm: Nordic Swan Ecolabel, Nordic Ecolabelling, SIS Miljömärkning.

'Putting out the green message' (2008) *Hotels.* January 2008: 34.

ScandicSIR (2007) *Scandic Sustainability Indicator Reporting.* Internal web site. (June 2008).

Scoviak, M. (2008) 'The year of green'. *Hotels* January 2008: 30–31.

Stipanuk, D.M. (2001) 'Energy management in 2001 and beyond: Operational options that reduce use and cost'. *The Cornell Hotel and Restaurant Administration Quarterly* 42(3): 57–71.

Stipanuk, D.M. (2003a) 'Benchmarks for hotel energy usage'. In *Proceedings of the World Energy Engineering Congress*. Association of Energy Engineers, Atlanta.

Stipanuk, D.M. (2003b) *Hotel Energy and Water Consumption Benchmarks*. Final Report. Washington DC: American Hotel & Lodging Association.

Sustainable Travel Report (2005) 'Selling eco- and sustainable tourism: the right mix'. *The Responsible Tourism e-Newsletter* 3(5).

Warnken, J., Bradley, M., and Guilding, C. (2005) 'Eco-resorts vs. mainstream accommodation providers. An investigation of the viability of benchmarking environmental performance'. *Tourism Management* 26: 367–379.

Wilson, B. (1999) 'Monitoring and targeting (at Accor)'. *Green Hotelier Magazine* 13: 2.

Wöber K.W. (2002) *Benchmarking in Tourism and Hospitality Industries*. Vienna: Vienna University of Economics and Business Administration, CABI International.

WTO (2004) *Indicators of Sustainable Development for Tourism Destinations–A Guidebook*. Madrid: World Tourism Organization.

8 Sustainable Ski Resort Principles
An Uphill Journey

Tania Del Matto and Daniel Scott

INTRODUCTION

As discussed throughout this volume, sustainability has become a stated goal for many tourism businesses and destinations. For ski resorts, there are a number of drivers behind the goal of sustainability (see also Chapter 2 in this volume). There exists a growing, though still incomplete understanding of the impacts ski resorts have on mountain environments (Hudson 2000; Clifford 2002; Rixen, Stoeckli, and Amman, 2003; Scott 2005; Wipe et al. 2005) and host communities (Clifford 2002). Ski areas engaged in the practice of Corporate Social Responsibility will move to minimise or eliminate these impacts. Indeed, the sustainable ski resort literature supports the argument that as corporate citizens, members of the ski resort industry have an ethical responsibility to act as stewards of the biophysical environment and as such pursue sustainability. Other firms may be driven to action in order to avoid regulation or further public scrutiny that might adversely affect demand or real estate investment. Ski resort owners and managers also have an inherent economic self-interest to pursue sustainability, as improved efficiency in resource use (e.g., energy and water) can provide substantial cost savings and real or perceived environmental performance can provide a marketing advantage within some market segments.

Despite these compelling factors and the many positive initiatives being undertaken in the name of sustainability, we are of the opinion that a sustainable ski resort[1] has not been adequately conceptualised. The current understanding of sustainable ski resorts is limiting because it encourages ski resort owners and operators to address sustainability challenges in a compartmentalised manner and in isolation of one another—an approach in an industry where sustainability has historically meant focusing on local or onsite resource efficiency (e.g., water, energy, solid waste) and ecosystem impacts (e.g., forest clearing, soil erosion, water levels) and thus fails to acknowledge the broader socio-ecological impacts of ski tourism (e.g., community social impacts, emissions related to transportation to ski areas).

This chapter aims to build on the current understanding of what constitutes a sustainable ski resort by developing sector-specific principles that operate on an integrated systems approach to sustainability. Sustainability principles describe what is 'sustainable' or simply 'more sustainable' and are meant to represent the 'ideals' of where we want ski tourism to be, they do not prescribe the methods for achieving sustainability, which are necessarily locally specific. The terms *integrated* and *systems* are distinct from one another in that integration is the way in which sustainability objectives are pursued, whereas systems is the way in which sustainability is studied. The integrated systems approach to sustainability avoids the inherent limitations of conventional approaches to sustainability, which tend to separate sustainability challenges into separate entities (as described in greater detail by Gibson 2002; Gibson et al. 2005; Gibson 2006; Lehtonen 2004; Pope, Annandale, and Morrison-Saunders 2004).

A SYSTEMS PERSPECTIVE OF A SKI RESORT

Viewing ski resorts from a systems perspective offers an alternative approach to investigating the essential requirements and outcomes for sustainable ski resorts in the context of whole systems, human and ecological subsystems, their state, changes in their state, and the linkages and interdependencies that exist between the systems. It is also important to acknowledge that ski resort activities operate within, and are influenced by, the larger societal and biophysical systems in which the ski resort is embedded.

The systems diagram in Figure 8.1 provides the conceptual framework of how a ski resort is nested within a complex set of biophysical and societal systems that include: the built system (Figure 8.2) consisting of buildings and physical infrastructure (i.e., ski trails, roads, water supply, and electricity) needed to house and service various types of ski resort activities; the human activity system (Figure 8.3) consisting of the host community (i.e., government, households, non-governmental organisations, and other stakeholders), businesses (i.e., suppliers, producers, and intermediaries), employment, real-estate, visiting guests, and public services (i.e., police, fire protection, public transportation, affordable housing, and land use planning); the biophysical system (Figure 8.4) consisting of plants, animals, and soils; and the environment consisting of the elements that provide life support for all living things (i.e., sun, energy, air, water, and minerals).

CONCEPTUALISING A SUSTAINABLE SKI RESORT

The eight sustainability principles proposed by Gibson et al. (2005) hereafter referred to as the 'Gibson principles' (Table 8.1) provide the foundation for the conceptualisation of a sustainable ski resort developed here. The Gibson principles are drawn from a broad and extensive body of literature by integrating considerations from various fields, including ecological

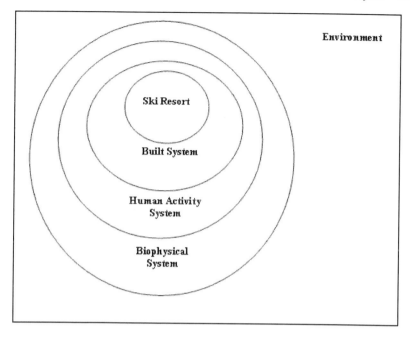

Figure 8.1 Systems diagram.

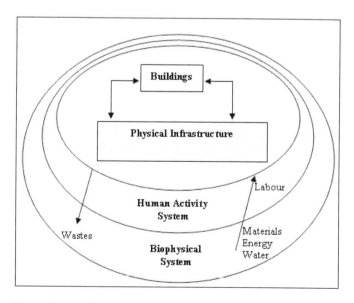

Figure 8.2 Built system.

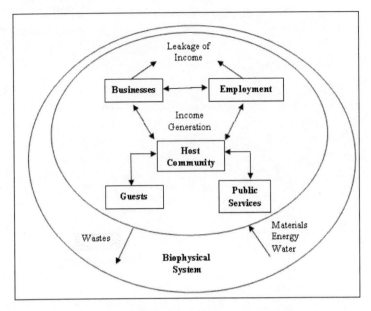

Figure 8.3   Human activity system.

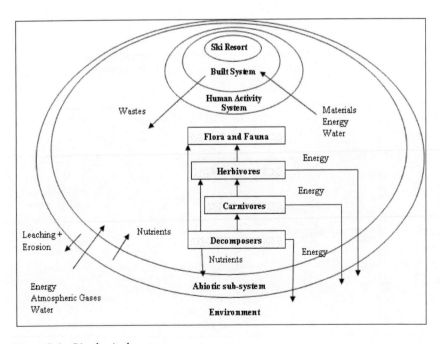

Figure 8.4   Biophysical system.

systems theory, corporate greening, social justice, ecological economics, community and growth management planning. The Gibson principles recognise the integrative aspects of sustainability as they represent a set of interdependent requirements that extend beyond the conventional categories (i.e., environmental, economic, and social).

The Gibson principles advocate that decision-making for sustainability makes a positive contribution to the well-being of biophysical and human systems, rather than traditional decision making for sustainability that advocates minimising negative impacts. Furthermore, the Gibson principles were created as core requirements for sustainability, serving as prerequisites for more specific sustainability based decision-making and with the

Table 8.1 Gibson Principles of Sustainability

Principle of socio-ecological system integrity: Build human-ecological relations that establish and maintain the long-term integrity of socio-biophysical systems and protect irreplaceable life support functions upon which human as well as ecological well being depends.

Principle of livelihood sufficiency and opportunity: Ensure that everyone and every community has enough for a decent life and opportunities to seek improvements in ways that do not compromise future generations' possibilities for sufficiency and opportunity.

Principle of intra-generational equity: Ensure that sufficiency and effective choices for all are pursued in ways that reduce dangerous gaps in sufficiency and opportunity (and health, security, social recognition, political influence, etc.) between the rich and the poor.

Principle of inter-generational equity: Favor present options and actions that are most likely to preserve or enhance the opportunities and capabilities of future generations to live.

Principle of resource maintenance and efficiency: Provide a larger base for ensuring sustainable livelihoods for all while reducing threats to the long-term integrity of socio-ecological systems by reducing extractive damage, avoiding waste and cutting overall material and energy use per unit of benefit.

Principle of socio-ecological civility and democratic governance: Build the capacity, motivation and habitual inclination of individuals, communities and other collective decision making bodies to apply sustainability principles through more open and better informed deliberations, greater attention to fostering reciprocal awareness and collective responsibility, and more integrated use of administrative, market, customary, collective and personal decision making practices.

Principle of precaution and adaptation: Respect uncertainty, avoid even poorly understood risks of serious or irreversible damage to the foundations for sustainability, plan to learn, design for surprise and manage for adaptation.

Principle of immediate and long-term integration: Attempt to meet all requirements for sustainability together as a set of interdependent parts, seeking mutually supportive benefits.

Source: Gibson et al. 2005

Table 8.2   Sustainable Ski Resort Principles

The **principle of socio-ecological system integrity**: recognises that there are limitations to quantitative growth and prescribes that ski resorts go beyond reducing human induced stresses on biophysical systems by pursuing opportunities that contribute to, rather than detract from, the integrity of both the biophysical environment and the host community it is nested with for the well-being of all residing ecosystems, residents, visitors and ski resort staff.

The **principle of livelihood sufficiency and opportunity**: recognises that ski resorts must decouple improvements in quality and service from further growth and consumption; create preferences amongst the skiing public for more sustainable consumption through education and marketing; and, enhance the capabilities of individuals and host communities to improve the quality of their lives through investments that support the local economy and benefit the wider host community interest.

The **principle of intra-generational equity**: calls upon ski resorts to deliver valued employment, community security, and opportunities for tranquility and fitness in a manner that enables one and all to fulfill their potential.

The **principle of inter-generational equity**: prescribes that ski resorts are morally obligated to make decisions with the needs of future stakeholders in mind, where the outcomes most likely preserve or enhance the opportunities of future generations to live while protecting the economic viability of the ski resort over an indefinite period.

The **principle of resource maintenance and efficiency**: directs ski resorts to reduce their net consumption of materials, energy and water across the life cycle chain of a ski resort, close the 'loop' on material flows and invest efficiency gains in areas that are deficient in natural and social capital.

The **principle of socio-ecological civility and democratic governance**: directs ski resorts to apply the sustainable ski resort principles through more open and better informed deliberations with host communities, by fostering social and ecological awareness and shared responsibility amongst internal and external stakeholders; and, using more integrative decision-making practices.

The **principle of precaution and adaptation**: prescribes that ski resorts respect uncertainty, avoid risks of serious or irreversible damage to the foundations of sustainability, invest in research and monitoring for greater understanding, design for surprise, manage for adaptation and adopt a long-term planning horizon.

The **principle of immediate and long-term integration**: states that ski resort owners and operators must pursue conformance with the sustainable ski resort principles as a whole, rather than in a compartmentalized way, thus seeking opportunities to contribute to all of them by arriving at decisions that strengthen the whole.

Source: Del Matto (2007)

intention that the principles be adapted for specific case use (Gibson et al. 2005), such as this application to ski resort based tourism.

Sustainability principles can inform a range of decisions from 'strategic' (i.e., corporate or government policy level) to 'site-specific' (i.e., project design or operational management level). Given the conceptual nature of defining a sustainable ski resort, here the Gibson principles were used as a framework for reviewing and incorporating insights from relevant literatures that discuss desirable characteristics of tourism businesses and destinations in sustainability terms. Emerging from this sector-specific adaptation process were a set of 'sustainable ski resort principles' (Table 8.2; Del Matto 2007), that represent key requirements of an idealised model of a sustainable ski resort and provide guidance for ski resorts to progress toward sustainability. As a package, the sustainable ski resort principles provide an introduction to what a sustainable ski resort should ideally strive for. The next task is to examine how the sustainable ski resort principles can be applied to ski resort operations, and this is done here using the ideal type analysis approach.

COMPARING THE SUSTAINABLE SKI RESORT PRINCIPLES TO THE CURRENT UNDERSTANDING OF SUSTAINABLE SKI RESORTS

Del Matto (2007) used ideal type analysis to compare and contrast the proposed sustainable ski resort principles (Table 8.2) with the present understanding of what constitutes a sustainable ski resort, as represented by viewpoints expressed in the scientific literature, non-government organisations (typically environmental NGOs that have examined the impacts of the ski industry), governments, and the ski resort industry. The intention of this comparison was to demonstrate whether the proposed sustainable ski resort principles advanced the debate beyond the existing understanding of what constitutes a sustainable ski resort. Table 8.3 summarises the insights acquired through this comparison.

The ideal type analysis revealed similarities and some key differences between the sustainability outcomes advocated in the current dialogue on sustainable ski resorts and those advocated by the sustainable ski resort principles. The five key differences outlined below demonstrate where the current understanding of sustainable ski resorts falls short of adequately defining a sustainable ski resort and where the sustainable ski resort principles advance the debate on what constitutes a sustainable ski resort (Del Matto 2007):

1. The sustainable ski resort principles clearly require that there be limits on quantitative growth and, as such, ski resorts must strive towards decoupling improvements in quality and service from further growth and consumption.
2. A sustainable ski resort contributes to narrowing the socio-economic gaps within the workplace and the host community while operating within a multigenerational timescale to ensure future generations are fairly represented.
3. A sustainable ski resort reduces its net consumption of materials and resources and invests these savings in areas that are deficient in natural and social capital.

Table 8.3   Comparison between the Sustainable Ski Resort Principles and the Current Understanding of Sustainable Ski Resorts

Sustainable Ski Resort Principles	Insights Acquired through the Current State of the Sustainable Ski Resort Discussion
Socio-ecological system integrity	• Ski resorts need to pursue opportunities that reduce human-induced stresses on biophysical systems (Todd and Williams 1996, NSAA 2000, NSAA 2005, SACC 2001, Schendler 2003, 2005, Eydal 2004, Lewis 2005)
	• Quantitative growth is acceptable provided measures are taken to minimize or mitigate impacts (NSAA 2005)
	• A limit must be set on quantitative growth (SACC 2001)
Livelihood sufficiency and opportunity	• Pursue education measures intended to influence a guest's consumption decisions while at the ski resort as well as at home (Todd and Williams 1996, Schendler 2003, Eydal 2004, Lewis 2005, NSAA 2005)
	• Debate on whether education measures will enhance the overall experience of guests or threaten to jeopardize a positive relaxing holiday experience (Lewis 2005)
	• Ski resort services should be provided in ways that do not compete directly with the local business community and in fact stimulate opportunities for local income generation (Hudson 2000, Castle 2004)
Intra-generational equity	• Ski resorts have a responsibility to ensure that employees are provided with affordable housing (Moore 2005)
	• Ski resorts should offer a wide range of accommodation in order to serve different economic classes of visitor (Hudson 2000)
Inter-generational equity	• There is acknowledgement from the ski resort industry to take action on sustainability in order to allow for skiing to be enjoyed by future generations (NSAA 2005) as well as an acknowledgement of the moral responsibility to take action for the well-being of future generations (Aspen Skiing Company 2006).
Resource maintenance and efficiency	• Use energy, water and materials in an efficient manner (SACC 2001, NSAA 2005)
	• Reduce the consumption of energy, water and materials across the life cycle chain using influence with suppliers (Todd and Williams 1996) and customers (Todd and Williams 1996, Schendler 2003, Eydal 2004, Lewis 2005, NSAA 2006b)
Socio-ecological civility and democratic governance	• Stakeholder involvement and collaboration plays a key role in a ski resort's journey towards sustainability (Flagestad and Hope 2001, Eydal 2004, Lewis 2005)
	• Decision-making on sustainability must be addressed in a systematic and integrated way (Eydal 2004)
	• Ski resorts should explore partnerships with stakeholders (NSAA 2005)

(continued)

Table 8.3 (continued)

Sustainable Ski Resort Principles	Insights Acquired through the Current State of the Sustainable Ski Resort Discussion
Precaution and adaptation	• Research is important to understanding the environmental and social processes that underpin the social and environmental carrying capacity related to the ski resort's operations (Eydal 2004)
	• Decision-making is to be flexible and subject to continuous improvement whereby feedback processes are in place to improve upon lessons learned and corrective action is taken where there is failure to achieve an intended outcome (Eydal 2004)
	• The need to adapt to changing climatic conditions is stressed as important (Scott 2005, Scott *et al.* 2003, 2007; Scott and McBoyle 2007, Bürki *et al.* 2003; Hennessy *et al.* 2003, OECD 2007).
Immediate and long-term integration	• Decision-making on sustainability must be addressed in a systematic and integrated way (Eydal 2004)

Source: Del Matto 2007

4. The sustainable ski resort principles require decision-making power to be shared amongst internal and external stakeholders.
5. Stakeholders must pursue opportunities to arrive at decisions that strengthen the well-being of both human and ecological systems through the integrated application of the sustainable ski resort principles.

APPLYING AN 'IDEALISED MODEL' OF A SUSTAINABLE SKI RESORT

The Case of Whistler Blackcomb (Canada)

The application of the sustainable ski resort principles serves to distinguish between sustainable outcomes and unsustainable outcomes at ski resorts and ultimately allows for this chapter to reflect upon the broader questions concerning the sustainability of the ski industry and tourism destinations in general. The application of the sustainable ski resort principles is not intended to quantitatively measure the progress that has been made towards sustainability within the selected case study, rather to qualitatively assess the state of sustainability at the case study and through the case study, a recognised leader in environmental initiatives in the sector, provide some insight into the state of sustainability in the ski industry as a whole.

The sustainable ski resort principles (Table 8.2) are made operational by applying them as the idealised model of a sustainable ski resort. Here ideal type analysis is conducted using the Whistler Blackcomb Mountain Resort (located in Whistler, British Columbia, and Canada) as a case study. Whistler Blackcomb is one of the three largest ski resorts in the world, with thirty-eight ski lifts providing access to over 200 ski trails covering 3,280 skiable hectares on two mountains and three glaciers (Whistler Blackcomb 2008). The resort hosts over two million skier visits annually and is the community's largest employer with 550 permanent full-time staff (Whistler Blackcomb 2008). The current conditions and trends of sustainability at Whistler Blackcomb are informed by its sustainability vision and policies, the frameworks it uses to guide decision-making toward more sustainable outcomes, and the plans and programmes it has established to address sustainability issues (De Jong-personal communication, October 2, 2007 and April, 2008). As evidenced by the many environmental awards Whistler Blackcomb has received, including the Golden Eagle Award for Overall Environmental Excellence in the North American Ski Industry (2005), the First Choice Responsible Tourism Award for Best in Mountain Environment (2006), and the British Columbia Award for Environmentally Responsible Tourism (2006), the only Canadian ski resort to receive an 'A' grade on the Ski Area Environmental Scorecard issued by Ski Area Citizens' Coalition, and large number of invitations to speak on the resort's sustainability initiatives (De Jong-personal communication 2007), Whistler Blackcomb is recognised as a leader in sustainability among ski resorts. Consequently, it was anticipated that Whistler Blackcomb would rate higher on the ideal type analysis than most ski resorts. While perhaps atypical in the ski industry, the Whistler Blackcomb case study is illustrative of initiatives that can be undertaken with regard to all eight of the proposed principles of sustainability, but also that even the most progressive tourism businesses have opportunities to become more sustainable.

The process of conducting ideal type analysis rests on viewing the current conditions and trends at Whistler Blackcomb through the lens of the sustainable ski resort principles. The analysis identifies whether the requirements of the sustainable ski resort principles are:

- Fully Present—Current conditions mirror the sustainable ski resort requirements;
- Strongly Realised—The sustainable ski resort requirements are well reflected, often in multiple dimensions, in current conditions;
- Partially Realised—Current conditions reflect some of the sustainable ski resort requirements; or,
- Absent—The sustainable ski resort requirements are absent.

The results of the ideal type analysis of Whistler Blackcomb are described below and summarised in Table 8.4.

Principle 1—Socio-ecological system integrity: Strongly Realised

Whistler Blackcomb has undertaken a number of initiatives to maintain the mountain's ecological resources. Over $1.5 million (since 1999) has been spent on mountain habitat restoration done in the ski area, including planting strategies that enhance wildlife habitat, utilisation of native species where possible, limit erosion within community watersheds to protect fish habitat (to-date seventy-two projects have been completed by a group of volunteers, organised by Whistler Blackcomb, and referred to as the Habitat Improvement Team. In 2006, a new chairlift was constructed in a sensitive sub-alpine zone. All construction was done over snow and with helicopters minimising the ecological impacts and Biologists, bear specialists and professional foresters were used in the planning to help reduce a footprint from 40% of the area down to only 5 per cent.

Whistler Blackcomb also contributes to programmes and engages its staff in initiatives that serve to strengthen the social fabric of the host community and the well being of the biophysical environment. Since 1992, the Whistler Blackcomb Foundation has raised funds ($3.8 million from 1992 to 2007) that support charitable social causes as well as contribute toward the preservation and restoration of biophysical environments, including support of the Protected Areas Network (a local government conservation programme), and the creation (in 2001) and ongoing support of the Whistler Blackcomb Foundation Environmental Fund.

Principle 2—Livelihood sufficiency and opportunity: Partially Realised

Whistler Blackcomb provides recreational and education opportunities to several disadvantaged groups. The 'Whistler Adaptive Ski Program' supports access to snow sports by physically and mentally challenged persons. The 'Zero Ceiling Society' offers snowboarding programmes to youth at risk from Vancouver. Similarly, the 'Whistler Blackcomb Outreach for Youth Program' provides mountain recreational experiences for inner city youth from Vancouver and child burn victims through the BC Firefighters Burn Kamp. First Nations communities also participate in numerous capacity building programmes with Whistler Blackcomb.

Whistler Blackcomb also offers its guests opportunities to learn and take positive action toward environmental stewardship. For instance, over forty-five interpretive sites have been established on the Whistler and Blackcomb mountains for visitors to learn about mountain ecosystems and understand what they are seeing.

Principle 3—Intra-generational equity: Strongly Realised

Whistler Blackcomb is not unique in terms of its challenges with respect to the need to provide affordable housing for ski resort employees in communities where real-estate prices have been driven up substantially by investment in recreational homes (Gill 1991; Gill 1997; Gill and Williams 1994; Laing 1998; Schendler 2005). The relatively low-end wages provided to Whistler Blackcomb seasonal staff, which is not atypical in the tourism sector, are addressed in part through the provision of employee housing on site (Hakala 2004).

Internationally, Whistler Blackcomb has raised funds and winter clothing, computers and books for projects in Romania, Nepal, Kashmir, and the Ukraine. Consulting support through staff time and travel has been provided to the UNEP for protecting UNESCO World Heritage Sites in mountainous regions and to help communities trying to promote tourism in their areas.

Principle 4—Inter-generational equity: Partially Realised

Efforts to reduce environmental impacts of current resort operations, particularly climate change focused strategies and ecosystems restoration (described under principle 1), could be considered investments for future generations. Such efforts conform to the Environmental Mission Statement of the National Ski Areas Association, which includes reference to providing future generations opportunities to ski, while overlooking many other dimensions of sustainability discussed in the eight principles: 'We are committed to improving environmental performance in all aspects of our operations and managing our areas to allow for their continued enjoyment by future generations' (NSAA 2005). Similarly, youth focused programmes (see principle 3) and environmental education programmes (see principle 6) also contribute to opportunities for future generations.

Principle 5—Resource maintenance and efficiency: Strongly Realised

Substantive progress has been made by Whistler Blackcomb to improve efficiency of energy, water and materials use. Water consumption in the resort has been reduced through the implementation of low-flush toilets in some resort facilities and waterless urinals in others. Filters have been put on drinking water taps at many locations to reduce the need to use bottled water. Slope operations have focused on reducing leaks through the replacement of snowmaking reservoir liners and the replacement of snowmaking pipe to also eliminate corrosion and the contribution of sediment to local water courses.

Whistler Blackcomb has also been successful at diverting 59 per cent of its waste from landfill (between 2001 and 2005) through an onsite recycling programme for than twenty-five types of materials, reuse of items and materials through a new 'Mountain Materials Exchange Program' (for ski equipment, desks and office equipment, dishes, furniture, and clothing), and composting food waste at a nearby facility (DeJong. personal communication, April 30,

2008). The resort has also reduced its material and resource consumption on a case-by-case basis by working with supply-chain providers, for example by purchasing reusable/recyclable products for food service and using recycled content paper products.

Energy efficiency gains and greenhouse gas emission reductions have been achieved through building retrofit programmes (lighting, insulation, energy systems controls, etc.) that combined reduced the resort's annual electricity consumption by 4,575,000 kWh. Vehicle fleets have been renewed with higher efficiency and in some cases hybrid models. Half of the snowmobile/grooming fleet has been replaced with low-emissions models as well. An employee carpooling programme is estimated to reduce carbon emissions by 617,000 kilograms annually. Whistler Blackcomb is also investing in the development of run-of-river micro-hydro projects that when completed would provide the annual electricity needs for its mountain operations (see Box 1).

Principle 6—Socio-ecological civility and democratic governance: Strongly Realised

Whistler Blackcomb has created a governance structure of four working groups that focus on sustainability planning in the areas of water, energy, transportation, and procurement (Szpala 2007). The resort also invests substantially in staff environmental training, with the approximately 50 sessions related to sustainability programmes offered per year reaching more than 2000 staff members. These initiatives foster a sense of shared responsibility amongst its staff—a key requirement of the principle of socio-ecological civility and democratic governance.

Whistler Blackcomb staff also contribute to sustainability focused education and governance off-site. Staff has established an Intrawest Sustainability Network to provide a forum of exchanging information

---

Box 8.1   On-site Renewable Energy Initiatives

Whistler Blackcomb has been producing renewable energy since 2006 when a half-megawatt run-of-river turbine was installed in Flute Creek (De Jong, personal communication 2008). Current construction is now underway to build a 6.6-megawatt run-of-river hydroelectric generation station in Fitzsimmons Creek between Whistler and Blackcomb mountains. This on-site renewable energy initiative has the potential to generate 32 gigawatt hours of electricity annually—an amount equivalent to what is consumed annually by Whistler Blackcomb as a ski operation (De Jong, personal communication 2008). Further, through its meteorological towers, Whistler Blackcomb continues to monitor wind patterns to help determine the viability of wind generation (Arthur De Jong, personal communication 2008).

on sustainability programmes and success stories with other company resorts. Staff also participates in a range of community events to showcase sustainability programmes, including Earth Day, Clean Air Day, Environment Week, the Commuter Challenge, and Mountain and Valley Clean up Days.

Principle 7—Precaution and adaptation: Strongly Realised

Whistler Blackcomb has pursued aspects of the principle of precaution and adaptation by investing in long-term environmental planning and more specifically, climate change mitigation and adaptation programmes. For instance, Whistler Blackcomb contributes to public understanding and the development of climate change mitigation and adaptation strategies by working with international organisations as well as the provincial government (SACC 2007). Further, Whistler Blackcomb has taken a leadership role by investing in on-site renewable energy generation, as described in Box 1, as a long-term planning measure that will contribute to the mitigation of climate change and the generation of a clean energy source. With stated goals for solid waste and energy use reduction, and many habitat restoration projects, the resort also invests in environmental research and monitoring.

Principle 8—Immediate and long-term integration: Partially Realised

With an Environmental Management System, a Mountain Planning and Environmental Resource Manager, two Environmental Coordinators, and a substantive investment in sustainability training of full- and part-time staff at the resort, sustainability has become part of the culture and represents an important mandate within many aspects of business planning at Whistler Blackcomb.

As demonstrated, Whistler Blackcomb has taken a leadership role among other ski resorts, through its actions on ecosystems restoration, employee housing, solid waste reduction, and its investments in on-site renewable energy generation. In addition, Whistler Blackcomb's governance structure forms a significant foundation to further advance towards sustainability. Can more be done? Despite these positive conditions, further efforts are needed to build Whistler Blackcomb's capacity to: decouple improvements in quality and service from further growth and resource consumption; redistribute efficiency gains in areas that are deficient in natural and social capital; proactively foster a shared sense of responsibility among external stakeholders (supply-chain, investors, customers); and invest in research and monitoring for greater understanding of the socio-ecological footprint which will serve to inform short-term and long-term targets to stimulate continual improvement. The following sustainability principles were found to be limited at Whistler Blackcomb, demonstrating how in some instances

Table 8.4 Application of Sustainable Ski Resort Principles to Whistler Blackcomb

Sustainable Ski Resort Principles	Existence of Sustainable Ski Resort Requirements at Whistler Blackcomb			
	Absent	Partially Realised	Strongly Realised	Fully Present
Socio-ecological system integrity			*	
Livelihood sufficiency and opportunity		*		
Intra-generational equity			*	
Inter-generational equity		*		
Resource maintenance and efficiency			*	
Socio-ecological civility and democratic governance			*	
Precaution and adaptation			*	
Immediate and long-term integration		*		

it falls short of decisions that contribute to the sustainable ski resort requirements as a whole:

Principle 2—Livelihood sufficiency and opportunity

Of the education initiatives that Whistler Blackcomb has created specifically for the skiing public, the emphasis is placed on creating a sense of ecological awareness through the establishment of interpretive sites and ecology tours throughout the resort area. These initiatives, while commendable, fall short of using education initiative to create preferences amongst the skiing public for more sustainable consumption as required by the principle of livelihood sufficiency and opportunity. Like other resort environments, an inherent contradiction exists when advocating the need for environmental stewardship as Whistler Blackcomb delivers experiences which often involve indulgence. Opportunities to address this concern may be found through 'buy local initiatives', which support the goals of sustainable consumption.

Principle 4—Inter-generational equity

Future generations are not formally acknowledged as a stakeholder by Whistler Blackcomb thereby the principle of intergenerational equity is partially realised. A similar gap is noted in the current state of the sustainable ski resort discussions in both academic literature and industry reporting, where there appears to be a missing linkage between acknowledgment of pursuing sustainability for the benefit of future generations and guidance on operating in the interests of future generations.

Principle 8—Immediate and long-term integration

While Whistler Blackcomb has demonstrated leadership in climate change mitigation, some of its activities in this area have come at the expense of other sustainable ski resort principles, therefore the principle of immediate and long-term integration is absent. For instance, the development of Whistler Blackcomb's high elevation ski terrain, as described in Box 2, was driven by the need to access the climate resources that would sustain ski operations into the mid-century under climate change scenarios (Honey, DeJong, and Schendler 2005; DeJong, personal communication, April 30, 2008). Consequently the ski terrain at Whistler Blackcomb has grown beyond its existing socio-ecological footprint thereby falling short of what is required by the principle of socio-ecological integrity. This point further emphasised issues of scale in integrated systems, because if the ski industry is seen holistically, and not at the ski resort level, then while Whistler Blackcomb and other resorts develop high elevation terrain to exploit needed climate resources, many other low lying ski areas are projected to become unprofitable and close ski runs (Bürki, Elsasser and Abegg 2003; Scott 2005; OECD 2007; Scott, McBoyle and Minogue 2007; Scott 2007); thus perhaps reducing the net footprint of the ski industry over time.

Box 8.2   High Elevation Terrain Expansion

The development of high elevation ski terrain is emerging as an adaptation strategy to climate change (Bürki et al. 2003; Scott and McBoyle 2007). Such development can be found amongst ski resorts in mountainous regions where at low elevations the impacts of climate change are evident in snow deficient winters. Climate models for Whistler Blackcomb indicate that the development of high elevation ski terrain (i.e., above 1800 metres) is needed in order to sustain the resort as a successful ski area into the mid century should an increase of 3 degrees Celsius be realised (Honey, DeJong and Schendler 2005; De Jong, personal communication, April 30 2008). In 2006, Whistler Blackcomb expanded its terrain by more than 1000 acres (SACC 2007), into a high elevation alpine area known as the Piccolo Peak/ Symphony Basin (Whistler Blackcomb 2008). Whistler Blackcomb developed a minimalist design to help reduce the footprint of this ski terrain expansion from the initial estimate of 40 per cent of the area down to only 5 per cent. Despite these measures, this expansion project affected fifty acres of old growth trees, which may have affected the habitat for the Coast Tailed Frog—a species at risk (SACC 2007).

Box 8.3 Snow Farming and Snow Making

Snow farming and snow making have been commonly practiced at ski resorts as a strategy to cope with climate variability for twenty-five years and will increasingly be relied on as climate change adaptation (Scott 2005). Whistler Blackcomb augments its natural snow with snowmaking at elevations below 1800 meters and relies on grooming equipment and helicopters to assist with snow farming effort). In addition, Whistler Blackcomb is adapting to less reliable snow levels by making its ski trails flatter—a practice referred to as trail grooming or land smoothing. The above practices allow ski resorts to open their trails earlier and with less snow. Although concerns have been raised about the environmental impacts of snowmaking, little scientific research has been conducted on its differential regional impacts on mountain vegetation, watersheds, or energy use and related GHG emissions; the latter of which are particularly affected by issues of scale. Snowmaking has been referred to as a perverse climate change adaptation because the large projected increase in snowmaking under climate change scenarios will further contribute to the very problem that requires this adaptation. While the large energy use for snowmaking is a very valid concern, both technical and regionally specific social factors must be considered in evaluating its sustainability. Technologically, snowmaking can be done with renewable energy and thus not contribute further to climate change. In the United States, sixty-one ski resorts now purchase renewable energy (primarily wind) for part of their operational energy use, and twenty-nine purchase 100 per cent of their energy needs from renewable sources (NSAA 2008b). Furthermore, the broader social context of regional tourism activity should be considered when determining the net impact of additional snowmaking activity on GHG emissions. If local ski areas are forced to close because of unreliable natural snow conditions, potentially thousands of skier visits may be displaced to other ski destinations or other forms of tourism. Alternate ski destinations may be nearby (less than 100 km away), but as was found in a study of potential ski area closures in the New England region of the USA, in many cases ski tourism may be transferred 500km or more (Scott 2007), with attendant emissions from car, coach, rail, or even air travel. If no alternate ski holiday is available regionally, skiers may opt for much more GHG intensive travel options.

Further, like other ski resorts around the world (Scott 2005), Whistler Blackcomb is investing heavily in the climate adaptation strategies of snow farming and snow making, as described in Box 3, which have sustainability impacts in terms of increase water and energy use as well as impacts to the ecological integrity of mountain vegetation. As with terrain expansion, the question of the sustainability of snowmaking is also dependent on issues of scale and become more complicated when the broader ski tourism industry is considered (see Box 3), however at the resort level the development of renewable energy onsite is at odds with current practice to promote and provide heli-ski tours, that when combined with any long-haul flight to visit the resort, are among the most energy intensive forms of tourism.

The above findings suggest that Whistler Blackcomb, and by extension the many other ski resorts not recognised as environmental leaders, do not presently address sustainability challenges in an integrated way as set out in the sustainability principles (Table 8.2). This is not surprising as the ideal type analysis performed in this chapter revealed that the current understanding of sustainable ski resorts falls short of the guidance provided by the sustainable ski resort principles in several key areas. Therefore, the use and integration of the sustainable ski resort principles into everyday decision-making would represent a fundamental advance towards ski resorts, including Whistler Blackcomb, becoming more sustainable.

REFLECTIONS AND CONCLUDING REMARKS

The sustainable ski resort principles (Table 8.2) offer a holistic interpretation of sustainability comprised of interdependent requirements, thereby broadening the scope of decision-making beyond the conventional categories (i.e., environmental, economic, and social). The integrated systems approach upon which this interpretation of sustainability is based makes it very difficult for ski resorts and other types of tourism destinations to fulfill all the requirements, which in reality may only be possible in the very long-term or perhaps not at all (Del Matto 2007). Despite these difficulties, the sustainable ski resort principles are successful at providing a vision of what an ideal sustainable ski resort should look like. They do not offer operational guidelines for applying the principles to both proposed and existing ski resort activities within all levels of decision-making. This is not considered to be a drawback, however, because the operational guidelines should be developed in collaboration with ski resort owners, operators, staff, and other stakeholders and so as to elicit their full commitment.

When applied, the sustainable ski resort principles serve to distinguish between sustainable outcomes and unsustainable outcomes and thus help to guide ski resorts on a path to sustainability. The integration of the proposed sustainable ski resort principles will infuse sustainability-based consideration into a wider range of decision-making. Without the sustainable ski resort principles, it is more likely that decisions made by ski resort owners and operators will lead to unsustainable outcomes.

In conclusion, this work opens opportunities for practitioners and stakeholders to use these principles and consider what fundamental changes need to take place for ski resorts to move towards sustainability. The sustainable ski resort principles provide a model for communities and the ski resort industry to evaluate current operations and decision-making processes, in order to determine what needs to take place to advance a ski resort further on the continuum of sustainability. The principles are also transferable to other areas of tourism and future work could interpret them for the specific application to other recreation and tourism destinations, such as coastal resorts.

NOTES

1. For this chapter, a 'ski resort' is considered to encompass all aspects of operations and infrastructure related to ski activities and tourism, including: ski slope operations (ski lifts, snowmaking, and grooming), hospitality and accommodations, and other retail-services (including equipment rentals, lessons, etc.). While highly integrated in North American markets, these various businesses can be largely independent in markets like Europe.

REFERENCES

Aspen Skiing Company (2006) *Sustainability Report,* Aspen, CO. Available at: http://www.aspensnowmass.com/environment

Bürki R., Elsasser, H., and Abegg, B. (2003) 'Climate change-impacts on the tourism industry in mountain areas'. Presentation given on April 9, 2003 at the *First International Conference on Climate Change and Tourism.* Djerba: Tunisia.

Castle, K. (2004) 'Vibrant villages or ghost towns?' *Ski Area Management* May 2004 43(3): 50.

Clifford, H. (2002) *Downhill Slide—Why the Corporate Ski Industry is Bad for Skiing, Ski Towns and the Environment.* San Francisco: Sierra Club Books.

DeJong, A. (2006) 'Pro-actively responding to climate change: Diversification, mitigation and adaptation strategies at Whistler Blackcomb'. Presentation at the *Alpine Resort Sustainability Forum.* Victoria, Australia. Available at: http://www.arcc.vic.gov.au/sustainabilityforum2006.htm

Del Matto, T. (2007) *Conceptualizing a Sustainable Ski Resort: A Case Study of Blue Mountain Resort in Ontario.* Master's Thesis. Waterloo, ON: University of Waterloo, Department of Environment and Resource Studies.

Eydal, G. (2004) *The Development of a Sustainability Management System for Ski Areas.* Master's Thesis. Burnaby, BC: Simon Fraser University/School of Resource and Environmental Management.

Flagestad, A. and Hope, C.A. (2001) 'Strategic success in winter sports destinations: A sustainable value creation perspective'. *Tourism Management* 22: 445–461.

Gibson, R.B. (2002) *Specification of Sustainability-Based Environmental Assessment Decision Criteria and Implications for Determining 'Significance' in environmental assessment.* Available at: http://www.ceaa-acee.gc.ca/0010/0001/0003/rbgibson_e.pdf

Gibson, R.B. (2006) 'Beyond pillars: Sustainability assessment as a framework for effective integration of social, economic and ecological considerations in significant decision-making'. *Journal of Environmental Assessment Policy and Management* 8(3), 259–280.

Gibson, R.B., Hassan, S., Holtz, S., Tansey, J., and Whitelaw, G. (2005) *Sustainability Assessment: Criteria, Processes and Applications.* London, UK: Earthscan.

Gill, A.M. (1991). "Issues and problems of community development in Whistler, British Columbia". In Gill, A. and Hartrnann, R. (eds.), *Mountain Resort*

Development: Proceedings of the Vail Conference, 27–31. Burnaby, BC: Simon Fraser University/Centre for Tourism Policy and Research.

Gill, A.M. (1997). 'Competition and the resort community: Towards an understanding of residents needs'. In Murphy, P.E. (ed.), *Quality Management in Urban Tourism.* West Sussex, England: John Wiley and Sons Ltd.

Gill, A.M. and Williams, P. (1994). 'Managing growth in mountain tourism communities'. *Tourism Management* 15(3): 212–220.

Hakala, K. (2004) *Affordable Housing in Resort Mountain Communities: Blue Mountain, Ontario.* Master's Thesis. Kingston, ON: Queen's University/School of Urban and Regional Planning.

Hennessy, K., Whetton, P., Smith, I., Bathols, J., Hutchinson, M., and Sharples, J. (2003) 'The impact of climate change on snow conditions in mainland Australia'. *CSIRO Atmospheric Research* 47.

Honey, M., DeJong, A., and Schendler, A. (2005) *Implementing Sustainable Winter and Summer Tourism in Northern and Central Montenegro: An Assessment of Current Strategies and Next Steps.* The International Ecotourism Society.

Hudson, S. (2000) *Snow Business: A Study of the International Ski Industry.* London: Wellington House.

Laing, C. D. (1998). *Developing Affordable Resident Housing in Ski Resorts: Municipal Programs and Policies for Whistler, British Columbia.* Master's Thesis. Winnipeg, MB: University of Manitoba/Department of City Planning.

Lehtonen, M. (2004) 'The environmental—social interface of sustainable development: Capabilities, social capital, institutions'. *Ecological Economics* 49(2): 199–214.

Lewis, J. (2005) *Sustainable Alpine Tourism: The British Ski Industry's Role in Developing Sustainability in the French Alps.* Master's Thesis. London: Imperial College, University of London.

Moore, S.R. (2005) *Corporate Social Responsibility and Stakeholder Engagement: A Case Study of Affordable Housing in Whistler.* Master's Thesis. Burnaby, BC: Simon Fraser University / School of Resource and Environmental Management.

NSAA (National Ski Areas Association) (2000) *Sustainable Slopes—The Environmental Charter for Ski Areas.* Available at: http://www.nsaa.org/nsaa/environment

NSAA (National Ski Areas Association) (2005) *Sustainable Slopes—The Environmental Charter for Ski Areas,* Lakewood, CO. Available at: http://www.nsaa. org/nsaa/environment

NSAA (National Ski Areas Association) (2006) *Sustainable Slopes Annual Report 2006,* Lakewood, CO. Available at: http://www.nsaa.org/nsaa/environment

NSAA (National Ski Areas Association) (2008a) *Keep Winter Cool Campaign.* Available at: http://www.nsaa.org/nsaa/environment/climate_change/keep_ winter_cool.asp

NSAA (National Ski Areas Association) (2008b) *Green Power Program Continues to Snowball.* Available at: http://www.nsaa.org/nsaa/press/0607/green-power.asp

OECD (2007) *Climate Change in the European Alps: Adapting Winter Tourism and Natural Hazards Management.* Paris: OECD. Available at: http://www. oecd.org/document/45/0,2340,en_2649_34361_37819437_1_1_1_1,00.html# (19 December 2006).

Pope, J., Annandale, D., and Morrison-Saunders, A. (2004) 'Conceptualizing sustainability assessment'. *Environmental Impact Assessment Review* 24: 595–616.

Rixen, C., Stoeckli, V., and Amman, W. (2003) 'Does artificial snow production affect soil and vegetation of ski pistes? A review'. *Perspectives in Plant Ecology, Evolution and Systematics* 5(4): 219–230.

SACC (Ski Area Citizens' Coalition) (2001) *Environmental Score Card Report 2001.* Available at: http://www.skiareacitizens.com

SACC (Ski Area Citizens' Coalition) (2007) *Welcome to the Ski Area Environmental Scorecard 2007–2008 Canadian Edition*. Available at: http://www.utsb.ca/research_ski_areas.html

Schendler, A. (2003) 'Applying the principles of industrial ecology to the guest-service sector'. *Journal of Industrial Ecology* 7(1): 127–138.

Schendler, A. (2005) *Sustainable Resort Development: A Report Prepared for The International Ecotourism Society and Rockefeller Brothers Fund*. Aspen, CO: Aspen Sustainability Associates.

Scott, D. (2005) 'Global environmental change and mountain tourism'. In Gössling, S. and Hall, M. (eds.) *Tourism and Global Environmental Change*, 54–75. London: Routledge.

Scott, D. (2007) 'Impacts on winter recreation'. In Frumhoff, P., McCarthy, J., Melillo, J., Moser, S., and Wuebbles, D. (eds.) *Confronting Climate Change in the US Northeast: Science, Impacts, and Solutions*, 81–90. Cambridge, MA: Union of Concerned Scientists.

Scott, D. and McBoyle, G. (2007) 'Climate change adaptation in the ski industry'. *Mitigation and Adaptation Strategies to Global Change* 12(8): 1411–1431.

Scott, D., McBoyle, G., and Mills, B. (2003) 'Climate change and the skiing industry in Southern Ontario: Exploring the importance of snowmaking as a technical adaptation'. *Climate Research*: 23: 171–181.

Scott, D., McBoyle, G., and Minogue, A. (2007) The implications of climate change for the Québec ski industry. *Global Environmental Change* 17: 181–190.

Szpala, M. (2007) *A Natural Step Case Study: Whistler Blackcomb*. The Natural Step Canada.

Todd, S.E. and Williams, P.W. (1996) 'From white to green: A proposed environmental management system framework for ski areas'. *Journal of Sustainable Tourism* 4(3): 147–173.

Whistler Blackcomb (2008) *Whistler Blackcomb—Official Ski Resort Website—Environment Page*. Available at: http://www.whistlerblackcomb.com/mountain/environment/index.htm

Wipe, S., Rixen, C., Fischer, M., Schmid, B., and Stoeckli, V. (2005) 'Effects of ski piste preparation on alpine vegetation'. *Journal of Applied Ecology* 42: 306–316.

9 Piloting a Carbon Emissions Audit for an International Arts Festival under Tight Resource Constraints
Methods, Issues and Results

Paul Upham, Philip Boucher, and Drew Hemment

INTRODUCTION

Arts and music festivals, shows, concerts, and similar events are all important leisure and tourist activities, not least in terms of the number of people involved. There are thousands of live performances in the UK each year, with a total of millions of people attending live music events (Bottrill et al. 2008). In 2007, some 85 major festivals took place in the UK alone, with a total audience of over 2 million people (ibid.). This chapter describes the method used to approximate the 'carbon footprint' of a relatively small, urban music festival.

It is relatively rare to find a detailed description of audit methodology in the open literature. It is also relatively rare to find a detailed account of the problems and decision processes involved, including the influence of a very small budget and limited data. Yet these problems and decisions are typical in carbon auditing where relevant data collection has not been on-going and where some of the relevant emission factors (explained below) are difficult to obtain. The main aim of the chapter, therefore, is to lay bare the first-time emissions auditing process of a leisure/tourist event. The report on the audit, together with a flyer and the prototype spreadsheet are freely downloadable from: http://www.futuresonic.com/07/eco2.html. It should be noted that this chapter is an updated version of some of the text in that report. While the issues and methods discussed here are applicable to any developed country, some of the emission factors are UK-specific.

The Festival and Partnership

The carbon emissions audit relates to *Futuresonic*, an annual three day international festival of electronic music and media arts in Manchester that attracts around 10,000 people. The festival began in 1996 and is presented by Future Everything, a not-for-profit community interest company and Regularly Funded Organisation client of Arts Council England. Future Everything established a partnership with Tyndall Centre Manchester, at

The University of Manchester, to help begin its work on measuring, managing and mitigating its contribution to climate change. The communications side of the partnership was assisted by Creative Concern Ltd, a sustainable development communications agency in Manchester.

Rationale for the Audit

The cultural sector is an integral part of the regional and national economy: The cultural sector accounts for 12 per cent of Gross Value Added (GVA), compared with, for example, 23 per cent of regional GVA by rural businesses and 6 per cent by the Financial and Business Service sector (Culture Northwest 2004). The sector is defined by the UK Government Department for Culture, Media and Sport (DCMS) as comprising seven domains and two sub-sectors, including tourism (DCMS 2007). The conceptualisation and categorisation of tourism alongside cultural events and practices emphasises the inter-relationships of travel and entertainment. The NW Regional Development Agency (NWRDA 2006) reports that the visitor economy is worth £10.9bn to the region, supporting 200,000 FTE jobs.

Effectively tackling the carbon emissions of the cultural sector will require supportive policy incentives for low greenhouse gas (GHG) technologies at the national level. Yet there remain political obstacles to this degree of change. Although quantitative, survey-type opinion polls in the UK repeatedly show substantial public concern about climate change (Anable, Lane, and Kelay 2006, 20), qualitative studies that ask people about their environmental concerns in the comparative context of other social and personal problems find that these other issues frequently take precedence (Poortinga and Pidgeon 2003 in Anable et al. 2006, 12). People are thus concerned about climate change and environmental protection, but these issues are still not top priorities for them. They also tend to expect, with some justification, that it is government that must take the lead on these issues.

Raising public awareness of the profound implications of climate change, and of the role of everyday consumption and behaviour, thus needs to remain a priority in the challenge of tackling climate change itself. Awareness and attitude change remain important precursors to effective emissions reduction. The act of conducting a carbon emissions audit and creatively communicating its results should help, in a small but real way, to bring climate issues to a higher level of salience. More specifically, it should make more obvious the fact that everyday leisure activity under present technological conditions nearly always has implications for climate change.

Previous Festival Emissions Audits

In the absence of compulsory emissions mitigation strategies, a number of festivals have undertaken emissions audits for the purpose of voluntary offsetting. Unfortunately, much of the public information on the offsetting of music events is in the form of short news articles, with little detail on methodology. Nevertheless, the information that is publicly available is summarised here.

Multi-venue Music Tours

The first offset activity for a music event tour seems to have been performed by The Levellers in 1998, including emissions resulting directly from the use of the venue, plus accommodation and transport for both fans and musicians. The audit was performed by the Edinburgh Centre for Carbon Management (ECCM), and the offset executed by Future Forests[1], which had formed the preceding year (Layne 1998). Since then, various other music events have been audited and offset with the same audit method, predominantly performed by ECCM and The CarbonNeutral Company (TCNC). Many of these have been high profile events, including the Foo Fighters, Pink Floyd, David Gray, and the Rolling Stones, for whom 90 per cent of emissions were reportedly caused by attendee travel (Masson 2003). The cost of forestry offsets for these were met by adding 15 pence to each ticket, paid for by tour sponsors (ibid.). Pearl Jam appears strongly committed to emissions mitigation activity (e.g., including the introduction of production trucks and tour buses running on 100 per cent biodiesel (The Ten Club 2006). Tour emissions are offset via donations to a portfolio of organisations involved in forestry protection and renewable energy (Van Schagen 2006). However, no detailed description of Pearl Jam's audit method is publicly available.

Music Festivals

T-In the Park, a major Scottish music festival held near Kinross has been carbon-audited by TCNC and its emissions offset. The scope of the audit included attendee and artist transport, accommodation, waste, and onsite energy use (VisitScotland Perthshire 2005). In 2007, the South by Southwest festival in Texas was carbon audited using projections from 2006 data, though only core business activity was included. Attendee and artist related emissions, which would have significantly increased the published 'footprint', were omitted. South by Southwest's offset comprised a mixture of native tree planting via a donation to the local parks department and the purchase of carbon credits from Austin's Green Mountain Energy Company (Dunn 2007; NME 2007).

Other Events

A carbon emissions audit was undertaken for The G8 summit of 2005, under UK Presidency in Gleneagles, Scotland, in conjunction with the UK Department for Environment, Food and Rural Affairs (DEFRA). The emissions audited were those resulting from all transport, accommodation, and venue usage. Offsetting followed via the purchase of 10,000 carbon credits with Gold Standard Clean Development Mechanism (CDM) certification, the projects for which took place in Cape Town, South Africa (DEFRA 2005a). Emissions from the 2006 FIFA World Cup, including significant attendee air travel, were also offset with Gold Standard CDM certified credits (MyClimate 2006). Again, detailed information on how these audits were performed is not readily available to the public.

AUDIT METHODS

General Approach

We developed and applied a practicable method with which to estimate a baseline for the festival's emissions, with the potential for future improvements in accuracy. Data collection was undertaken by supervised undergraduate students and without a research budget[2]. This limited the extent to which we could survey visitors regarding their origins and destinations, and also limited our access to venue energy consumption data. The report and audit spreadsheet (which could be developed further) also had to be finalised in a tight timescale to allow for dissemination of the results in time for the festival itself. These constraints may well be similar to those under which other events organisers would operate. The research process has thus been of the action research type and undertaken in far less than ideal conditions. Nonetheless, all assumptions are kept explicit.

Scope

The scope of the audit is limited primarily (but not wholly) to significant, first-order carbon dioxide emissions. These are defined as CO_2 emissions that are directly attributable to the festival and that would not have otherwise occurred. By 'directly' is meant 'at the point of consumption or activity' and not over the life cycle of products used or activities undertaken. This is typical of such audits. In future, where life cycle data is readily accessible, these may be noted as 'second-order plus' emissions. Judgment has been used to select those emission categories likely to be most significant. In future years, the scope of the information collated could be expanded. This could also extend to other gaseous, solid, and liquid wastes and to resource inputs to the festival.

Transport, visitor and hence emissions data relate to the previous year, as data for 2007 was not available at the time of writing, nor in time to produce results that could be communicated to the 2007 festival audience. However, commitment to the audit process has already begun to improve monitoring processes. The central ticket booking website now asks purchasers for information on their point of origin, and volunteers will ask a sample of festival participants, across venues, for their originating postcode and mode of travel.

The emissions categories for 2007, using 2006 data, are:

- transport of artists to and from the festival (split by distance).
- transport of visitors and delegates (the festival includes a conference) to and from venues (split by distance).
- venue 'background' energy consumption.
- artists' equipment electrical consumption.
- hotel energy consumption per visitor-night.

These categories were selected by the research team on the basis of what were measurable and likely to be significant, first order emissions sources. The indicators and their metrics are those that are commonly used in relation to energy and emissions accounting. The following sections relate in turn to the emissions factors (explained in the following) used within the emissions categories.

Emission Factors

Emissions factors are usually derived via an averaging process to more-or-less accurately represent the emissions of types of engine under a range of conditions (e.g., DEFRA 2007—note that these factors are more recent than those used in the present study). The emission factor is multiplied by a quantity reflecting use: e.g., distance or duration. The accuracy of the emissions estimate obtained via the use of emissions factors can be improved by, for example, using vehicle-specific factors rather than a factor that averages across all vehicle of that type. The intention is in this project, where possible, is to work towards use of vehicle-specific factors in future—for example, representative aircraft types for specific routes.

Aircraft Emission Factors

Aircraft emissions estimates are affected by assumptions relating to aircraft load factor, type, age, operational procedures, and capacity, in addition to the most obvious variable—flight distance. As it is impossible to account for these variables on a per-flight basis, most UK-based offsetting organisations use DEFRA's (2005) average factors (now updated as [DEFRA 2007]). These are gross averages from which the emissions of an individual aircraft may depart substantially, and are presented on the basis of a kg per passenger kilometer metric (i.e., kg of CO_2 one passenger being transported one kilometer). The factors used are:

- Aircraft (Short Haul) 0.18kg per pkm
- Aircraft (Long Haul) 0.11kg per pkm
- Aircraft (for air passengers of unknown origin) 0.145kg per pkm

In addition to the limitations of using gross average factors is the uncertainty and debate over how best to account for the additional warming effects of non-CO_2 aircraft emissions, relative to CO_2. Research under the European Commission-funded TRADEOFF project (Sausen et al. 2005), which provided updated estimates of the atmospheric effects of aviation emissions, estimated the mean, net effects of these to be 1.9 times higher than CO_2. However, this does not include contrail-induced cirrus cloud. Including cirrus may substantially increase the warming effects of high altitude aircraft emissions, relative to surface emissions of CO_2, to a theoretical maximum of 5.1. However, it should be noted that not only is an accurate multiplier unknown, but the calculation appropriate for indicating these warming effects is keenly debated, due to their region-specific effects and, as with other greenhouse gases, their substantially differing lifetimes (e.g., Shine et al. 2005). Furthermore, the above multipliers are strongly influenced by the warming effects of historic CO_2 emissions by aircraft (due to the long life of CO_2), and do not simply reflect the impact of GHG emitted now for future generations (Wit et al. 2005; Forster, Shine, and Stuber 2006; Peeters, Gössling, and Williams 2006).

Gössling et al. (2007) show that, of thirty-five offsetting organisations with online calculators, half currently use no multiplier to account for non-CO_2 effects, one quarter use a factor of 2.0, and 5 companies use 2.7 or 3.0. Four other companies allow the customer to decide to include a higher/lower radiative forcing index (RFI, sometimes called 'uplift factor'). The use of different multipliers will heavily influence the calculation of the amount of CO_2-e generated by a given flight. In this project, we provided aircraft emissions estimates with and without the application of a multiplier of three. Given the range 1.9 to a possible 5.1, a value of three is plausible. Nonetheless it is important to emphasise that such 'uplifted' CO_2 values need to be treated as highly indicative. RFI is defined in relation to aviation as a sector, operating in a defined time period in the past, or in future scenarios. Technically, it cannot be used in relation to individual flights because the different gases emitted and atmospheric effects caused by aircraft vary by geographical location and lifetime. They are thus qualitatively different to GHGs that become well-mixed in the atmosphere and hence are comparable via the Global Warming Potential metric (Peeters, Williams, and Gössling 2007). However, while uplift or RFI should not be used in emissions accounting in relation to individual flights, it arguably has an important role to play in communicating to travelers the additional warming effects of high altitude aviation emissions. For this reason it is used here alongside CO_2-only values. Finally, it should be noted that we did not take into account transportation to and from airports by car, bus, train, etc., as this information was not available to us.

Surface Transport Emission Factors

As we did not know the specific types or classes of cars used by visitors, the emission factor for cars is derived from the application of a weighted average reflecting the proportions of diesel and petrol-fuelled cars in use in the UK, to DEFRA's (2005b) emissions factors for average cars of those types. That is: $(0.206 \times 0.17 + 0.792 \times 0.18) = 0.178$kg CO_2/km. These values are themselves derived from the following information:

Total number of private cars licensed in the UK in 2005: 26,207,700 (DfT 2006a)
Number of private petrol cars licensed in the UK in 2005: 20,762,200 (DfT 2006a)
Number of private diesel cars licensed in the UK in 2005: 5,399,400 (DfT 2006a)
per cent of private petrol cars licensed in the UK in 2005: 79.2 per cent
per cent of private diesel cars licensed in the UK in 2005: 20.6 per cent
Emission factor for average petrol car in the UK in 2003: 0.18kg CO_2/km (DEFRA 2005b)
Emission factor for average diesel car in the UK in 2003: 0.17kg CO_2/km (DEFRA 2005b)

The bus emissions factor is given by the National Atmospheric Emissions Inventory (NAEI 2005) as 0.374kgCO_2/km per bus. The rail emissions factor is given by DEFRA (2005b) as 0.04kg CO_2/km per person, and takes into account the proportion of electric to diesel train kilometres in the UK in 2003 (DEFRA 2005b). While the rail emissions factor can be simply multiplied by the assumed number of people travelling by train, the car and bus factors require assumptions relating to load factor, i.e., the number of people in the vehicle. Bus load factors will vary at different times of day and on different routes. An average UK value of twelve people per urban bus is used here (after Barrett, Scott, and Vallack 2001). Car load factor is assumed to be two, but this will need to be validated in future rounds. The tram per person emission factor was derived from local information as 0.034kgCO_2/passenger km. DEFRA (2007) give an average per passenger km emission factor for light rail as about twice this, at 0.065 kgCO_2/passenger km, possibly reflecting the older tram systems of Sheffield and Newcastle.

Hotel Emission Factors

The energy intensity of different types of tourist accommodation varies considerably, as do the results of different studies involving similar accommodation types. For example, Becken, Frampton, and Simmons' (2001) values for campsite emissions are half those of Gössling et al.'s (2002) values for the same kind of venue; Becken et al.'s (2001) values for pensions/beds are well over four times those of Gössling (2002) values for the same. The variation may be caused by differences in the type of energy supply as well as different levels of consumption. More specifically, they may result from the varying level of services (such as swimming pools) and facilities (such as air conditioning) offered, as well as a result of variation in energy supply type in different geographic areas (Gössling, 2002). The factors used here are taken from Gössling (2002) but UK-specific values should be obtained for future iterations. The values used here are 20.6kgCO$_2$ per hotel person-night and 19kgCO$_2$ per hotel person-night for self-catering/Youth Hostel.

Venue Emission Factors and Estimation

The Futuresonic festival takes place in some twenty venues across the city of Manchester, some of which are open air, and one of which in 2007 is in a through-route within a large shopping mall. If the festival were not to take place, these venues would be open anyway, some booked to other artists, others catering for drinking or shopping clients. Nevertheless, the fact remains that the festival does require these venues to be open in order to function. Here, venue emissions are defined as the sum of two source categories. The first category is the 'background' emissions of the venue without the festival. These are the emissions associated with the venue simply being open. They are caused by all of the fossil energy consuming appliances that are functioning in the venue but which are not solely associated with the festival. For example, fixed lighting (i.e., lighting that is not installed temporarily for the venue or that is only used for performances), heating, cooling, cash tills, fridges, computers, etc.

The second emissions source category relates to the festival-only emissions at the venue. This consists of all fossil energy consuming appliances installed or used temporarily for the duration of the Futuresonic event. The category includes stage lighting, amplifiers that are temporarily brought in: in fact the entire 'kit list' of the artists. Where possible, the audit may in future years add additional detail, such as emissions associated with beverage and food consumption at the venues.

Different options are possible for estimating background venue emissions. A calculator could allow a user to enter (a) an annual bill (i.e., sum of money), or (b) annual kWh consumption, or (c) floorspace in m^2. These could then be converted to an emissions value for an assumed nine hours total of rehearsing and performing, a duration based on the experience of Futuresonic staff. For example, for emissions associated with background electrical consumption, if an annual value for electrical consumption was available, this could be multiplied by 9/8760 (there are 8760 hours in a 365 day year), and the product multiplied by DEFRA's electrical grid factor of 0.43kgCO$_2$/kWh. Further data entry options could be offered, such as for electrical consumption in May and detail relating to gas consumption. For the time being, option (c)—a floorspace factor—has been used to estimate venue emissions. This is necessarily a coarse estimate, as the value is based on hotel energy consumption and is not specific to music events venues in general or to specific venues. Subsequent work by students (Robson et al. 2007), using electricity bill and meter data not available for use in the results presented here, derived more refined emission factors for electrical consumption in very approximate size bands of venue. These are large: 0.0028 per m² floorspace per hour, and small: 0.0205 per m² floorspace per hour.

For electrical consumption associated with the kit-list, the power consumption of typical kit-list equipment has been established via internet-based music equipment catalogues and retailers, in kW units, and this multiplied by the performance time, followed by multiplication by DEFRA's electrical grid factor. The kit list used here is partial and needs refining in future. Finally, although the objective should be to minimise the artists' consumption, it should be noted that kit list emissions should not be added to metered or billed values, as double counting may result.

Estimation of Participant Numbers

Participant numbers have a strong influence on the audit results because they drive the per person emissions factors. In areas where those factors are high, such as air travel, minor inaccuracies over participant numbers are likely to have a relatively large effect on the final emissions total. The precise number of Futuresonic participants is not known because not all of the events are ticketed and because a central ticket register is not kept (this may change in future). The 2007 festival also includes a presentation at a shopping mall, to which many thousands of people will be briefly exposed. Nevertheless, the number (if not the origins) of artists and conference delegates is known precisely, which is fortunate, as these participants travel the furthest.

A further complication is that the festival issues a total of about 500 three-day passes that allow attendance at several events. Of the approximately 10,000 people in total who experience Futuresonic, the event organisers assume that the other 9,500 participate in an average of two events per trip into the city, which gives 19,000 attendances. However, it should not be assumed that these people travel into the city more than once when they have not done so, so the number of return trips associated with the 9,500 people is 4,750. The 500 people with three day passes are assumed to attend an average of 6 events each, with one return trip per day. This entails a further 3,000 participant events and 1500 return trips. The total number of venue attendances is therefore assumed to be 22,000 and the total number of return trips in and out of the city is assumed to be 6,250.

Estimation of Transport Distance and Mode

Estimation of transport distances ideally requires knowledge of the journey departure point of every participant. While it is not feasible to obtain this information for everyone, airport origins are known for most of the artists and conference delegates. Nevertheless, transport distance and mode are probably the most significant sources of potential error, as the values can be relatively large in scale and are summative. Whereas venue emissions are constant and the per-person contribution to them declines with increasing numbers of participants, each additional trip adds considerably to the emissions total, and the assumed distance and mode of travel will substantially affect that contribution. These source categories are therefore priorities for obtaining more accurate information in future years.

Aircraft Journeys

To our knowledge, it is almost exclusively artists and conference delegates who use aircraft to access the festival. Relatively good, but still partial, information is available for their origins and the Futuresonic booking website now has a facility by which those booking a ticket can enter their point of departure, to assist with the 2007 audit. Here, the distance calculator used is Great Circle Mapper (Swartz 2008). This entails a simplification of journey length but is readily available and provides a distance that can be multiplied by the long haul emissions factor given previously, plus the multiplier to account for additional warming. We have applied a detour factor of 1.05, following Gössling (2006), to account for deviation from the great circle routes due to airspace issues, preferential take-off and landing routes, etc. Trips to and from the airport are excluded due to a lack of relevant information.

Local, Regional, and National Journeys

Data on local, regional, and national journeys, undertaken by non-artists, were not readily available for this pilot version of the audit and are thus a substantial, potential source of error. In future, modal split should be entered by the user and guide distances to example locations offered.

For the present audit, transport mode, number of passengers and approximate origins are known for local artists. However, for non-artists, given the lack of monitored data, we have modified and applied percentage values for modal split as obtained from the 2005 National Travel Survey (DfT 2005). The particular modal split chosen is for leisure travel (DfT 2005, Table 7.1). However, this is likely to over-represent the use of walking and cycling as transport modes (17.7 per cent and 18 per cent respectively), these being popular leisure activities in themselves. Accordingly, car usage has been retained at 70 per cent (this merges DfT's passenger and driver categories), while the remaining 30 per cent has been allocated on a best-guess basis of 13 per cent to bus, 13 per cent to tram and 4 per cent to walking and cycling jointly. These percentages are then applied to the assumed 8,000 attendees travelling from Manchester (i.e., 80 per cent of the assumed 10,000 participants).

Regarding the approximately 15.5 per cent assumed to be coming from the English regions (1550 people), one return trip has been assumed (as stated), with an assumed 70 per cent travelling by car (based on DfT 2005) and 30 per cent by train. Again, this needs validation. Similarly, without origins data we have split the regional attendees into five groups and have assumed that each group travels from a different major English city (London, Birmingham, Sheffield, Leeds, and Liverpool). Regarding the participants with 3 day passes, their assumed 1,500 return trips are also assumed to use a modal split as just described for local attendees. That is, 70 per cent by car, 13 per cent by bus, 13 per cent by tram and 4 per cent by walking and cycling jointly.

Travel *distance* is equally problematic to estimate and we emphasise that this, again, remains a likely source of significant error at this stage. We assume one notional 16 km return trip for all local attendees. Clearly this is a simple average that needs to be improved upon via survey data. Regional/national journey distances from the assumed five cities have been calculated using an internet-based route calculator, but again the core assumption of travel from those cities is highly approximate.

RESULTS SUMMARY

Please see the full report for results details (Upham et al. 2007).

Table 9.1   Summary of Emission Sources and Values

Emission sources	CO_2 emissions (kg)
International attendee transport (uplifted)	702,794
International attendee transport (not uplifted)	234,265
Artists' non-local transport (aircraft uplifted)	47,052
Regional attendee transport	29,915
Artists' non-local transport (aircraft not uplifted)	21,456
Local (Manchester) 1 day attendee transport emissions	8,587
Artist accommodation	2,391
Venues	1,679?[3]
Artists' local transport	487
Artist rehearsal and performance	257[4]
Local (Manchester) 3 day pass attendee transport	101
Total (aircraft uplifted)	791,584
Total (aircraft not uplifted)	297,459

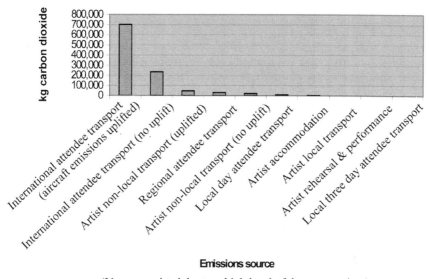

(Venues omitted due to a high level of data uncertainty)

Figure 9.1 Summary of emission sources and values.

RESULTS AND RECOMMENDATIONS

Table 9.1 summarises the results in terms of emission sources, showing their relative magnitudes. Figures 9.1 and 9.2 show the same in graphic form, with the latter excluding aviation emissions so that the relative size of the smaller sources can be seen. We estimate that the 2006 Futuresonic international music festival generated first order carbon dioxide emissions

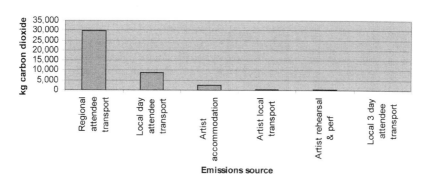

(Venues omitted due to a relatively high level of data uncertainty)

Figure 9.2 Summary excluding source categories with aircraft emissions.

of some 297–791 tonnes, with the higher value including 'uplifted' aviation emissions for communication purposes only. As the organisers estimate 10,000 attendees in total, this works out as 29.7–79.1 $kgCO_2$ per attendee. For comparison, Bottrill et al. (2008), in a UK-wide study of live music performances, estimate (as here, with explicit admission of major data access problems) similar festival per-attendee emissions of 25 kg CO_2e (no uplift).

The emissions total of a relatively small festival to which many artists and some attendees fly, has been and will continue to be strongly influenced by aviation emissions, such that greater accuracy with respect to international visitor origins will significantly influence the results in any future iteration. Greater accuracy in surface transport emissions would have the next most significant influence on the results, and gaining a more accurate estimate of venue emissions should be another priority.

Carbon offsetting is the most obvious form of potential mitigation, and Gössling et al. (2007) review offsetting issues in relation to aviation. Arguably, Futuresonic should commit to offsetting its carbon emissions, choose the offset firm carefully, and view offsetting as a minimum expression of its climate responsibility. There is no wholly objective way of recommending offsetting firms, but *Ethical Consumer* magazine has recommended *Atmosfair* and *Equiclimate*, with *Global Cool* and *Pure* also recommended (Welch 2007).

Offsetting aside, the main options for facilitating the reduction or mitigation of transport emissions are to encourage both the supply and use of lower emission transport modes. At the supplier level, there may be scope for discussion with the City Council and bus companies regarding use of biodiesel in dedicated buses or minibuses. However, biodiesel can be environmentally damaging at the production stage in its life cycle, depending on the source of the feedstock (Department for Transport 2006b, EC 2006; see also Biofuelwatch, 2008). Other bus fuel/propulsion options may be feasible but will depend on bus company decisions: Biodiesel is suggested here only because it is a relatively straightforward option. Environmentally, the best source of biodiesel would be that made from waste cooking oil (Zah et al. 2007).

The prospects for influencing transport mode at the consumer/visitor level are also small or modest but are nevertheless worth pursuing. The UK National Atmospheric Emissions Inventory (NAEI 2005) (http://www.naei. org.uk/index.php) 2005 emission factor for carbon (for CO_2 multiply by 44/12) emissions by the average diesel bus (urban driving) is 0.374kg per vehicle km, which can be compared to the average diesel car (0.036kg) and average petrol car (0.047kg), all for carbon per vehicle km, urban driving. This means that an urban bus only needs to carry nine passengers to emit less carbon dioxide per person than an average car (if we average the diesel and petrol car to give 0.0415kg), assuming single occupancy. Although the bus should be taken rather than an additional car added to congested roads, in simple per passenger terms the CO_2 benefits obviously decline as the car is filled with people. If the car has two occupants, the bus needs to carry 18 passengers to perform better, and so on. For comparison, DEFRA's (2005) average rail emissions factor is a (surprisingly high) 0.04 $kgCO_2$ per passenger km, which the average car could improve upon with each additional passenger. DEFRA's (2007) factor, not used here, is 0.0602 $kgCO_2$ per passenger km. A task for future iterations is to investigate NAEI emission factors and region-specific load factors to improve the accuracy of the emissions estimates. Given this information, one option is to give Futuresonic participants incentives to travel by bus, train, or in a full car, for example via a reduction in entry price. More controversially, to reduce aviation emissions, Futuresonic could consider giving priority to European artists. While this raises many issues, it would probably be the single most effective act that the organisers could take to reduce the climate impact of the festival.

In terms of methodological improvements, implementing data collection systems is a clear priority: particularly data on attendee and artist origin, destination, and transport mode. For attendees, this can be achieved via sampling with questionnaire surveys. However, while this need not be repeated annually, it does require some level of resource. Realistically the most likely option in this particular case is an occasional visitor survey staffed by students as part of environmental project work, though late night events may need to be omitted for safety reasons. For venue energy consumption, the average values obtained by Bottrill et al. (2008) may prove useful. At the time of writing, there will be no 2008 carbon audit of Futuresonic because the teaching situation of one of the authors has changed and we no longer have access to students.

One final point: Estimating emissions does not in itself reduce them. If an organisation has highly limited time and financial resources, it may be preferable to target these at developing an incentive system for attendee car-sharing and use of public transport; for greenfield festivals, if affordable, using solar, wind, and biodiesel from waste cooking oil for power and fuel supply; and choosing artists who are already on tour in the region.

ACKNOWLEDGMENTS

Thanks to the following people for data, advice, and/or student supervision: Carly Mclachlan of Tyndall Centre Manchester and Manchester Business School; Drew Hemment of imagination@lancaster, Lancaster University and Future Everything; Gina Hewitt and Malcolm Duffin at Future Everything; Steve Connor and Holly Waterman at Creative Concern; Jessica Symons of Krata. Thanks to those venues that supplied electricity consumption or floorspace data, especially *Kro* and Manchester Academy. Many thanks to the North West Regional Development Agency for seed-funding this project. Last but not least, thanks to the following Manchester

Business School students, who undertook much of the primary research for this report: Andrew Appleton, Jessica Bryan, Patrick Coyle, Nana Parry Sarah Robson, Matt Rooney, David Shoebottom, and Becki Tate.

NOTES

1. Renamed in 2005 as The Carbon Neutral Company (TCNC).
2. The NW Regional Development Agency has covered the audit-related overhead costs of Future Everything and Creative Concern, and materials costs for dissemination of the results.
3. Robson et al (2007), using metered and billed data, derived a more accurate value of $7,597kgCO_2$, which is 4.5 times higher than the estimate in Table 9.1. A nine hour consumption time was assumed, for three days. Metered and billed data for large and small venues was extrapolated to the venues as a whole, based on their approximate size.
4. Similarly, Robson et al. (2007) derived a more accurate value of $324kgCO_2$, using a more accurate kit list and electrical consumption values, which is 1.3 times higher than the estimate in Table 9.1. Both improved estimates in footnote 4 and 5 remain very small relative to the hundreds of thousands of kg of CO_2 released by the transport modes.

REFERENCES

Anable, J., Lane, B., and Kelay, T. (2006) *An Evidence Based Review of Public Attitudes to Climate Change and Transport Behaviour*. For the Department for Transport, London. Available at: http://www.dft.gov.uk/pgr/sustainable/climatechange/areviewofpublicattitudestocl5730 (22 May 2008).

Barrett, J., Scott, A., and Vallack, H. (2001) 'The ecological footprint of passenger transport in Merseyside'. Stockholm Environment Institute, University of York, York. Available at: http://www.merseytravel.org/pdf/EFofPassengerTransport.pdf (22 May 2008).

Becken, S., Frampton, C., and Simmons, D., (2001) 'Energy consumption patterns in the accommodation sector—the New Zealand case'. Analysis, *Ecological Economics* 39: 371–386.

Biofuelwatch (2008) *Biofuelwatch*. Available at: http://www.biofuelwatch.org.uk/index.php (22 May 2008).

Bottrill, C., Liverman, D., Boykoff, M., and Lye, G. (2008) *First Step, UK Music Industry Greenhouse Gas Emissions for 2007*. Environmental Change Centre, Oxford University. Available at: http://www.juliesbicycle.com/wp-content/uploads/2008/05/jb-first-step-e-report.pdf (22 May 2008).

Culture Northwest (2004) *A Snapshot of the Creative Industries in England's North West*. Prepared by Culture Northwest for the Department of Culture, Media & Sport, Manchester. Available at: http://www.northwestcultureobservatory.co.uk/document.asp?action=view&id=127 (22 May 2008).

DCMS (2007) '1.01 Definitions'. *Reference Library*. Department for Culture, Media and Sport, London.

DEFRA (2005a) *Offsetting UK Presidency of G8 and EU*. Defra, UK. Available at: http://www.defra.gov.uk/environment/climatechange/uk/carbonoffset/presidency.htm (22 May 2008).

DEFRA (2005b) *Environmental Reporting—Guidelines for Company Reporting on Greenhouse Gas Emissions*. Department for Environment, Food & Rural Affairs, London. Available at: http://www.defra.gov.uk/environment/business/envrp/pdf/envrpgas-annexes.pdf (22 May 2008).

DEFRA (2007) *Passenger Transport Emissions Factors. Methodology Paper*. Department for Environment, Food & Rural Affairs, London. Available at: http://www.defra.gov.uk/environment/business/envrp/pdf/passenger-transport.pdf (22 May 2008).

Department for Transport (2006a) 'Transport statistics for Great Britain: 2006 edition' (Section 9; Table 9.4). Available at: http://www.dft.gov.uk/pgr/statistics/datatablespublications/personal/mainresults/ (22 May 2008).

Department for Transport (2006b) *Towards a UK Strategy for Biofuels—Public Consultation*. Department for Transport, London. Available at: http://www.dft.gov.uk/consultations/archive/2004/tuksb/towardsaukstrategyforbiofuel2047 (22 May 2008).

Dunn, C., (2007) *South by Southwest Hops on Carbon Neutral Bandwagon, Treehugger, Culture and Celebrity*. Available at: http://www.treehugger.com/files/2007/01/south_by_southw.php (22 May 2008).

EC (2006) *An EU Strategy for Biofuels*. Communication from the Commission, 8.2.2006 (EC, Brussels). Available at: http://ec.europa.eu/agriculture/biomass/biofuel/com2006_34_en.pdf (22 May 2008).

Forster, P.M.D.F., Shine, K.P., and Stuber, N. (2006) 'It is premature toinclude non-CO_2 effects of aviation in emission trading schemes'. *Atmospheric Environment* (1994) 40(6): 1117–1121.

Gössling, S. (2002) 'Global environmental consequences of tourism'. *Global Environmental Change* 12: 283–302.

Gössling, S. (2006) *Analysis of Tourist Flows, Greenhouse Gas Emissions, and Options for Carbon Offsetting Schemes in the Context of Linnéjubileet 2007*. Report prepared for Linnédelegationen, Stockholm, Sweden..

Gössling, S., Broderick, J., Upham, P., Ceron, J.P., Dubois, G., Peeters, P., and Strasdas, W. (2007) 'Voluntary carbon offsetting schemes for aviation: Efficiency and credibility'. *Journal of Sustainable Tourism* 15(3): 223–248.

Layne, A. (1998) *Levellers Want You to Breath Clean Air*. Rolling Stone, 15 May 1998. Available at: http://www.rollingstone.com/artists/thelevellers/articles/story/5927289/levellers_want_you_to_breathe_clean_air.html (22 May 2008).

Masson, G. (2003) *Sympathy for the Environment at Stones' UK Shows*. Reuters News Service, Published by Planet Ark, viewed 5 January 2006. Available at: http://www.planetark.org/dailynewsstory.cfm/newsid/22049/story.htm (22 May 2008).

MyClimate (2006) *FIFA World Cup 2006 is Climate Neutral*, MyClimate, Switzerland. Available at: http://www.myclimate.org/download/060407_myclimate_mm_WM_klimaneutral_en.pdf (22 May 2008).

NAEI (2005) *National Atmospheric Emissions Inventory*. AEA Energy and Environment. Available at: http://www.naei.org.uk/index.php (22 May 2008).

NME, (2007) *South by Southwest Goes Green*. NME news, IPC Media. Available at: http://www.nme.com/news/iggy-and-the-stooges/25916 (22 May 2008).

NWRDA (2006) *Northwest Regional Economic Strategy—2006 Baseline Report*. North West Regional Development Agency, Manchester. Available at: http://www.nwda.co.uk/what-we-do/policy-and-strategy/regional-economic-strategy.aspx (22 May 2008).

Peeters, P., Gössling, S., and Williams, V. (2006) 'Air transport greenhouse gas emission factors'. Presentation made during the *Tourism and Climate Change Mitigation Conference*, 11–14 June 2006. Westelbeers, The Netherlands.

Peeters, P., Williams, V., and Gössling, S. (2007) 'Air transport greenhouse gas emissions'. In Peeters, P.M. (ed.) *Tourism and Climate Change Mitigation, Methods, Greenhouse Gas Reductions and Policies,* 29–50. Breda, The Netherlands: NHTV Academic Studies.

Poortinga, W. and Pidgeon, N. (2003) *Public Perceptions of Risk, Science and Governance: Main Findings of a British Survey of Five Risk Cases.* Centre for Environmental Risk, University of East Anglia, Norwich, U.K.

Robson, S., Coyle, P., Tate, R., and Rooney, M. (2007) *Venues Audit for Futuresonic Festival with a Carbon Footprint Analysis.* Unpublished undergraduate research project, Manchester Business School, The University of Manchester, Manchester, U.K.

Sausen, R., Isaksen, I., Grewe, V., Hauglustaine, D., Lee, D.S., Myhre, G., Köhler, M.O., Pitari, G., Schumann, U., Stordal, F., and Zerefos, C. (2005) 'Aviation radiative forcing in 2000: An Update on IPCC (1999)'. *Meteorologische Zeitschrift* 14(4): 555–561.

Shine, K.P., Berntsen, T.K., Fuglestvedt, J.S., and Sausen, R. (2005) 'Scientific issues in the design of metrics for inclusion of oxides of nitrogen in global climate agreements'. *PNAS* 102: 15768–15773.

Swartz, K.L. (2008) *Great Circle Mapper.* Available at: http://gc.kls2.com/ Breda, The Netherlands.

The Ten Club (2006) *Pearl Jam's Continued Efforts to Offset Carbon Emissions'.* The Ten Club, USA. Available at: http://www.pearljam.com/activism/carbon. php (11 April 2008).

Upham, P., Boucher, P., Connor, S., Hemment, D., and Waterman, H. (2007) *Futuresonic: Pilot Carbon Audit April 2007.* Tyndall Centre Manchester, The University of Manchester. Available at: http://www.futuresonic.com/07/eco2. html (22 May 2008).

Van Schagen, S. (2006) *Jam Session—Pearl Jam Guitarist Stone Gossard Chats About the Band's Environmental Ethos.* Grist Magazine Inc, USA. Available at: http://www.grist.org/news/maindish/2006/07/21/vanschagen/index.html (22 May 2008).

VisitScotland (2005) *T in the Park Festival Now Carbon Neutral.* VisitScotland, UK. Available at: http://www.perthshire.co.uk/index.asp?pg=382 (22 May 2008).

Welch, D. (2007) 'Carbon offsets—Enron environmentalism or bridge to the low carbon economy?' *Ethical Consumer* 106 (May/June).

Wit, R.C.N., Boon, B.H., van Velzen, A., Cames, M., Deuber, O., and Lee, D.S. (2005) *Giving Wings to Emission Trading. Inclusion of Aviation under the European Emission Trading System (ETS): Design and Impacts.* 05.7789.20 Delft: CE.

Zah, R., Böni, H., Gauch, M., Hischier, R., Lehmann, M., and Wäger, P. (2007) 'Ökobilanz von Energieprodukten: Ökologische Bewertung von Biotreibstoffen'. (A Life Cycle Assessment of Energy Products: Environmental Impact Assessment of Biofuels). EMPA für die Bundesampt für Energie, die Bundesampt für Umwelt, und die Bundesampt für Landwirtschaft: Bern, 2007. Available at: www.empa.ch/plugin/template/empa/3/60542/—-/l=2 and http://www. theoildrum.com/node/2976#more (22 May 2008).

10 Voluntary Carbon Offsets
A Contribution to Sustainable Tourism?

John Broderick

INTRODUCTION

In recent years tour operators, airlines, concerned travellers, and high profile celebrities have turned to carbon offsetting to mitigate the impact of air transport. The promise is that for a small increase in ticket price, the atmospheric consequences of your holiday or business trip can be 'neutralised' by dealings in the carbon markets. This may be sweetened by adding the prospect of non-climate benefits to those locally involved with the offset project; an Indian community may gain access to a regular mango crop through carbon forestry or UK pensioners enjoy warmer housing as the result of an insulation programme. Enthusiasts for voluntary offsetting argue that it engages polluters outside of existing regulation in issues of environmental sustainability and provides investment in a range of normatively worthwhile causes, biodiversity conservation, low carbon technology, and human development for example. Nevertheless, there are justifiable reasons to be wary of the prospects of carbon offsetting to deliver a sustainable tourism industry. Indeed, governments, academics and journalists urged caution as they witnessed very rapid growth and innovation in the retail voluntary offset market during 2006 and 2007 (Adams 2006).

Although it is not the only source of greenhouse gas emissions that has been associated with voluntary offsetting, the issue is of particular relevance to the tourism industry. Chapters 5, 7, 8, and 14 of this volume demonstrate the impact of aviation emissions from tourism and cannot be dismissed. Without rehearsing the arguments developed there, two aspects of the analysis are especially pertinent; (i) the significant proportion of a European emissions budget, commensurate with avoiding 2°C warming from pre-industrial times, that would be occupied by the aviation sector under current growth expectations; and (ii) the limited possibilities for technological improvements in aircraft operational efficiency to mitigate this (Bows and Anderson 2007). Together these conclusions indicate that emissions offsetting, modal shift away from aviation, or outright travel demand reduction will be required to progress towards environmental sustainability. Whilst the latter two deserve detailed consideration (see also Chapter 12, this volume), the intention of this chapter is to provide an overview of the carbon markets and answer a number of questions relevant to the tourism industry.[1] Is this a credible mitigation strategy? What is it that is being traded in these markets? By whom? With what implications?

Broadly, carbon offsetting may be defined as indirect emissions reduction through a market mechanism; GHG reductions made by one actor, be that a nation, industry, corporation, or individual, are sold to an unrelated actor to defray their expectation or commitment to emissions reduction. The nascent carbon markets enable these transactions by providing a set of institutions and commodities, tonnes of carbon dioxide equivalent (tCO_2e) represented by credits and allowances, to structure this exchange. Often, the terms *carbon offsetting* and *carbon neutrality* are used to represent the voluntary purchase and retirement, cancellation, of project derived emissions credits in equivalent quantity to the emissions from a specified activity. However this is only one possible form of transaction and not exclusively the case. The purchase of emissions permits from another agent in a 'cap and trade' regulatory system could be understood as offsetting in a broad sense. As the aviation and tourism industries seem likely to continue to grow their absolute contribution to climate change, it is almost inevitable that they will employ offsetting of some sort to justify this expansion.

The discussion of this issue is organised into four parts. The first will consider carbon credits in their various guises and the means by which they are created, bought, and sold. The variety and terminology can at first seem bewildering but it is useful to consider voluntary carbon markets systematically and in relation to allied environmental policies and regimes. The second part then looks at how their purchase has become associated with and promoted by a range of actors in the tourism industry, primarily considering the interactions of the aviation industry and the carbon markets. From there, it will offer a brief outline of the criticisms directed at the voluntary carbon market and recent efforts to address these failings, primarily in its governance. Finally, it will briefly consider the structural implications of offsetting and how this is relevant to the tourism industry in the medium to long term.

THE CARBON MARKETS

Regulatory Markets

The regulatory carbon markets are a series of environmental policy initiatives that seek to reduce aggregate greenhouse gas (GHG) emissions from a group of actors[2] by creating tradable financial instruments. Each operates over a specific scale and jurisdiction, with carefully specified assets, reduction commitments, and legal entities. Conceptually, it is useful to distinguish between allowance-based and credit-based systems. Allowance based 'cap and trade' systems divide responsibility for a fixed, total quantity of tolerable pollution amongst the actors to be regulated. At the end of a defined period, each actor must surrender allowances equal to the amount of pollution attributable to them. In a sense, the allowance becomes a factor of production, comparable to labour or steel, and essential for the economic activity regulated. It is important to note that the distribution of permits, the 'rights to pollute', at the outset is not necessarily uniform, nor tied to the obligation to surrender them. Distribution may be decided by 'grandfathering' (allocation by historical precedent), on the basis of an industrial baseline or standard, or as the outcome of an auction, thus generating revenue for the regulator. If they have a financial exchange value amongst regulated actors, emissions rights become assets and their distribution represents a transfer of wealth.

The European Union Emissions Trading Scheme is the most prominent example of this type of system, incorporating approximately 45 per cent of EU CO_2 emissions. Presently, it covers 12,000 industrial installations in twenty-seven countries, which each year must monitor their activity and surrender permits, European Union Allowances (EUAs), equivalent to their CO_2 emissions. If an installation does not have sufficient permits, it must pay a fine of €100 per additional tonne CO_2, considerably more than the current EUA trading price of €21.[3] The total quantity and distribution of permits is set in the member states' National Allocation Plans (NAPs), which must be approved by the European Commission. Allocations are controversial; in the UK's case 93 per cent of permits are distributed free to installations on the basis of their historic emissions and only 7 per cent are made available through auction. The extent to which Phase I (2005–2008) had an over-allocation of allowances and how much it contributed to reductions in emissions is debatable (Kettner et al. 2008; Ellerman, 2005; Ellerman and Buchner, in press). However, the cap was widely acknowledged as being not terribly stringent and a widely reported 'price collapse', from approximately €32 to €9 per tonne, followed the publication of installation emissions data in April 2006. Phase II of the EU ETS runs from 2008 to 2012 and is expected to meaningfully contribute to the EU's effort to meet its Kyoto obligations.

Credit based systems reward actors for environmental performance that is quantifiable and in some way an improvement on existing norms. The regime authority awards credits based on a particular set of criteria, which may then be traded with other entities to meet their regulatory obligations. In this way, a financial incentive is created for environmental improvement. For example, an iron foundry installs a new, efficient furnace and is awarded credits equivalent to the difference in emissions if 'business as usual' had continued. It does not require all of the credits generated to meet its own regulatory commitments and so sells the surplus to another installation that has not invested in improvements. In the case of stock pollutants such as greenhouse gases then there is also the possibility of credits being awarded for the removal of pollution from the atmosphere, for example through afforestation projects producing a biological carbon 'sink'.

Credit systems are typically associated with project-by-project assessment and crediting, however this need not be the case. Standards for whole industrial sectors may be defined with automatic crediting for entities operating above a performance threshold or alternatively a diverse group of interventions may be packaged together and awarded credits as one and indeed these modalities are currently being explored in the regulated carbon markets. The source of the authority to issue the credit is key and in this sense it is worth noting that credit-based systems align more closely to traditional 'command and control' policies than allowance systems (Ellerman 2005). Actor level abatement decisions are constrained by the terms and conditions of the regime authority rather than being subject to the logics of the market and the decisions of each actor. In fact, no measurement of emissions need occur in a credit regime, the regulator may choose to grant credits on the basis of standalone policy or investment so long as confidence is maintained in the regime (Ellerman 2005). This feature is at the heart of debates about the effectiveness of project-based credits and will be discussed in Section 4.

Currently, regulatory carbon markets exhibit substantial overlap in geography, actors, and instruments. The global scale of trading systems for greenhouse gases is justified on the basis of their being long lived and well mixed in the atmosphere. The offset mantra runs that 'a tonne, is a tonne, is a tonne' (Pease 2005; Bristow 2006; Rosen 2007; Grabl and Brockhagn 2007) wherever and however it is removed or prevented from being released to the atmosphere. Under the Kyoto Protocol, industrialised nation states listed under Annex 1 are committed to emissions caps on a 'basket of 6' greenhouse gases—carbon dioxide (CO_2), methane (CH_4), nitrous oxide (N_2O), hydrofluorocarbons (HFCs), perfluorocarbons (PFCs), and sulphur hexafluoride (SF_6). For example, the EU 15 must collectively reduce emissions by 8 per cent by 2012 from a 1990 baseline. Assigned Amount Units (AAUs) are distributed to represent each nation's total allowable emissions, but may be exchanged between governments to meet the final treaty targets. The Clean Development Mechanism (CDM) and Joint Implementation (JI) are credit-based schemes that are intended to support technology transfer and broader sustainable development objectives in developing countries, the primary distinction being that JI projects take place in nation states with emissions caps whilst CDM projects do not. The credits they respectively generate, Certified Emissions Reductions (CERs) and Emissions Reduction Units (ERUs), can be used in lieu of AAUs for national governments to meet their treaty commitments. Similarly, the EU Linking Directive (2004/101/EC) allows CERs and ERUs entry to the EU ETS subject to certain quantity restrictions defined by the NAPs and exclusion of those credits generated from biological sequestration, nuclear, and large hydroelectricity projects greater than 20 MW. As a result, much of the activity in the CDM can be attributed to the EU ETS (Hepburn 2007), which is the world's largest carbon market, by both volume and value (Capoor and Ambrosi 2007).

Although substantially smaller than the EU ETS, the CDM is the largest framework for the supply of carbon credits from emissions reduction projects. The approval, validation, and exchange of credits are overseen by the UNFCCC secretariat, which is made up of the CDM Executive Board (EB) and a series of expert panels and working groups. Emissions reductions projects may be implemented in any of the Non-Annex 1 countries (i.e., those that do not have their own binding targets for the 2008–2012 period). The majority of these projects are located in India, China, Brazil, and Mexico with very few in sub-Saharan Africa.[4] By number registered, typical projects are electricity generation from biomass or other renewable sources, the capture of methane from landfill sites, or decomposition of agricultural waste. The quantification of credits is complex, subjective, and contentious. A hypothesised 'business as usual' (BAU) scenario is decided upon and defines an emissions baseline against which the measured emission of the project can be compared. The project and hence the reductions are said to be 'additional' if they occur as a result of specific carbon financing and intervention and would not have happened otherwise. This issue will be discussed further in Section 4 as it has particular resonance in the voluntary markets. Once issued, the project developer can sell on CERs to national governments, private business, or any other legal entity as with other commodities and financial instruments.

The development of a CDM project consists of both a physical implementation and along side it a bureaucratic process from initial design to issuance of the credits themselves (Figure 10.1). Contracts between project developers and CER purchasers are usually drawn up in advance of the final issuance of the credits, the price reflecting the project implementation cost, administrative overheads and the risk of non-delivery of the final credits. Because of the complexity of the CDM process and requirement for legal, financial, and technical expertise to complete registration, there is a substantial transaction cost and hence a bias towards projects with larger returns. CDM specific costs, those related to documentation, validation, and registration can be between US$38,000 and $US610,000 depending upon project type, scale, novelty, and complexity, in addition to the project implementation capital and operational costs (CD4CDM and EcoSecurities (2007). Whilst there are simplified regulations and procedures for small scale projects,[5] the registration process may still take up to eighteen months to complete. Concerns about approval bottlenecks and understaffing of the secretariat having been raised by developers, however, they are now thought to be overstated (Røine et al 2008).

In all, 949 projects have been registered and 128 million CERs issued, each being notionally equivalent to a tonne of CO_2 prevented from release to the atmosphere.[6] It is estimated that a total of 2700 million CERs will be issued by the end of 2012 and the close of the current commitments under Kyoto. Including both forward contracts and secondary sales, the World Bank estimates the sum value of trading in this new commodity to have been €3.8bn during 2006 with an average price of €8.40 per CER. This represents very rapid growth given that the first project, a landfill gas capture project in Brazil, was registered with the CDM Executive Board as recently as 2004. Long-term expansion of the CDM is heavily dependent upon the ongoing UNFCCC negotiations for a post-Kyoto climate treaty and the particular measures it will specify.

Box 10.1 Process of CDM Project Development

1. *Planning* and *design*—a Project Design Document (PDD) is drawn up, specifying the technical details, participants, management, and organization of the project. This is then approved by the Designated National Authorities (DNA) of all states involved.
2. *Validation*—the PDD is formally checked against the CDM approved methodologies (project archetypes outlining means of implementation and calculation of credits) by a Designated Operational Entity (DOE, a third party organisation approved by the Executive Board [EB] of the CDM providing technical and audit services). If a new methodology is to be used, this must first be submitted to and approved by the EB.
3. *Registration*—the PDD and validated documentation is examined by the EB before formal registration of the project. A fee to the CDM is paid at this stage.
4. *Monitoring*—project becomes operational and data is collected by the project participants in accordance with the PDD.
5. *Verification*—the project and monitoring data are inspected and assessed for GHG reductions by an independent DOE, *ex post* (i.e., after they have occurred). The DOE then formally certifies the quantity of CERs that the project has earned.
6. *Issuance*—the Executive Board conducts a final review, CERs are created in the CDM Registry and allocated to the project participants' accounts. An administration fee is deducted by the CDM and 2 per cent of issued CERs are assigned to a fund for climate change adaptation in vulnerable countries.

Voluntary Carbon Markets

On the opening day of the 2004 Buenos Aires session of the UNFCCC, the international bank HSBC publicly declared it would 'go carbon neutral' with an internal carbon management strategy and commitment to offset the remainder through carbon credit purchases (HSBC 2004). Initiatives like this are becoming more and more prevalent as climate change becomes a significant issue of public concern. In addition to managing emissions at source, businesses, governments, and individuals are joining the carbon markets voluntarily, buying credits and allowances over and above mandatory requirements. The majority of this voluntary activity is the purchase of carbon credits generated by project activities similar to the CDM. For instance, as part of their Green Goal initiative, FIFA spent €1.2m to offset emissions from the operation of the Germany 2006 World Cup by commissioning 100,000 tCO_2e from emissions reductions projects in South Africa and India (Öko Institut 2006). Similar pledges of 'carbon neutrality' have been made by the UK Government office estate, The Rolling Stones, and the media group BSkyB. These initiatives have been echoed by individual travellers and tourism operators sensitive to environmental issues in a parallel retail market.

There are a range of possible motivations for voluntary action such as the discharge of moral obligation, desire to promote transition to a low carbon economy, pursuit of a 'green' image, pre-emption of regulation, or the identification of new opportunities for financial gain. Due to the informal, opaque, and fragmented nature of this market it is very difficult to reliably estimate the volume and value of credits sold as voluntary offsets and distinguish demand from corporate, government, and consumer buyers. In aggregate, the World Bank estimates at least 10 $MtCO_2e$, worth US$100m. were traded in 2006 (Capoor and Ambrosi 2007), whilst a more recent survey put the volume traded at 13.4 $MtCO_2e$ representing growth of over 80 per cent on the previous year (Hamilton et al. 2007). These figures are estimates of transactions and not the quantity of credits that were retired (i.e., finally purchased and removed from circulation, equivalent to the surrender of permits in a regulatory system, which the latter survey estimated at 23 $MtCO_2e$ in total to date). Despite the remarkable growth rates, this is a very small amount if we consider that current anthropogenic GHG emissions were estimated to be 49 $GtCO_2e$ in 2004, up 24 per cent since 1990, and are understood to have to decrease globally by at least 50 per cent by 2050 to avoid temperature rises greater that 2°C above pre-industrial times (Metz et al. 2007). However, this is to dismiss the social, political, cultural, and economic significance of the development of these market mechanisms, and as noted previously, the nature of emissions growth from the tourism industry justifies consideration in this volume.

It should also be noted that not all voluntary carbon trading is conducted as bilateral or 'over the counter' deals between large organisations or as retail sales. In the USA, the Chicago Climate Exchange (CCX) is a hybrid system, that is, 'cap and trade' with credit generating projects, with voluntary membership but legally binding commitments. There are currently 130 Full Members representing a wide variety of organisations including electricity generators, primary and secondary industry, municipalities, states, and educational institutions. There are also Associate Members, with smaller output committed to monitoring and offsetting their emissions, and Participant Members who trade in the market or commission projects but are not compliant with the reductions programme. The full Kyoto 'basket of 6' gases are included and the trading unit is the Carbon Financial Instrument (CFI) representing 100 tCO_2e. Although none of the formal members are associated with the tourism industry, there are currently a number of organisations retailing credits certified through the CCX for voluntary retirement as offsets.

While the first voluntary offset agencies were initiated in the early 1990s, there has been rapid expansion in the sector in the last five years with consumer access to the carbon markets aided by the founding of many new retail agencies. Considering air travel offsets specifically, Gössling et al (2007) identified forty anglophone agencies offering the sale of carbon credits online (Figure 10.1) and the website Carbon Catalog[7] currently lists seventy-four organisations retailing credits to the general public. Registered charities, not-for-profit, and for-profit companies are present in the market. They are predominantly based in Western Europe, although increasingly in North

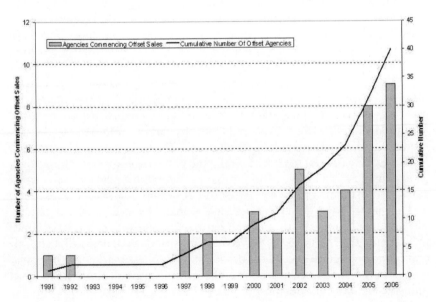

Figure 10.1 Year of inception for offset agencies (reproduced from Gössling et al 2007).

America and Australia). A few purchase, retail, and retire CERs, ERUs, and EUAs from the regulatory market supply chain, however the majority manage projects in-house, work with local NGOs, or commission projects from independent developers. In some cases, offsets have been seen by NGOs as additional revenue streams for existing activities, most frequently projects associated with biodiversity conservation and tropical forestry.

The acronym VER, for Verified, or Voluntary, Emissions Reductions, has been coined for carbon credits produced outside of the regulatory systems. Although such credits notionally represent 'real and verifiable' reductions the use of the term VER does not imply that they are awarded by a common, authoritative, and transparent institution. In reality, there is a great diversity in accepted project types, calculative procedures, interpretations of additionality, degrees of third party oversight, and the stringency of emission reduction verification. Whilst there are a series of open and proprietary standards in use by project developers these are not mandatory and act only to reassure consumers and differentiate products.

There is similar diversity in the projects funded to generate carbon credits. Table 10.1 presents a typology and nonexhaustive list of VER generating projects, their proposed nonclimate benefits, and notes on other indirect consequences. Proponents of voluntary offsetting cite the diversity of projects as evidence of innovation that is stifled by the bureaucratic requirements of the CDM and emphasise the role of consumer choice in promoting projects with wider 'sustainable development'[8] benefits (Bayon et al. 2006). A majority of the organisations surveyed in Gössling et al 2007 held forestry projects in their portfolio and for half this was the sole source of their carbon credits. Forestry also forms a major segment of the voluntary market by volume, Hamilton et al. (2007) estimating this to be 36 per cent in 2006. For example, avoided deforestation marks a significant distinction between the voluntary and regulatory markets. Whilst afforestation and reforestation are permitted in the CDM as valid offset methodologies under the Marrakesh Accords (2001) the protection of existing forest is not. Such projects are restricted to the voluntary market and have received strong support from a number of environmental NGOs, notably Conservation International and Wildlife Conservation Society.

As an alternative to investing in credited projects, there are a small number of voluntary offset agencies that use donations to purchase and retire credits from cap and trade systems such as the EU ETS.[9] The logic is that under a total output cap, all permits will be used by the participants and so by removing these credits that would otherwise be used by to meet industrial commitments, total emissions are reduced. The artificial shortfall should also drive up the allowance price and encourage low carbon investment. The UK Government's Code of Best Practice for voluntary offset providers endorses this mode as it avoids key problems of credit systems; (a) the possibility of wrongly awarding credit (i.e., for reductions that would have occurred regardless of intervention), and (b) indirect increases in emissions,

Table 10.1   Voluntary Offset Project Typology

Project type	Example	Promoted 'sustainable development' benefits	Notes	Source
Avoided Deforestation	Makira Forest, Madagascar	Endemic species conservation. Transfer of some aspects of land management to local communities to improve welfare.		http://www.conservation.org/learn/forests/Pages/project_makira.aspx
Afforestation & Reforestation	EBEX21 Forest Regeneration, New Zealand	Land owner income, reduced soil erosion, native biodiversity protection and enhancement, recreation		http://www.ebex21.co.nz/ebex_members.asp
Fugitive gases: Methane capture from cattle manure	'Farm Power' anaerobic digesters, USA	Nutrient cycling and reduced nitrogen run off	USA has no Kyoto target so projects are considered additional in terms of international climate policy	http://www.terrapass.com/projects/farm-power.html#holsum
Fugitive gases: Coal mine methane	Rhine-Ruhr coal mine methane capture, Germany		Although within Annex 1 nation, methane capture from mining is currently beyond legal requirements	http://www.carbonneutral.com/projects/projects.asp?id=941
Fossil fuel substitution: Solar	Solar thermal rice cookers, India	Local employment opportunities, improved air quality	Mirrored solar collectors replacing diesel fuel and firewood in hospitals and school canteens	http://www.atmosfair.de/index.php?id=159&L=3
Fossil fuel substitution: Wind turbine	Farmer owned distributed wind turbines, USA	Reduced energy expenditure, promote distributed energy infrastructure	Credits calculated against the average emissions intensity of electricity supplied from grid	http://www.nativeenergy.com/pages/our_projects/14.php

Fossil fuel substitution: Human labour	Treadle pumps, India	Reduced cost and increased frequency of irrigation, improved production and diet, reduced labour migration, reduced soil erosion	Baseline of hired diesel pump used in quantification of credits. Presented in UK press as a juxtaposition of Western leisure travel and child labour in impoverished communities [Nelson 2007].	http://www.climatecare.org/projects/countries/treadle-india'
Fossil fuel substitution: Biomass	MPPL Biomass Power Plant, India	Local employment and revenues, reduced ground-water pollution	First CDM project to generate Gold Standard Credits	http://www.myclimate.org/index.php?lang=en&m=project&uum=overview&uum=k arnataka
Fuel switch	Truck stop electrification, USA	Improved local air quality and reduced noise impact, better nights sleep for truckers improving road safety	Substitute engine idling for grid electricity with lower emissions intensity	http://www.climatetrust.org/offset_truckstop.php
Energy efficiency: Heat	Improved wood burning stoves, Madagascar	Reduced respiratory disease, reduced ecosystem degradation, reduced time/expense spent collecting wood fuel	Efficient cooking stoves reduce fuel use	http://www.bvco.org.uk/BVCO_Projects_Madagascar.htm
Energy efficiency: Electricity	Distribution of energy efficient lightbulbs, South Africa	Improved access to lighting and reduced cost for beneficiaries	Project has not yet delivered reductions through CDM since announcement [EAC 2007]	http://www.defra.gov.uk/environment/climatechange/uk/carbonoffset/presidency.htm

often termed *leakage*, if the project activity displaces one emitting activity or stimulates another. However, retirement of EUAs has received strong criticism given that allowances are almost entirely freely allocated by 'grandfathering' rights, and hence purchase would represent a donation on the basis of historically polluting behaviour, and that the overall allocation of Phase I contained so much 'hot air' that any voluntary retirement would have a negligible effect. As a consequence, this mode of voluntary offsetting has received very little attention and support in comparison to credit purchase and retirement.

Having discussed the supply side of retail voluntary offsetting, it is briefly worth considering the demand side. In the regulatory markets, international commitments and national legislation formally define the environmental responsibilities of the relevant actors. However, the voluntary market must somehow create that expectation and provide its own tools to quantify emissions prior to their exchange. Consequently, calculators are a key feature of retail offset websites enabling individuals to purchase the appropriate quantity of credits to 'neutralise' an activity of their choice. Almost all websites include air travel but many also offer estimates for emissions from household heating, electrical appliances, road, rail, and ferry transport and even nappies. These are necessarily estimates, usually taking fuel consumption and a standardised emissions factor as a proxy, as it would be practically impossible to directly measure tail pipe emissions for individual activities. Some offset agencies use relatively sophisticated calculation models, including multiple flight phases, estimates of occupancy rates, and fuel efficiencies for aircraft known to operate on particular routes, whilst others offer crude approximations of short and long haul flights, a typical year's driving, national average household emissions, or even an arbitrary fixed sum or given quantity.

There are two important points to raise here; (a) one of the oft asserted strengths of voluntary offsetting, that it imparts knowledge of GHG emission sources and relative quantities to participants, hinges on the process of data collection, calculation and representation, and (b) that air transport raises issues of the environmental commensurability of different emissions types. Taking the first point, while many offset companies and independent organisations amenable to carbon trading (Ceppi 2006; Sustainable Development Commission 2008) argue that offsetting is the final stage of a complete carbon management strategy, it is clear that isolated donations through 'quick fix' offsets and automatic services[10] do little to engage the customer with the impact of their individual behaviour, nor do they contextualise a given activity. Silverjet and Eurostar[11] have gone so far as to incorporate offsets routinely into their tickets, without requiring a charge or individual intention. Without the putative socio-cultural influence of the process of calculation prior to credit purchase, added emphasis must be placed on the veracity of emissions reduction credits. Arguments that offsetting provides a shadow carbon price, if indeed a shadow price is appropriate for decision making when considering this problem, are also questionable given their very low price. Credits typically sell in the voluntary market for an order of magnitude less (Box 2) than estimates of the marginal social cost of carbon[12] and substantially below long term estimates of marginal abatement costs for climate stabilisation at less than 2°C above pre-industrial temperatures (Enkvist et al. 2007).

On the second point, uncertainties and inconsistencies remain within the understanding and representation of the multiple non-CO_2 climate impacts of aviation. As well as creating contrails and altering cloud formation, aircraft emit oxides of nitrogen (NO_x), which subsequently influence ozone and methane concentrations in the atmosphere. These warming and cooling effects vary over different timescales and their strength may be influenced by location, altitude, season, and time of day (Peeters et al. 2007). From latest IPCC AR4 figures it can be estimated that aviation causes a radiative forcing between 1.9 and 4.7 times as much as that due to CO_2 alone (Graßl and Brockhagen 2007).

This ratio refers to the radiative forcing of historic (long-lived) CO_2 emissions versus current (short lived) non-CO_2 impacts, while for the purpose of offsetting the future radiative forcing of 1t of GHG emitted would ideally be accounted for (Forster et al. 2006; Wit et al. 2005). Half of the online calculators surveyed by Gössling et al (2007) use 'uplift factors' of between 1.9 and three and multiply the quantity of CO_2 emissions by this amount. While these are not strictly justified by the physics of the situation, there is undoubtedly a warming effect over and above CO_2 and for simplicity in the voluntary market inclusion may be warranted. Ultimately, the problem is with the assumption of equivalence between emissions, which is demanded by the conceptual framework of offsetting. Fundamentally different phenomena are made equivalent in order to be transacted in a simple commodity exchange. This is a theme emphasised by critics of offsetting as a whole (Lohmann 2005).

INTERACTIONS BETWEEN CARBON MARKETS AND THE TOURISM INDUSTRY

> There is now wide agreement that emissions trading is the most effective way of tackling aviation's contribution to climate change. Other measures, such as carbon offsetting, also have an important role to play.
>
> Rt Hon Tony Blair Prime Minister,
> Foreword to Sustainable Aviation (Sustainable Aviation 2005).

For certain members of western society 'hypermobility' has become the norm (Gössling and Hall 2005) and the regulation of air transport emissions has been met with some resistance and a 'psychology of denial' in policy discussions and Western media (Stoll-Kleemann et al. 2001; Gössling et al. 2001; and Peeters 2007). Aviation and shipping are currently excluded from international environmental regimes and national accounting systems prompting conspicuous campaigns by environmental NGOs and vocal criticism from a number of journalists. These sectors will not remain excluded from regulation indefinitely, the Kyoto Protocol itself requires Annex I Parties to ' . . . pursue limitation or reduction of (GHG emissions of from aviation), working through the International Civil Aviation Organization . . . ' (UNFCCC (1997, [Article 2.2]). In its 2004 Assembly, the ICAO declined to adopt a separate global legal instrument but instead focus on voluntary measures and incorporation of aviation into existing states' trading schemes but progress since has been slow.

Within Europe, inclusion of aircraft emissions into the EU ETS is supported in general terms, by the European Commission, member states, airlines, and industry associations, as an economically efficient mechanism for achieving climate stability (House of Lords 2006; Sustainable Aviation (2005). It is likely that aviation will enter in some form in 2011 or 2012, although the details of attribution of emissions, emissions rights allocations, and trading mechanisms were not finalised at the time of writing.[13] EU proposals have lacked support from the ICAO, and at the September 2007 assembly a controversial counter resolution was passed requiring member states to sign agreements with third parties operating in their airspace before introducing market-based measures. The feeling from the EC is that the organisation is impeding progress, indeed in a strongly worded press release Luis Fonseca de Almeida, Director General of Civil Aviation for Portugal, said

> We are disappointed by the outcome and believe ICAO has abdicated the leadership role given to it in the Kyoto Protocol. That is a very great failing that should concern us all.
>
> (European Commission 2007)

Despite being passed by the majority of delegates, the EU has declared that it does not feel bound to respect this 'mutual consent' resolution having formally registered its reservations (*ibid*), effectively foreshadowing a legal challenge by the United States.

Upon inclusion in the EU ETS initial analysis suggests that the aviation industry will meet its obligations primarily by purchasing credits from other trading entities with over-allocations or lower marginal abatement costs. As defined earlier, the aviation industry would be 'offsetting' its mitigation requirement by buying substantially from other sectors. In the short term, the EU ETS will likely have a limited effect on aircraft emissions, ticket prices, or overall demand for air transport (House of Lords 2006). Whilst the effect on freight is uncertain, it seems likely that airlines will on average pass through 100 per cent of permit costs to travellers, whether directly incurred in purchase or as an opportunity cost if freely allocated (Gibbs and Retallack 2006; Vivid Economics 2007). Tyndall Centre estimates of the impact of high carbon prices on ticket prices have been made on a range of allocation, baseline, passenger growth, and fleet efficiency assumptions (Anderson, Bows, and Footitt 2007). Even in scenarios where the price is assumed to be an order of magnitude higher than present, at €300 per tCO_2e in 2017, per passenger supplements are found to be just €15-€40, €70-€155 and €140-€310 for short, medium, and long haul flights respectively. Economic modelling of UK industry suggests that border tariffs or multilateral sectoral agreements may be required to protect a few industrial sectors unable to pass through the cost of permits to consumers because of international competition (Hourcade et al. 2007). These estimates were based upon the full auctioning of EUAs at €20 per tonne CO_2, plausible conditions for Phase III of EU ETS (2012–2016) and it is not clear what implications a very high carbon price, stimulated by demand from aviation under a stringent cap, would have for other energy intensive sectors in the long term.

The import of credits from the CDM has been suggested as one way in which prices may be essentially 'capped' by exploiting lower marginal abatement costs in developing nations. In the short term there are regulated limits on the proportion of reduction commitment that can be met by CERs for each individual trading entity. Current proposals suggest that a uniform level will apply, equivalent to the average proportion of CERs in each of the National Allocation Plans.[14] Without agreement of a successor to Kyoto post-2012 this possibility cannot be assumed to continue and questions of equity, integrity and international politics will undoubtedly have influence.

In the interim period before legislative action there are those who wish to be seen to be mitigating the impacts of their emissions. Credit purchase and retirement over and above legal requirements, often simply referred to as 'voluntary offsetting', has received considerable attention by airlines and tour operators as a means to maintain consumer demand for high emissions activities. Co-ordinated promotion and support from ICAO is unlikely, their 2007 report on voluntary emissions trading being vague, and whilst speculating about the possibility of unilateral offsetting schemes (ICAO 2007, see 3.6) distances itself from direct involvement or specific proposals (*ibid*, see 3.7). Table 10.2 summarises a variety of such initiatives and demonstrates the different ways the tourism industry has become involved with offsetting. In some cases the tourism industry has been at the forefront of the voluntary offsetting market. For example, the Forum Anders Reisen association of independent tour operators was instrumental in founding a complete offset retailer, Atmosfair. Some tour operators, such as Blue Ventures, have directly commissioned offset projects and a bespoke crediting scheme, TICOS, is seeking to make emissions reductions projects more tangible for the tourist by co-locating them in the holiday destination. Considering airlines, there is again variety, from one off promotions, Quantas and Delta have both gone 'carbon neutral' for a single day, to integrated purchase of credits during ticketing as EasyJet has done.

Table 10.2 Tourism Industry Offset Initiatives

Lead Actor	Implementation	Source of offset credits	Examples
Offset retailer	Direct sale through website. Calculator includes specific tourism component (typically just transport but sometimes hotel nights)	Retailer may develop in house projects or purchase credits from external developer or open market	Atmosfair, My Climate, CO2 Balance, Climate Care, The Carbon Neutral Company
Travel press, guide books	Travel sections of print media, online equivalents and guide books provide links and references to offset retailer	Retailer may develop in house projects or purchase credits from developer or open market. Standards vary.	Guardian offsets all emissions from travel section journalists, Lonely Planet likewise. Rough Guide includes offset retailer's calculator on its website.
Travel agency	Opt-out offset in booking procedure	Mixed portfolio: VCS, GS VERs and CERs	Lastminute.com with Climate Care
	Retail offsets alongside holidays	CCX and proprietary standard credits	Expedia
		Mixed portfolio: VCS, GS VERs and CERs	Travelcare
	Encourage travellers to use external offset retailer		Responsible Travel
Trade association	Initiated offsetting agency	GS CERs and VERs	Forum Anders Reisen
		Proprietary standards	TICOS
Tour operator	Manage own offsetting projects and retail credits to tourists	Proprietary standards	Blue Ventures

Hotel	Offset emissions from all operations	Green-e Certified credits	Ecoventura with Native Energy
	Offset emissions from all operations	Mixed portfolio: Proprietary and GS VERs	Rufflets Hotel, Scotland; URBN Hotel, China
Airline	Offset included in all ticket prices	Proprietary standard credits	Silverjet
	Option at point of sale, calculator automatic within booking site	Proprietary standard credits—Afforestation, Reforestation, Avoided Deforestation	Delta Air Lines with The Conservation Fund
		CDM CER	British Airways, EasyJet, Flybe
		Greenhouse Friendly standard (Aus)—Afforestation, Reforestation, Avoided Deforestation	JetStar, Virgin Blue
	Option following sale, calculator linked to booking site	Mixed portfolio: proprietary and GS CERs	Continental Airlines with Sustainable Travel International
		Voluntary Carbon Standard	Cathay Pacific
		Proprietary standard credits	SAS with The Carbon Neutral Company
		GS CERs and VERs	Virgin Atlantic
		ERUs	Air New Zealand
	Promotion of external retailer on website, no specific calculator	Proprietary standard credits—Afforestation, Reforestation, Avoided Deforestation	Air Canada with Zerofootprint

(continued)

Table 10.2 (continued)

Lead Actor	Implementation	Source of offset credits	Examples
	Credits for sale on board transport	Greenhouse Friendly standard (Aus)—Afforestation, Reforestation, Avoided Deforestation	Quantas
		GS CERs and VERs	Virgin Atlantic
	Credits purchased to cover all emissions from aircraft on a single day		Quantas 'Fly Carbon Neutral Day' 19th September 2007 Delta Air Lines, Earth Day promotion, 22nd April 2007
Rail operator	Offset included in all ticket prices	Mixed portfolio: VCS, GS VERs and CERs	Eurostar
	Option following sale, calculator linked to booking site	Proprietary standard credits—Afforestation, Reforestation, Avoided Deforestation	Amtrak

The table is not comprehensive but offers indicative examples of relationships between tourism actors engaged in carbon offsetting.

Motivation for these various actors to engage with the offset industry appears to be primarily to maintain reputation of environmental sensitivity to both consumers and regulators. Whether this has been effective is debatable. In 2006, British Airways were castigated by the UK Parliament's Environmental Audit Committee for the poor profile of their offset scheme and the subsequent 'risible' uptake. In the case of Air New Zealand reputation issues extend beyond the airline itself to encompass the national tourism industry as a whole. Speaking on TV news, Prime Minister Helen Clark directly associated the Air NZ offset programme with maintenance of long haul inbound tourism and associated revenues: 'If they know that we have got a scheme running through our national airline to offset the carbon emissions that's fantastic, first world consumers are looking for that' (3 News (2008)).

However, close association with offsetting is a risky reputation strategy following substantial criticism of the voluntary offset industry in the mainstream Anglophone press and broadcast media (Muir 2004; Dhillon and Harnden 2006; Chapman 2006; Elgin 2007; Faris 2007; Clarke 2007; Harvey 2007; Harvey and Fidler 2007; Davies 2007b; Johnston 2007). The consequences may be reflected in the poor uptake for consumer opt-in at point of booking although this could also be a result of low awareness of climate change, apathy, or cost sensitivity. BA explained that in the first two years of their scheme, they sold only 1600 tonnes of offsets per year, equivalent to just four return flights of a 777 to New York (EAC 2007). Even considering estimates of the volume of retail voluntary offset purchases as a whole, and assuming that emissions are offset one to one by the credits purchased, the quantities are very limited in comparison to emissions arising from the aviation sector. Globally, the best estimate of voluntary credit retirement in 2006 was 4.2 Mt. (Hamilton et al. 2007) whereas annual emissions from aviation are were estimated to be 553 Mt CO_2 in 2002 and rising rapidly since (Eyers, et al. 2004). Not surprisingly, absolute growth in emissions from air transport is not mentioned in industry press releases, however, offsetting being 'only part of the solution' is emphasised.[15] Ongoing improvements in fuel efficiency are regularly cited although these are more likely driven by the sustained high price of oil. It seems reasonable to believe that in the EU, voluntary offset schemes will play a minor role in the environmental management of the aviation industry in comparison to formal regulation. However, their discursive impact and influence elsewhere, notably the United States, may be more pronounced.

CRITIQUES AND RESPONSES IN THE VOLUNTARY CARBON MARKETS

I had a debate with a member of the cabinet, who I won't shame by naming; he was saying there's no problem with a third runway (at

Heathrow), because as more people fly in to Britain we'll do carbon offsetting somewhere else. Complete bollocks.

Ken Livingston
Mayor of London (Siegle 2008).

In counterpoint to the obvious enthusiasm of some for the use of offsetting as a climate change mitigation strategy, there has been substantial and sustained criticism. This ranges from the humorous but superficial, the Cheatneutral website offers to compensate for your sexual peccadilloes 'Helping you because you can't help yourself',[16] to the trenchant and sophisticated arguments developed by environmental and social justice NGOs (Lohmann 2006; Smith 2007). It should be noted that discussions are not restricted to those external to the market, indeed both buyers and sellers of offsets often seek to distinguish their mitigation strategy or product from competitors. This section will aim to summarise the main points of criticism and discuss the various means proposed to address them, considering them in terms of particular crediting systems, problems associated with specific projects or project types, and finally 'in principle' and systemic arguments. The focus will be on credit-based systems rather than offset by retirement of allowances from a capped system.

Crediting Systems

Critiques of credit based systems, both voluntary and regulatory, often centre on the 'real' and 'verifiable' nature of the commodity. Often this is used as a distinction between those with and those without formal verification procedures, through the CDM or otherwise. This serves to mask the fact that *all* credits are generated as counterfactual constructions and social conventions within trading institutions. As Dan Welch of the Ethical Consumer magazine puts it 'Offsets are an imaginary commodity created by deducting what you hope happens from what you guess would have happened' (Welch 2007). That is not to say that there is no differentiation in the likelihood of an individual project to achieve emissions savings, nor that their governance structures are similarly arranged. However, it is important to retain the notion that all such credits are the result of particular institutional arrangements and are not unproblematic or tangible entities.

The exchange relationships in an offset transaction hinge around the means of calculation and the ability to equate emissions in one circumstance with reductions in another. We must consider (a) what quantity of emissions 'reductions' can be reliably associated with a given project, without being double counted in an existing commitment, (b) whether the 'reductions' attributed to the project the exclusive result of offset revenue (i.e., are they 'additional'?), and (c) how the credit stream is registered and assigned to buyers exclusively to prevent double selling. These are not trivial issues and as emphasised before there are no 'hard and fast' scientific means of assessing different systems.

The CDM utilises a rigid crediting procedure (Box 1) that includes submission of a calculation methodology and baseline for review by an independent board, audit of the project's documentation, and subsequent operation by private consultants, in a sense 'carbon accountants', and finally the disbursement of uniquely registered credits *ex post*, i.e., after the operation of the project. As there is no Kyoto cap on non-Annex 1 states, reductions are not double counted in international commitments, but both the host country and project developer have an incentive to inflate the quantity of credits to be claimed by a given project. The methodology and baseline can be very complex and may for example involve the determination of economic and technological parameters for an industry, region, or country, or emissions factors for a particular electricity system. Additionality is then argued with reference to the baseline. The case may be made in different ways; is the project reliant on the credit revenue stream for existence, are there national and international policies that incentivise or disincentivise the project, are there local institutional or capacity barriers to a project's implementation, is the project different to 'common-practice'? Leakage, the increase in emissions that may occur as a result of the project in another location or at a later date should also be accounted for. For instance, the imposition of a forest reserve in one locale may simply displace timber extraction activities to a neighbouring site or to take the example of an efficiency rebound effect, financial savings in a household budget from reduced energy expenditure may be used to purchase a motor vehicle or further appliances. In some circumstances sophisticated macroeconomic tools may be employed (Shrestha and Timilsina 2002) but ultimately this is again matter of judgement as to where and when boundaries are drawn in calculation. The DOE that reviews the project and ultimately the CDM Executive Board must make a decision to accept or reject it such that undeserved credits are not rewarded and the effectiveness of the Kyoto Protocol diminished.[17] However, the subjective nature of this judgement results in a tension between the desire for 'flexibility' and low cost versus the environmental integrity of regime. For example, it is in fact possible to *retrospectively* apply for registration if it can be argued that 'incentive from the CDM was seriously considered in the decision to proceed with the project activity'.[18]

There have been mixed reports from the operation of the CDM arguing that despite their depth, the procedures are not being rigorously enforced. A recent review of Indian projects by Axel Michaelowa, a world authority on baseline and calculation methodology and member of CDM Registration and Issuance Team, noted that one third of the fifty-two projects inspected failed to adequately demonstrate additionality (Davies 2007a) by for example claiming that new investment in industrial hardware would not have been profitable without CDM revenue. Many of the hydro-electricity projects registered under the CDM are also criticised for not being additional, especially given the maturity of the technology and existing avenues of support from international financiers and development banks. Haya (2007) further argues that 'common practice' is weakly defined; all projects must overcome 'barriers' to implementation and financial calculations of return on investment are vulnerable to manipulation because of information asymmetry between developer and regulator. This is illustrated with reference to Chinese large hydro projects where new projects registered in the CDM represent 5.1GW a substantial proportion of total installation of 9GW in 2007 against previous annual installations of around 7.7GW. In this case, political and economic conditions are manifestly supportive raising significant questions of any specific claims of additionality. These examples do not argue against the CDM's effectiveness *in toto* but neither can they be dismissed as failings unique and specific to individual projects or auditors. Instead, it is argued here that they point to inevitable, systemic weaknesses that can only be addressed to a certain degree. The trade off between 'flexibility' and climatological security is persistent and must be recognised in operating CDM as a central plank of international climate policy.

These same issues in the commodification process are encountered in the unregulated voluntary market but amplified by lack of uniformity, coordination or transparency. *Caveat emptor,* the principle of buyer beware, applies even more so especially in the retail market where it is often very unclear at the point of sale as to what is being funded, where projects are located, how the credit will be cancelled and when the carbon reduction is claimed to have occurred or is expected to occur. Without a central institution, credit purchasers, individual or corporate, have had to rely upon the reputation of the company and their individual auditing procedures. As very few purchasers have the relevant knowledge, expertise, time or access to information, the potential for fraud is high. This is not denied by those within the market, for example Jonathan Shopley, chief executive of The Carbon Neutral Company was recently quoted as saying that

> There are credibility issues and there are cowboys around. It is probably to be expected for an industry at this stage but we need a set of standards and outside verification so that self-regulation can engender trust and integrity in the market.
>
> (MacAlister 2007)

Lohmann develops this further arguing that the voluntary market for credits resembles a 'lemons market' where the quality of goods being bought and sold cannot be proved or discriminated (Lohmann 2005). As such, buyers may be unwilling to pay a premium for a more reliable product, prices decline, and there is pressure to further reduce quality.

In response to this phenomenon there has been a rush of voluntary standards and audit procedures from different quarters. The UK Department for International Development supported the introduction of the Plan Vivo programme for small scale community forestry, a group of environmental NGOs presented the Gold Standard first for CDM projects and then also voluntary, a number of emissions traders and their industry body the International Emissions Trading Association developed the Voluntary Carbon Standard and the ISO itself has produced ISO 14064–2. Kollmuss, Zink and Polycarp present a helpful comparative analysis of these and other standards (Kollmuss et al. 2008) for the interested reader. The closest that the market has come to formal regulation is in the UK where the Department for the Environment Food and Rural Affairs (DEFRA) has proposed a Code of Best Practice for offset retailers. At present this will only endorse the sale of CERs from the CDM, ERUs from JI and EUAs from the EU ETS but has left the door open for voluntary standards should they demonstrate rigorous project selection, monitoring and successful operation of a credit registry. Importantly, it also specifies how credits are to be managed by a retailer and cancelled within a specific timescale.[19] It is not yet clear how, and with what consequences, these standards will be negotiated between buyers, sellers, and governments elsewhere in the world.

Specific Project and Project Type Criticisms

It would be naïve to assume that project interventions will only have consequences for GHG emissions and otherwise be neutral. Credit generating projects bring with them site specific and generic consequences that are both intentional and incidental. Although some failures are conspicuous, the withered mango trees funded by the pop group Coldplay for example (Dhillon and Harnden 2006), the voluntary market's haphazard development makes it very difficult to generalise about the consequences of its growth for those involved in the projects.

Problems with specific projects are often more conspicuous than those associated with crediting regimes but there is a relationship between the two, the characteristics of the regimes defining those project types that are acceptable for crediting. In the CDM, forestry is accepted only where it is afforestation or reforestation but in the voluntary market there is a strong presence of credits from avoided deforestation. Carbon sequestered in forestry is not permanent, trees being vulnerable to fire, drought, and pests, and there has been warranted concern about its inclusion in a market based climate policy. It is also clear that investment in forestry does nothing to move economies away from dependence upon fossil fuels or promote innovation into new technologies. However, forestry projects are attractive to a retail voluntary market as there are often the co-benefits, real or presumed, of watershed protection and biodiversity conservation, generic positive associations between trees and pro-environmental behaviour and easy comprehension by Western consumers.

In contrast, retail voluntary market exchanges rarely involve credits produced by industrial projects, such as the destruction of refrigerant by-product hydrofluorocarbons (HFCs), whilst they constitute a large proportion of the CDM volume given their very high global warming potential (GWP_{100}, an estimation of warming impact over 100 years used to provide equivalence between gases). Industrial investments have received criticism for their lack of sustainable development benefits to local communities (Öko Institut 2006; Olsen 2007) and in some cases encountered popular resistance as a consequence of exacerbating environmental injustice (Smith 2007). The CDM has been accused of politically legitimising and financially supporting 'bad neighbour' industries such as sponge iron smelters or waste landfill sites (Lohmann 2006).

Industrial monoculture plantations have also drawn substantial and justifiable criticism from indigenous peoples, conservation and development organisations (Carrere and González 2000; Bachram 2004). Although there is support for small scale, bundled community forestry from some quarters (Boyd et al. 2007), it is not clear how tensions between local welfare and self determination are to be reconciled with international, institutional requirements for low cost, security, and permanence. The CDM has also been criticised for geographic bias, with two thirds of credits being supplied by three countries, China, India, and Brazil, with consequently very little investment in Least Developed Countries (Cosbey et al. 2006). Reports on voluntary credit supply appear to suggest it is somewhat more diverse, however, in the absence of substantial emissions reduction legislation, the USA voluntary offset market is developing a large number of indigenous offset projects through the CCX and other standards (Hamilton et al. 2007). This signals a desire for progress from certain individuals and organisations but is contributing nothing by way of technological or capital transfer North to South.

Voluntary and regulatory standards that include project screening and environmental and social impact assessment prior to initiation may serve to mitigate many of the negative impacts and promote those projects with greater benefit to their host communities. However, fundamentally the governance of the voluntary market is such that the terms of exchange weigh in favour of the credit consumer who is by definition distant and isolated from its impacts, a feature the carbon trade has in common with other instances of neoliberalisation (Bumpus and Liverman 2008). In the absence of a robust governance regime, the power of the purchaser may overwhelm both rigorous science and the public good.

Criticisms of Offsetting in Principle

Aside from the means of generating carbon credits, offsetting relationship may also be criticised in principle from both ethical and pragmatic positions.

Pragmatically, a relatively low cost of carbon and ease of engaging in off-setting acts as a disincentive to structural change in the consumption of energy which is necessary given the profound and rapid reductions required to avoid dangerous climate change (Metz et al. 2007). Market mechanisms may yet prove to be very successful at gathering 'low hanging fruit' but in doing so are delaying inevitable changes and in the process misdirecting attention from existing arrangements that subsidise fossil fuel extraction and consumption. Infrastructure lock-in through a prolonged period of low incentive to change, for example the purchase of a new fleet of long haul aircraft and construction of new terminal facilities, could lead either to a write-off of resources or ultimately avoidance of change altogether. Credit regimes also have the potential to influence the offset host's policy development by creating a situation of moral hazard; actors may delay taking action to curb emissions in the short term for gain under a future regime (Hepburn 2007).

A series of forceful ethical positions against offsetting have also come from proponents of social justice. Offsetting is predicated upon the profound difference in per capita emissions between industrialised and industrialising countries and it assumes that an equal price should be paid by all for the mitigation of climate change by 'outsourcing' reductions to economies where savings are, currently, low cost. 'Carbon colonialism' ignores historic responsibilities and in the case of voluntary retail offsets focuses attention on the individual not the society (Bachram 2004), reducing what might be seen as a political problem to a financial transaction. The agenda for development and acceptable degrees of climate impact are set by technical, financial and political elites in the 'business as usual' baselines of credits systems and not by those most at risk or through democratic means (Lohmann 2001). This marks the distinction between a New Zealand land owner who is paid for not farming land, to permit regrowth of forest, and the residents of Tuvalu who will face some of the most immediate impacts of climate change with little compensation even though they have been *not* flying to Europe on holiday for generations. Of course, this comparison sounds ridiculous, but highlights the essentially arbitrary nature of what is deemed an emission to be offset and what is an activity to be rewarded. Where there are opportunities for substantial wealth creation in the development of such markets the benefits accrue to the already wealthy (Bumpus and Liverman 2008) and as John O'Neill eloquently argues, to invoke a market solution to environmental disputes is to entrench the same social order that is responsible for production of those disputes in the first instance (O'Neill 2007). Considering forestry, offset projects are a long term appropriation of land and carbon cycling capacity from tropical to temperate nations on the terms of those in the rich North (Lohmann 2001), whilst the greatest impacts of environmental change will be felt by tropical, least developed countries.

CONCLUSION

The aviation industry and those components of tourism industry dependent upon it are presented with a substantial challenge by the profound and pressing need to reduce GHG emissions from industrialised economies. Current growth rates are incompatible with a carbon-constrained future. Carbon offsetting, generically stated as the reallocation of mitigation effort between actors, offers some promise for an industry with limited technological or organisational possibilities to improve efficiency. How society addresses this issue is currently being negotiated at a range of scales but market based mechanisms are increasingly favoured as a means of coordinating action. They are predicated upon the equivalence of different sources of emissions in principle and practice and a great deal of effort, both political and technical is being invested in their progression. The EU ETS cap and trade system is likely to include aviation emissions from 2012. Current proposals do not look likely to have a great deal of impact on the tourism industry but may lead to conflict with other industrial sectors.

In the meantime, a small but increasing number of tourists are offsetting the 'unavoidable' emissions from their transport and accommodation by voluntarily purchasing project generated credits. The credits are created by private actors using a variety of procedures, from the large scale UN FCCC Clean Development Mechanism to opaque proprietary standards. All are essentially based upon justifying that a given project would not have happened under 'business as usual' circumstances (i.e., without a carbon broker intervening), and subsequently quantifying an amount of emissions saved by the intervention. Because of the frequently contentious nature of these claims, there have been serious questions raised as to the credibility and effectiveness of these programmes and subsequently efforts to respond to these from the private sector and government. Given that the validity of any given credit cannot be unequivocally defined, and that there are a number of incentives for developers to overstate the project benefits, good governance and enforcement in the systems that generate them is paramount to reduce fraud. There are also initiatives such as the Gold Standard intended to ensure that projects funded also generate positive outcomes for those communities that host them. These protective measures are contested as barriers to efficiency and heavy-handed burdens on the system and are unlikely to be widely adopted.

In the long term and at global scales, the tourism industry will have to alter its structure in response to legislative and commercial pressure. For better or worse, carbon offsetting looks likely to slow the pace of this transition by reducing the financial and political incentives for change, unless the claimed benefits of communicating climate change to new audiences translates into an enabling environment for broader emissions reduction and social and technical innovation.

NOTES

1. Empirical content of this chapter is drawn primarily from the survey voluntary carbon credit retailers reported in Gössling et al. (2007), updated and extended as of February 2008.
2. As defined by the regime—nation states, corporations, and individual industrial installations for example.
3. Source: Point Carbon http://www.pointcarbon.com/ accessed 27/02/08, over the counter price for EUA delivery 01/12/08.
4. Further detail can be found on the UNFCCC website http://cdm.unfccc.int/ Statistics/index.html
5. Depending on project type, small scale is defined as projects generating less than 60,000 CERs per annum, those based on installation of equipment rated at <15MW or resulting in efficiency savings of <60GWh per annum (CDM document reference: FCCC/KP/CMP/2006/10/Add.1)
6. Figures from UNFCCC website as of 03/03/2008 http://cdm.unfccc.int/ index.html
7. http://www.carboncatalog.org/providers/ (accessed 06/03/08).
8. Inverted commas highlight the contested nature of sustainable development, with particular resonance in this context. One of the cornerstone principles of the CDM is that it enables sustainable development by providing economically efficient climate change mitigation and the transfer of technology and expertise to developing nations. How it achieves this is questionable both in terms of the specific projects supported and also the larger principle of shifting the burden of mitigation.
9. Including but not limited to the PURE Clean Planet Trust, Carbon Clarity and Equiclimate.
10. See for example Blue Ventures Carbon Offset http://www.bvco.org.uk/ BVCO_fix.htm, ReduceMyFootprint http://www.reducemyfootprint.travel/ individuals/calculators/quick.cfm or the national per capita average subscription service of Carbon Planet http://www.carbonplanet.com/store/subscriptions (all sites accessed 06/03/08).
11. See http://www.flysilverjet.com/Press-release-2006–11–26.aspx and http://www.eurostar.com/UK/uk/leisure/about_eurostar/environment/off_setting.jsp (both accessed 06/03/08).
12. In year 2000 prices Stern estimates the MSCC to be $85/tCO$_2$e $\approx$ £53/tCO$_2$e $\approx$ £194/tC although the calculation is in itself problematic. A recent DEFRA review of the academic literature reports a wide variety of estimates for the marginal social cost of carbon emissions from less than $0/tCO$_2$e to over $400/tCO$_2$e (year 2000 prices; Downing and Watkiss 2005), highly dependent upon the pure rate of time preference, 'business as usual' assumptions and consideration of system dynamics in the long term (Stern 2006, 25 and Box 13.1).
13. Readers are directed to the European Commission website for latest details on progress http://ec.europa.eu/environment/climat/aviation_en.htm
14. The equivalent proportion inferred from Phase II NAPs is 13.4 per cent (approximately 1.3 GtCO$_2$e over the whole period) http://europa.eu/rapid/ pressReleasesAction.do?reference=IP/07/1614
15. Flybe 04.06.2006: 'Flybe launches world's first aircraft eco-labelling', Virgin Atlantic 07.11.2007 'Tea? Coffee? Carbon offset? Virgin Atlantic launches first ever onboard carbon offset scheme', British Airways 15/01/2008: 'Green initiative launched', Delta Air Lines 18/04/2007 'Delta to Launch Worldwide Carbon Offset Program for Customers This Summer.', EasyJet.
16. http://www.cheatneutral.com/ What is cheat offsetting? (accessed 28/03/2008).

17. For every CER issued there will almost certainly be an increase in emissions elsewhere as they are surrendered to meet EU ETS and/or Kyoto net emissions commitments.
18. CDM document reference: Guidelines for Completing the PDD Version 6.2, p. 11.
19. This effectively excludes *ex-ante* systems whereby a credit is claimed 'up front' on the expectation of a successful reduction in the future. For example, energy efficient light bulbs distributed in 2007 may only save an appreciable quantity of carbon in a ten-year period. However an offset transaction may be arranged to supply the capital to fund the project on the expectation of the savings. This is clearly a precarious state of affairs.

REFERENCES

3 News (2008) *Air NZ Launch Scheme to Allow Passengers to Offset Carbon Footprints.* Available at: http://www.tv3.co.nz/News/Story/tabid/209/article ID/50650/cat/41/Default.aspx (27 March 2008).

Adam, D. (2006) 'Can planting trees really give you a clear carbon conscience?' *The Guardian.* Available at: http://environment.guardian.co.uk/climatechange/story/0,,1889830,00.html (7 October 2006).

Anderson, K., Bows, A., and Footitt, A. (2007) *Aviation in a Low Carbon EU.* Technical report, Tyndall Centre, prepared for Friends of the Earth, September 2007.

Bachram, H. (2004) 'Climate fraud and carbon colonialism: The new trade in greenhouse gases'. *Capitalism Nature Socialism* 15(4): 5–20.

Bayon, R., Hawn, A., and Hamilton, K. (2006) *Voluntary Carbon Markets: An International Business Guide to What They Are and How They Work.* Earthscan, 2006.

Bows, A. and Anderson, K.L. (2007) 'Policy clash: Can projected aviation growth be reconciled with the UK government's 60 per cent carbon-reduction target?' *Transport Policy* 14(2): 103–110.

Boyd, E., Gutierrez, M., and Chang, M.Y. (2007) 'Small-scale forest carbon projects: Adapting CDM to low-income communities'. *Global Environmental Change—Human and Policy Dimensions* 17(2): 250–259.

Bristow, S. (2006) *Going Carbon Neutral—The Right Path for Business?* The Climate Group, 2006.

Bumpus, A. and Liverman, D. (2008) 'Accumulation by decarbonisation and the governance of carbon offsets'. *Economic Geography,* 84(2).

Capoor, K. and Ambrosi, P. (2007) *State and Trends of the Carbon Market.* Technical report, IETA & World Bank, Washington D.C.

Carrere, R. and González, A. (2000) (eds.) *Climate Change Convention: Sinks that stink.* World Rainforest Movement. Available at: http://www.wrm.org.uy/actors/CCC/sinks.html

CD4CDM and EcoSecurities (2007) *Guidebook to Financing CDM Projects.* Available at: http://cd4cdm.org/Publications/FinanceCDMprojectsGuidebook.pdf

Ceppi, P. (2006) *The Carbon Trust Three Stage Approach to Developing a Robust Offsetting Strategy.* The Carbon Trust, UK, 2006.

Chapman, M. (2006) 'Green government plan "a fiasco"'. *BBC Radio Five Live.* Available at: http://news.bbc.co.uk/1/hi/business/6092460.stm (29 October 2006).

Clarke, T. (2007) 'Great green smokescreen'. *Channel 4: Dispatches.* Available at: http://www.channel4.com/news/articles/dispatches/the+great+green+smokescreen/589267 (16 July 2007).

Cosbey, A., Murphy, D., Drexhage, J., and Balint, J. (2006) *Making Development Work in the CDM Phase II of the Development Dividend Project*. International Institute for Sustainable Development.

Davies, N. (2007a) 'Abuse and incompetence in fight against global warming'. *The Guardian*. Available at: http://business.guardian.co.uk/story/02093836,00. html (2 June 2007).

Davies, N. (2007b) 'The inconvenient truth about the carbon offset industry'. *The Guardian*. Available at: http://environment.guardian.co.uk/climatechange/ story/0,,2104395,00.html (17 June 2007).

DEFRA (2007) *Consultation on Establishing a Voluntary Code of Best Practice for the Provision of Carbon Offsetting to UK Customers*. Technical report, Department for Environment, Food and Rural Affairs.

Dhillon, A. and Harnden, T. (2006) 'How coldplay's green hopes died in the arid soil of India'. *The Telegraph*. Available at: http://www.telegraph.co.uk/news/ main.jhtml?xml=/news/2006/04/30/ngreen30.xml&sSheet=/news/2006/04/30/ ixhome.html (29 April 2006).

Downing, T. and Watkiss, P. (2005) *Overview: The Marginal Social Costs of Carbon in Policy Making: Applications, Uncertainty and a Possible Risk Based Approach*. DEFRA. Available at: http://www.defra.gov.uk/environment/climatechange/research/carboncost/pdf/downing_watkiss.pdf

Economist (2007, March 17) 'Ripping off would be greens?' *The Economist* 382(8520): 36.

Elgin, B. (2007) 'Another inconvenient truth: Behind the feel-good hype of carbon offsets, some of the deals don't deliver'. *Business Week*. Available at: http:// www.businessweek.com/magazine/content/07_13/b4027057.htm (last accessed 26 March 2007).

Ellerman, A. (2005) 'A note on tradeable permits'. *Environmental and Resource Economics* 31(2): 123–131.

Ellerman, A. and Buchner, B. (in press) 'Over-allocation or abatement? A preliminary analysis of the EU ETS based on the 2005–06 emissions data'. *Environmental and Resource Economics*.

Enkvist, P.-A., Naucler, T., and Rosander, J. (2007) 'A cost curve for greenhouse gas reduction'. *The McKinsey Quarterly* 1.

Environmental Audit Committee (2007) *The Voluntary Carbon Offset Market*. Sixth Report of Session 2006–2007. London: The Stationery Office.

European Commission (2007) *Europe Stands Firm on Ambitious Action to Cut Aviation Emissions*. Press Release, 28 September 2007.

Eyers, C., Eyers, J., Norman, P., Middel, J., Plorh, M., Michot, S., and Atkinson, K. (2004) *AERO2k Global Aviation Emissions Inventories for 2002 and 2025*. QinetiQ.

Faris, S. (2007) 'The other side of carbon trading'. *Fortune Magazine*. Available at: http://money.cnn.com/2007/08/27/news/international/uganda_carbon_trading.fortune/index.htm?postversion=2007082911 (30 August 2007).

Forster, P.M.D.F., Shine, K.P., and Stuber, N. (2006) 'It is premature to include non-CO_2 effects of aviation in emission trading schemes'. *Atmospheric Environment* 40(6): 1117–1121.

Gibbs, T. and Retallack, S. (2006) *Trading Up: Reforming the European Union's Emissions Trading Scheme*. London: IPPR.

Gössling, S., Broderick, J., Upham, P., Ceron, J.-P., Dubois, G., Peeters, P., and Strasdas, W. (2007) 'Voluntary carbon offsetting schemes for aviation: Efficiency, credibility and sustainable tourism'. *Journal of Sustainable Tourism* 15(3): 223–248.

Gössling, S. and Hall, C.M. (2005) *Tourism and Global Environmental Change: Ecological, Social, Economic and Political Interrelationships*. London: Routledge.

198   *John Broderick*

Gössling, S. and Peeters, P. (2007) 'It does not harm the environment! An analysis of industry discourses on tourism, air travel and the environment'. *Journal of Sustainable Tourism* 15(4): 402–417.

Graßl, H. and Brockhagen, D. (2007) 'Climate forcing of aviation emissions in high altitudes and comparison of metrics'. Max-Planck-Institut fuer Meteorologie, Hamburg.

Hamilton, K., Bayon, R., Turner, G., and Higgins, D. (2007) 'State of the voluntary carbon markets 2007: Picking up steam'. *The Ecosystem Marketplace*, Washington DC. Available at: ecosystemmarketplace.com/documents/acrobat/StateoftheVoluntaryCarbonMarket18July_Final.pdf (15 March 2008).

Harvey, F. (2007) 'Beware the carbon offsetting cowboys'. *Financial Times*, 25 April 2007. Available at: http://www.ft.com/cms/s/dcdefef6-f350-11db-9845-000b5df10621,dwp_uuid=3c093daa-edc1-11db-8584-000b5df10621.html (29 March 2008).

Harvey, F. and Fidler, S. (2007) 'Industry caught in "carbon credit" smokescreen'. *Financial Times*, 26 April 2007. Available at: http://www.ft.com/cms/s/6d52acc6-f392-11db-9845-000b5df10621.html (29 March 2008).

Haya, B. (2007) *Failed Mechanism: How the CDM is Subsidizing Hydro Developments and Harming the Kyoto Protocol.* Berkley, CA: International Rivers.

Hepburn, C. (2007) 'Carbon trading: A review of the Kyoto mechanisms'. *Annual Review Of Environment And Resources* 32: 375–393.

Hourcade, J.C., Neuhoff, K., Demailly, D., and Sato, M. (2007) *Differentiation and Dynamics of EU ETS Industrial Competitiveness Impacts.* Cambridge, UK: Climate Strategies.

House of Lords (2006) *Including the Aviation Sector in the European Union Emissions Trading Scheme.* Number 21 in European Union Committee, Session 2005–2006, 21st Report of Session 2005–2006.

HSBC (2004) *HSBC World's First Major Bank to Go Carbon Neutral.* Press release, 6 December 2004. Available at: http://www.hsbc.com/1/2/newsroom/news/news-archive-2004/hsbc-worlds-first-major-bank-to-go-carbon-neutral (29 March 2008).

ICAO (2007) *Report on Voluntary Emissions Trading for Aviation.* April 2007. Available at: http://www.icao.int/icao/en/env/vets_report.pdf (29 March 2008).

Johnston, I. (2007) 'A gift from Scotland to Brazil: Drought and despair'. *The Scotsman*, 7 July 2007. Available at: http://news.scotsman.com/index.cfm?id=1060072007 (29 March 2008).

Kettner, C., Koppl, A., Schleicher, S.P., and Theninus, G. (2008) 'Stringency and distribution in the EU emissions trading scheme: First evidence'. *Climate Policy* 8(1): 41–61.

Kollmuss, A., Zink, H., and Polycarp, C. (2008) *Making Sense of the Voluntary Carbon Market: A Comparison of Carbon Offset Standards.* Frankfurt: WWF Germany.

Lohmann, L. (2001) 'The dyson effect: Carbon "offset" forestry and the privatization of the atmosphere'. *International Journal of Environment and Pollution* 15(1): 51–78.

Lohmann, L. (2005) 'Marketing and making carbon dumps: Commodification, calculation and counterfactuals in climate change mitigation'. *Science as Culture* 14(3): 203–235.

Lohmann, L. (2006) *Carbon Trading: A Critical Conversation on Climate Change, Privatisation and Power.* Uppsaka: Dag Hammarskjold Foundation.

MacAlister, T. (2007) 'Offsetting chief warns of carbon cowboys'. *The Guardian*, 18 June 2007. Available at: http://business.guardian.co.uk/print/0,,330042880-108725,00.html (29 March 2008).

Metz, B., Davidson, O.R., Bosch, P.R., Dave, R., and Meyer, R.A. (2007) (eds.) *Climate Change 2007: Mitigation. Contribution of Working Group III to the*

Fourth Assessment Report of the Intergovernmental Panel on Climate Change. Cambridge University Press, Cambridge, UK: IPCC.

Muir, H. (2004) 'Trees, the eco-investment of choice. but now campaigners question forests firm'. *The Guardian*, 30 September 2004. Available at: http://www.guardian.co.uk/waste/story/0,12188,1316050,00.html (29 March 2008).

Öko Institut (2006) *Green Goal Legacy Report.* Organizing Committee FIFA World Cup, November 2006. Available at: www.oeko.de/oekodoc/292/2006–011-en.pdf (29 March 2008).

Olsen, K.H. (2007) 'The clean development mechanism's contribution to sustainable development: a review of the literature'. *Climatic Change* 84(1): 59–73.

O'Neill, J. (2007) *Markets, Deliberation and Environment.* London: Routledge.

Pease, R. (2005) 'Request for forest carbon credits'. *BBC News*, 17 July 2005. Available at:http://news.bbc.co.uk/1/hi/sci/tech/4557541.stm (29 March 2008).

Peeters, P., Williams, V., and Gössling, S. (2007) 'Air transport greenhouse gas emissions'. *Tourism and Climate Change Mitigation*, 29–50.

Røine, K., Tvinnereim, E., and Hasselknippe, H. (2008) *Carbon 2008—Post-2012 is Now.* Point Carbon.

Rosen, M. (2007) 'Carbon credit report'. *The Monthly*, August 2007. Available at: http://www.themonthly.com/feature-08-07.html (29 March 2008).

Shrestha, R.M. and Timilsina, G.R. (2002) 'The additionality criterion for identifying clean development mechanism projects under the Kyoto protocol'. *Energy Policy* 30(1): 73–79.

Siegle, L. (2008) 'When Thom met Ken'. *The Observer Magazine* March 28: 23.

Smith, K. (2007) *The Carbon Neutral Myth: Offset Indulgences for Your Climate Sins.* Transnational Institute.

Stern, N. (2006) *Stern Review: The Economics of Climate Change.* Cambridge University Press: Cambridge, UK.

Stoll-Kleemann, S., O'Riordan, T., and Jaeger, C.C. (2001) 'The psychology of denial concerning climate mitigation measures: Evidence from Swiss focus groups'. *Global Environmental Change—Human and Policy Dimensions* 11(2): 107–117.

Sustainable Aviation (2005) '*Technical Report*'. AOA, BATA, SBAC, NATS, 2005. Available at: http://www.sustainableaviation.co.uk/

Sustainable Development Commission (2008) '*Carbon Neutrality and Offsetting*'. Available at: http://www.sd-commission.org.uk/pages/carbon_neutrality.html (29 March 2008).

UNFCCC (1997) *Kyoto Protocol to the United Nations Framework Convention on Climate Change.* United Nations, Dec 1997. Available at: http://unfccc.int/kyoto_protocol/items/2830.php (29 March 2008).

Vivid Economics (2007) *A Study to Estimate Ticket Price Changes for Aviation in the EU ETS.* Technical report, DEFRA.

Welch, D. (2007) 'Enron environmentalism or bridge to the low carbon economy?' *Ethical Consumer* 106: 12–20.

Wit, R.C.N., Boon, B.H., van Velzen, A., Cames, M., Deuber, O., and Lee, D.S. (2005) *Giving Wings to Emission Trading.* Technical report, CE Delft.

Part III

Innovation

Sustainable Tourism Futures

11 The Implementation of Sustainable Tourism
A Project-Based Perspective

Tim Coles

INTRODUCTION: SITUATING SUSTAINABLE TOURISM KNOWLEDGES

Sustainable tourism remains one of the most enigmatic concepts in tourism management. Although an agreed meta-statement remains elusive, a *de facto* understanding has emerged among practitioners and scholars on the basic conceptual building blocks as well as the guiding ethos (Butler 1999). Whatever the precise definition used, it is clear that our understanding of the concept, as well as its potentials and pitfalls for future management and governance, have been advanced by numerous insights from a multitude of investigations of programmes, initiatives, schemes, plans, ventures, and projects that purport to deliver sustainable tourism. In effect, knowledge production about sustainable tourism has frequently been in the 'context of application', or through Mode II to adopt Gibbons' et al. (1994) terminology. Just how far our understanding has been advanced is debatable. One of the principal criticisms of work on sustainable tourism is that it has been limited by singular, inward-looking, ring-fenced, and often unconnected case studies. Not only do the specificities of these cases defy straightforward generalisation and wider transferability, but the proliferation of Mode II tourism studies, especially those sponsored by business interests, may also frustrate the generation of deeper understanding and theory building (Tribe 2004; Coles 2009).

This chapter examines the extent to which advances from organisational science that advocate 'the project' as the principal unit of analysis are able to contribute to our understanding of current and future practices of sustainable tourism. According to Sydow and Staber (2002, 216), a project is simply a series of time-delimited tasks that are bundled together to yield a particular outcome (see also Lundin and Söderholm 1995). Grabher (2004a, 104) views projects as a 'temporary organizational arena in which knowledge is combined from a variety of sources to accomplish a specific task'. Project-based approaches focus on the changing characteristics of a project over time, including the tasks involved, the participants, the knowledge applied and created, the outcomes, and their evaluation.

In this respect, there are some limited overlaps with existing analyses of sustainable tourism, for instance in the way in which tourism collaborations function over time (Jamal and Getz 1995) and their life-cycle (Caffyn 2000). Treuren and Lane (2003) present a view of tourism policy-making as a project. Significantly, however, project-based perspectives have developed a distinctive conceptual apparatus of their own and they force us to think 'beyond the project'. The external project milieux, in addition to the relationships and networking between projects, as it were, 'upstream' and 'downstream', are just as important as the internal operation of individual projects. Most importantly, though, such approaches reveal that knowledge in a number of sectors of economic activity is produced by, resides in, and circulates through projects and their participants. Indeed, their strongest advocates argue that people 'think in projects' and as such projects are emblematic of a wider shift in the way in which society and economy are organised (Sydow, Lindkvist, and DeFillippi 2004).

PROJECTS, PROJECT-BASED ORGANISATIONS, AND PROJECT ECOLOGY

Although projects have become 'cool' in recent times (Grabher 2002a), the idea of the project as an organisational unit in business has existed for some time (Burke 2003). As early as the 1950s, distinctive bodies of knowledge started to emerge on project management (Gaddis 1959). These focused on how to introduce and manage change, above all, in the engineering, construction, and manufacturing sectors. At its simplest, project management is '. . . the application of knowledge, skills, tools and techniques to project activities in order to meet stakeholder's needs and expectations from a project' (in Burke 2003, 3). Given the complexity of projects and the wide array of potential participants, the study of project management has tended to concentrate on concepts, technologies, and frameworks that optimise the delivery of desired outcomes. Engwall (2003) has noted that, for some, the vast and contested bodies of knowledge that have been generated to this end signify that 'project management' should be considered a science in its own right, while Pollack (2007) identifies the emergence of two distinctive paradigms in the 'discipline'. The 'hard paradigm' is rooted in positivist/realist epistemology and concentrates on problem-solving and is more predominant in the practice and academic study of project management; in contrast, the 'soft paradigm' which deploys interpretivist theory to structure problems is less common but no less insightful concentrating as it does on people and intangibles such as learning, participation, and social processes.

The proliferation of project management as a construct would suggest that economic, social, and political activity is increasingly being organised around a series of connecting, time-limited projects, each of which has its own lifespan. As Dornisch (2002, 308) points out with respect to economic organisation, the project contrasts with . . . more stable forms of organization, in that it is created most fundamentally as a provisional, temporary vehicle for joint action which a set of loosely tied actors construct to achieve a common aim or carry out a pressing task.

Rather than embedding enterprises in strong, lasting, durable relationships, projects represent relatively weak, short-term, and flexible connections for the purpose of delivering economic outcomes. Examples abound in the building and construction industry, research and development, finance, film, architecture, pharmaceuticals, and the car industry (Midler 1995; DeFillippi and Arthur 1998; Hartman, Ashrafi, and Jergeas 1998; Ekinsmyth 2002; Grabher 2001; 2002b, 2004a; Sydow and Staber 2002). There are notable differences in the nature of projects across and even within different sectors. For instance, the engineering and construction industries are identified as exemplars of established, orthodox approaches to project management, whereas the arts—through media such as exhibitions, productions, concerts and films—feature much shorter-term, more dynamic, and fluid project-based arrangements (Hartman et al. 1998; Lindgren and Packendorff 2007). Winter et al. (2006, 700) note that, since the original work stemming from engineering, there has been a qualitative shift from product- to value-centric views such that 'business projects involving IT' have replaced 'IT projects'; that is, tasks are no longer exclusively focused on creating an asset, but rather creating value and benefit for stakeholder groups.

Projects are therefore common forms of socio-economic organisation and of self-regulation among the multiple communities of practice that understand and use the principles, practices, and technologies of project management (Hodgson 2002). Projects exist in several forms which may vary over time and space depending, not least, on the nature of the task, the historical background, the political positioning, as well as the actors involved, their social relations, and their motives for participation (Engwall 2003). Hobday (2000) differentiates project-based organisations in which most of their internal and external activity is conducted through projects, from project-led organisations whereby projects may be frequently conducted but their activities are predominantly either volume- and/or operations-based (Sydow et al. 2004). Grabher (2002b, 246) identifies distinctive socio-spatial forms coalescing around projects which he terms 'project ecologies' or 'a *heterarchic* form of social organization . . . that, despite dense patterns of interaction, is less systemic and less coherent than the more established territorial innovation models' such as industrial districts, clusters, or learning regions.

Project ecologies comprise sets of temporary, spatially specific alliances and collaborations of actors blending their competencies and expertises in order to deliver particular tasks and outcomes. These are not necessarily enduring or persistent relationships with long-term linkages. Rather, they are often driven by rivalry and they preserve the identities of the actors (Grabher 2002b, 246). Project ecologies rely on proximity and a thickness of organisations because they 'hinge on a dense fabric of lasting ties and networks that provide key resources of expertise, reputation, and legitimization' (Grabher 2004a, 104). The integration of individuals and organisations (or not) in projects is based on such features as social ties with other participants and/or their reputation from previous projects. The architecture of project ecologies—and hence a potential analytical focal point—is based on the identification of the following key components: namely, the core team as the main learning arena of the project; the firm, or where benefits from project-based learning reside; epistemic communities, or all participants who in some way contribute to knowledge creation; and finally, the personal networks, which are vital framework to allow projects to form in the future.

Projects and project ecologies are important in driving innovation and dissemination. It is unusual to find project actors involved in just single projects; rather, their skills, knowledge, expertises, and experiences are routinely put to work in several projects simultaneously as well as sequentially. Where there are overlaps between projects, common actors become vital agents in knowledge transfer (Prencipe and Tell 2001; Grabher 2001). Particular concepts may be exchanged while views and opinions may be articulated about certain prospective projects, project managers, contractors, and regions in which to work. This socialisation of knowledge plays a role in the type of team and accordingly the competencies that can be assembled in order to deliver a particular initiative.

Project-based learning points to the considerable importance in contemporary economy and society of knowledge creation in the context of application (cf. Gibbons et al. 1994). However, not all Mode II knowledge creation functions in the same manner. Grabher (2004b) has identified two distinctive forms of knowledge creation in Mode II project ecologies, which he describes as cumulative and disruptive learning. These ecologies exhibit different architectural traits (Table 11.1). They are not polar opposites nor are they intended to be ideal types. Rather, they suggest the predominant characteristics of each ecology based on their architectures. Such ecologies remain largely untested beyond Grabher's empirical research on the software and advertising 'industries'. They do, though, raise the question, if Mode II is as argued increasingly relevant, of how best to ensure that knowledge is effectively and appropriately 'sedimented'? For instance, it is not beyond the realm of possibility that knowledge produced in the context of application by a project team may be dispersed, diminuted, or even lost altogether when the team is dissolved. As actors move from project to project, there is the risk that they may not remember, nor have the chance to reflect on, or use knowledge produced in previous projects (Grabher 2004b, 1492). This risk would appear to be higher among disruptive ecologies.

Table 11.1 Characteristics of Contrasting Types of Project Ecology

Cumulative Learning	Disruptive Learning
Modularization of knowledge	Originality of knowledge
Learning by repetition	Learning by switching ties
Stable teams	Switching teams
Benefits of recurring ties	Benefits of reconfiguring relationships
Move from one-offs to repeatable solutions	Demand for difference
Create conventions	Defy conventions
Orchestration	Improvisation
Task-centred	Client-centred
Projects with clients	Projects for clients
Technical lock-in	Personal lock-in
Experience	Know-whom
e.g., Software ecology, Munich	e.g., Advertising ecology, London

Source: adapted from Grabher (2004)

RECOGNISING VERSUS RESEARCHING TOURISM PROJECTS: INSIGHTS FROM OTHER SECTORS

Projects are clearly a significant feature of contemporary organisation in other sectors of the economy and tourism is no different in this respect. Short-term initiatives are implemented by a number of actors in the private, public, and voluntary sectors to achieve particular outcomes in a number of areas of tourism management. For instance, Hawkins and Mann (2007, 359) observe that since the mid-1960s the World Bank has financed over 300 projects of a directly tourism or tourism-related nature. The World Bank, they assert, no longer funds tourism projects and this decision was taken on the basis of the application of standard approaches to project appraisal and evaluation. The projectification of tourism is also evident in O'Sullivan et al.'s (2003) discussion of the 'Network of Excellence for Action in Tourism' (NEAT), a European Union (EU) funded intervention to improve the contribution of tourism to the economy in South Wales. Their conclusions are drawn from an evaluation study commissioned by the project manager to meet EU funding regulations.

While project-based organisations and the guiding principles of project management are relatively straightforward to identify in the tourism sector, to date there has been practically no discussion about the epistemological and methodological challenges they raise here as there has been in other fields of study. For instance, Smyth and Morriss (2007) comment on the apparent tensions between generality and particularity elsewhere in studies of projects. On the one hand the dominant trend is to identify general trends but this marginalises the particular; on the other hand, if the particular experience is emphasised too greatly, then we run the risk that we may not be able to learn more about common patterns, shared meanings, and experiences, nor may we be able to make normative recommendations. This is, in many respects, a standard dilemma in case study research, tourism research included. As Grabher (2004b) argues, the case study is the principal modus operandum of project-based analysis per se (Engwall and Westling 2004). A combination of primary and secondary quantitative data may allow reasonably detailed mapping of projects or project ecologies, but it is almost impossible to know the background population of (all) relevant projects. Detailed insights on social relations are best revealed through qualitative research, in particular through in-depth interviewing often conducted in several rounds over time. In some instances, visual methodologies also offer important clues (Engwall and Westling 2004). There are therefore significant costs associated with such approaches to knowing and researching projects. Clearly, such research is facilitated by research grants and/or permissions to access the key actors (Grabher 2004b).

Interviews allow consensus (and dissonance) on critical issues to be revealed and reached, and to a degree obviate the background population problem. They are not however unproblematic and they raise questions regarding the ethics of data capture and reporting, as well as how to make complex, even 'fuzzy' concepts from academe understandable to those in practice (Coles, Liasidou, and Shaw 2008). By virtue of the nature of the close nature of many (professional) communities (of practice) and the need to ensure confidentiality and anonymity, it is often necessary to use indirect, reported speech. Access to key information sources—whether key actors or documents—may be restricted. Both are standard issues but the latter is all the more frustrating because many projects by virtue of their public funding or remit are required to have extensive documented evidence for reasons of transparency and these paper trails should be accessible to members of the public as well as research workers more specifically. Even with a Freedom of Information Act in force in the United Kingdom, it is not always possible to access documents because one has to know what one is looking for, and how to frame a request accurately and appropriately.

SUSTAINABLE TOURISM AS A POLICY PRIORITY IN SOUTH WEST ENGLAND

The epistemological and methodological tensions between first recognising and then researching projects and project ecologies in the tourism sector have been encountered in the empirical work on which this chapter is based. This is drawn from a series of research projects over a period from 2002 to 2007 in which sustainable tourism is featured. In effect, this chapter is informed by knowledge that has been produced in the context of application and derived from such exercises as consultations, interviews, project appraisals, and identification of best practice in the South West of England. As one of England's most popular tourism destinations, the South West has promoted itself in recent times on the richness of its natural and cultural landscapes. Active place marketing campaigns use the slogan 'It's in our nature' to denote a high quality of life and standard of living based on the regional environment. Not surprisingly, the principles of sustainable development have been at the forefront of the policy agenda.

Since 1999, the South West of England Regional Development Agency (SWRDA) has been responsible for long-term economic development through such activities as stimulating investment, job creation, boosting entrepreneurship and business growth, addressing skills and knowledge gaps, and promoting the principles of sustainable development (see Goodwin, Jones, and Jones 2005 on UK governance structures). As a non-departmental public body, it has recognised tourism as a key sector to support. In light of the central importance of tourism to society, economy, and environment in the region, the SWRDA was invited by the Department for Culture Media and Sport (DCMS)—the UK government department responsible—to lead inter-regional dialogue and debate on tourism policy in England. Other English regions have lost or are in the process of losing their regional tourist boards (RTBs)—in some cases, the role of the RTB has been taken on directly by the Regional Development Agency (RDA). In contrast, despite its financial difficulties and support from the SWRDA, at the time of this writing, South West Tourism (SWT) continues to provide leadership on regional tourism matters from marketing to human resources, and planning to destination management. As part of its most recent strategy, 'Towards 2015', SWT has made the further implementation of sustainable tourism in the region as one of its three long-term priorities (SWT 2007). Arguably, SWT is also at the forefront of international efforts to encourage more widespread sustainable tourism. With three partners, it is the author of a new volume in the highly successful *Dummies Guide* brand franchise devoted to the subject (Tourism South East, South West Tourism and West Sussex County Council 2008). It is within this policy context that the empirical evidence discussed in the next section was collected.

PROJECTS AND THE IMPLEMENTATION OF SUSTAINABLE TOURISM

Sustainable tourism projects are, it seems, also 'cool' in South West England. Despite the methodological and epistemological challenges, project-based analyses of sustainable tourism contribute a number of important insights. The first is that the use of the term *project* (and hence the conceptual fiat) is far from trivial in tourism governance. The word 'project' may all too often be used as a catch-all, 'umbrella term' in both research and practitioner circles to describe a wide array of programmes, initiatives, schemes, and the like.

Perhaps the most high profile use of the word 'project' for both domestic and international audiences is in the name of an iconic regional attractions, the Eden *Project* (author emphasis) in Cornwall (www.edenproject. com). Internationally recognisable for its distinctive geodesic domes that are reminiscent of golf balls, here sustainable tourism is *the* project. As a visitor attraction, the Eden Project has set out to be a platform from which to broadcast the messages of sustainable development through the practices and medium of tourism. Its humid tropics and warm temperate biomes housed in the domes explore aspects of human–environment relationships primarily through plants (Hultman and Gössling 2008), and with no little success. Eden was designed for 750,000 visitors per year when it opened in 2001, but by 2006 visitor numbers had stabilised at 1.2 million visitors (Paterson 2006, 1). The Eden Project has generated a large public debate about its effects on local and regional tourism patterns. In its defence, a range of studies have been conducted that suggest the Eden Project has played a vital role in enhancing the sustainable development of tourism in the county (DTZ 2005; Paterson 2006): Visitor numbers and spend have increased; jobs have been created; the tourism sector performs more efficiently; the economic benefits have been widely felt; local purchasing and contracts have been encouraged; Eden has widened visitation through the calender and taken pressure off the traditional summer 'high season'; and Eden has generated repeat visits.

The exact magnitude of the Eden effect is contested. Here it is more important to reflect that its enduring success is based on its highly projectified operations and interactions with its visitors (Jackson 2000). Originally, it was deliberately named for the connotations of the word 'project'. Tim Smit (2001), the visionary force behind Eden, describes how it got its name, 'So, it was game on and every game needs a name. Thinking up names is a pain [. . .] I like the word Project. It's got dynamism, direction and a sense of evolution about it.'

For Smit, projects are ideologically vital in for the delivery of particular outcomes and outputs. The word was used deliberately to galvanise support and to clarify the thinking of the stakeholders involved. As he recounts, 'Eden was never about plants and architecture; it was always about harnessing people to a dream and exploring what they are capable of' (Smit 2001).

Thus, the term had multiple meanings to key constituencies, which could be exploited. For local residents and business people, the term denoted a particular opportunity ring-fenced in time but not location. For those contractors and sub-contractors involved in the physical realisation of the attraction from the building trade, architects, surveyors, and land agents, the term 'project' was a familiar and comfortable tool in their conceptual armoury. It helped to manage their expectations of what they were about to become involved in. For investors, the term 'project' implied a degree of prudence and propriety by establishing clear sets of outputs and outcomes, which formed a basis for the monitoring and evaluation of the scheme in which they had a stake. As Mark Paterson (2006, 3), the Tertiary Education Co-ordinator at Eden, comments, the concept of the 'project' remains a major driver for expansion and rejuvenation of the product offer several years after its development and opening:

> The use of the word "project" is as much a title as a reality. Eden is far from a finished article and remains a dynamic project of ongoing development. Among the many prospects is the potential for a third covered biome—the Dry Tropics Biome—highlighting the importance of water in arid regions. Eden will remain a venue of being many things to many people.

Among the 'many things' Eden has been the venue for live entertainment projects such as a Live8 concert and the so-called 'Eden Sessions', evening concerts with major international artists. Since opening further construction projects have taken place on site, including the opening of a new education centre, 'The Core'. In late 2007, Eden was however unsuccessful in its attempts to obtain national lottery funding for 'The Edge', its new concept for the third biome refined to focus on issues of climate change and the human–environment interface.

Eden is far from the only sustainable tourism initiative implemented in this manner in either the county of Cornwall or the South West region. Between 2001 and 2006 there were at least 37 other projects conducted in Cornwall with total investment of £71 million, or £1.9 million per project (Table 11.2). These were funded by the EU Objective One programme,

Table 11.2 The Funding of Tourism Projects Through the EU Objective One Programme in Cornwall, 2001–2006

District Focus of Project	Number of Projects	Objective One Share (£)	Total Project Investment (£)	Average Project Investment (£)
All Cornwall	11	20,376,906	52,273,805	4,752,164
Caradon	1	797,725	1,828,613	1,828,613
Carrick	5	525,190	1,676,777	335,355
Isles of Scilly	2	356,406	947,545	473,773
Kerrier	2	449,913	928,004	464,002
North Cornwall	3	224,836	839,110	279,703
Penwith	5	2,133,655	5,205,564	1,041,113
Restormel	8	2,552,212	7,367,180	920,898
Total	37	27,416,843	71,066,598	1,920,719

Source: calculated from Objective One (2007)

which throughout this period intended to increase prosperity, create sustainable communities, and capitalise on the economic opportunities of the distinctiveness of the region. Environmental sustainability is explicitly raised as one of the cross-cutting themes (Objective One 2005). This array of projects included the next stage in Eden's expansion ('The Core'), as well as schemes for protecting fragile environments, reducing visitor pressures in time and space, investing in local community employability and infrastructure, improving access for the disabled, new product development, feasibility studies, a local tourism forum, and marketing campaigns. Not included are projects with other funding arrangements, from more modest private sector initiatives to high profile iconic attractions such as the National Maritime Museum in Falmouth. Of course, the precise degree to which these projects facilitate or embody sustainable tourism is debatable. Notwithstanding, were an inclusive approach to be employed, they progress sustainable development through tourism in several different ways (Table 11.3). Regional distinctiveness is prioritised (measures 5.1, 5.2, and 5.5) and a closer reading of the Programme Complement (Objective One 2005, 84ff) indicates they advocate making full and best use of public resources but not at the expense of the future.

From the recent history of tourism development in Cornwall, sustainable tourism would appear to be nurtured practically on a highly focused, project-by-project basis rather than through the implementation of a wider, over-arching macro-policy, or detailed plan from the regional or national level. While there is a regional policy for tourism (Towards 2015), this sets out a vision and aspirations rather than an intricate delivery plan on a county-by-county or destination-by-destination case. Regional and local policy and planning certainly does not include provision for such specific

Table 11.3   The Strategic Priorities of Objective One Tourism Projects in Cornwall, 2001–2006

Measure	Purpose	Number of Projects
1.3	Developing competitive business	4
2.1	Strategic sites and premises	2
2.4	Strategic regional infrastructure	1
3.2	Learning for competitive business and enterprise	1
4.6	Promoting the adaptation and development of rural areas	8
5.1	Securing the benefits from the arts, culture and heritage	8
5.2	Enhancing and developing the public product	10
5.5	Improving and developing the public realm	3

Source: adapted from Object One (2005, 2007)

projects or initiatives. These may in many cases be the end result of public–private partnerships. However, they are driven in the first instance by private sector operators who respond to more general calls for funding ideas or invitations to tender. RTBs and RDAs do not have statutory responsibility for planning at the local or county levels in England (Goodwin et al. 2005). Clearly, they exercise considerable power and influence in public–private partnerships and in the allocation of public funding. The apparently ad hoc or even responsive nature of the implementation of sustainable tourism is a clear manifestation of the principles and practices of project management facilitating opportunities rather than a strategic vision per se driving change. For instance, to secure funding, applicants had to refer to the Programme Complement, which was, in effect, a set of project management regulations (Objective One 2005). Similar manuals of practice (such as HM Treasury's Green Book and White Book, as well as the Single Programme Appraisal Guidelines–SPAG) have governed applications to the SWRDA for capital projects, such as the HMS Scylla artificial diving reef and the expansion of the Living Coasts (Torbay, Devon) attraction. Targets, performance indicators, funding arrangements, project administration, and governance structures were all set out in great detail. In turn, a distinctive, loose, and plainly problematic, consumptionist view is sometimes articulated whereby tourism is vital to the sustaining of particular levels of revenue, employment, or investment first and foremost. Bricolage is in effect a means of delivering or sustaining geographies of competitive advantage through tactical investment (DTZ 2005) and uneven development results (Table 11.2). For instance, at a sub-regional level, as the home of Newquay, one of the UK's most popular resort towns as well as the Eden Project, Restormel [Borough] has fared relatively well from the allocation of project-based funding, whereas the Isles of Scilly and Caradon [District Council] have been much less successful.

There are distinctive project ecologies of sustainable tourism associated with Objective One. Considerable knowledge about sustainable tourism has been generated within, but is not restricted to, Cornwall. Rather, this has formed a significant resource for others within the region to access by building relationships that stretch over space rather than necessarily just rely on geographical proximity. Other project ecologies of sustainable tourism are evident in the region, not least in the Bristol locality. Closer to London than Penzance in the west of the region, Bristol is home to many of the central government functions devolved to the South West (through Government Office for the South West—GOSW). Bristol is also home to many regional bodies and regional head offices, and hence it has a high level of local institutional thickness.

The organisation 'our south west' (2007) describes itself as '*the* focus on sustainability in the South West. . . . ' (emphasis original) which is 'managed, developed and co-ordinated by the Government Office for the South West [GOSW, in Bristol]. . . . to support the strategic work of GOSW and regional partners in moving the region towards a more sustainable future'. 'our south west' has recently been championing the importance of the climate change agenda in tourism (and other sectors) through the South West Climate Change Impact Partnership (SWCCIP), which is funded in part by the GOSW and SWRDA, and in which South West Tourism participates.

The nature of the inter-connections among the community of practice focused on sustainable tourism are far more intricate than one may even expect upon first inspection. The first hotel in Cornwall to take up GOSW's so-called 'Greener Events Guide' was also the venue for the Cornwall Sustainable Tourism Project (CoaST) so-called 'Building on Distinction' event at which twenty-three ambassadors of good practice were recognised. Funded by Objective One, this event was allied to a £96k project provided information and guidance to small- and medium-sized enterprises (SMEs) on how to increase markets, marketing opportunities, cost savings, and competitiveness through the application of the principles of sustainable tourism (Objective One 2007).

More linkages can be mapped out that serve to demonstrate the density of ties. Located in offices literally facing GOSW in Bristol, Sustainability South West (SSW) has its headquarters. SSW sets out its credentials as 'the independent Champion Body for sustainable development in the South West of England. We are an awareness raising and advisory charity supporting action on sustainability' (SSW 2007). SSW's flagship tourism project is called 'Future Footprints' which is a 'sustainable tourism campaign in the South West of England that works with tourism businesses to help their customers and visitors "keep the South West special."' Its current campaigns in this respect are focused on the taglines of 'Try Local, Buy Local', 'See what's on your doorstep', and 'Look for tourism that cares' (FF 2007).

Typical of the public–private partnership genre, SSW claims to 'coordinate Future Footprints on behalf of a partnership of organisations who input via a steering group'. The partners include: CoaST; Devon County Council, the Eden Project, Natural England, Penwith District Council, Salt Media, South Hams District Council, South West Tourism, and the National Trust. A list of the demonstration projects it has appropriated as examples of best practice in sustainable tourism appear in Table 11.4. Kynance Cove is a National Trust property, which benefited from an Objective One grant of £308k in 2002 towards a £616k project to improve visitor facilitates and reduce visitor pressures. Similarly, the activities of CoaST are featured, while CoaST reciprocates by recognising the work of SSW on its web site (CoaST 2007).

These connections may appear intricate and complex, even a confusing 'alphabet soup' of acronyms. In part, this is the purpose for presenting the data in such a manner. They begin to represent fully the manner in which knowledge about sustainable tourism is transferred by key interlocking projects and participants. Quite often these connections are highly subtle and not at all easy to disentangle, and in some cases

Table 11.4 Demonstration Projects Highlighted by Sustainability South West

Project	Scope
The Barn	B&B in Glastonbury dedicated to highest environmental standards
The Local Food Company	Green principles and local sourcing of producers and produce
Golowan Festival	10-day celebration of Cornish identity structured around the Feast of St. John in Penzance, Cornwall
South Hams District Council	Pioneer and advocate of green business schemes and held up as international example of best practice
Otterton Mill	Tourist attraction in coastal Devon driven by commitment to local culture, tradition and environment
Kynance Cove	Popular National Trust property on Lizard Peninsula in Cornwall that has just undergone extensive renovation to manage visitor pressures
Exploring West Cornwall by 'Boot, Bus, and Branch-line'	Initiative designed to encourage visitors to use integrate public transport in the most peripheral parts of the county
Blackdown Hills Area of Outstanding Natural Beauty	Rural partnership among community in Devon and Cornwall aimed at supporting tourism ventures that are sensitive to the needs of local environment, economy and people

Source: adapted from FF (2007)

the relationships are based on serendipity or happenstance as much as strategy and planning or reputation and track record. Notwithstanding, certain key actors are common to many projects of differing scope and they become vital agents in the dissemination of knowledge, experience, skills, and best practice examples. GOSW is routinely involved because it is deputed to administer EU and UK government funding initiatives, while the RDA is recognised as the lead agency for economic development and another channel for public funding (Goodwin et al. 2005). SWT is present because, in its capacity as the RTB, it provides leadership in a key regional economic sector and the RDA is a major patron. Several interviews carried out between 2001 and 2007 have repeatedly confirmed the view that there has become almost a tacit list of *de rigeur* agencies that have to be part of any successful tourism initiative. This includes the above as well as other non-departmental public bodies (NPDBs) such as the Environment Agency, English Nature and the Countryside Agency (now Natural England), and English Heritage. Almost invariably, the latter two would be required where a scheme would impact on the natural or cultural landscape.

This tactical 'badging' of key partners is common practice across the region. The same may be said of soliciting and receiving endorsements through short 'soundbites' in order to build a project's or an organisation's credentials and legitimacy. For instance, on SSW's web site, the Major Project Manager (Policy) for Natural England opines, 'Our support for Future Footprints contributes to more sustainable use of the countryside whilst meeting our aim or [sic] helping everyone respect, protect and enjoy the South West' (FF 2007).

On the same site another supporter, the Eden Project, offers strong endorsement for SSW's work because, 'We have a long term view that there is no such thing as sustainable tourism. We need to move as quickly as possible to a point where all tourism is sustainable' (FF 2007).

SSW is routinely consulted because of its apparent ability to add to the credentials of a proposal and it is perceived to have access to the highest echelons of central government. Its president is Jonathan Porritt, the high profile environmental commentator and Chair of the Sustainable Development Commission (SDC), the UK government's independent watchdog. Not to be outdone, CoaST has been 'invited to be a member of the Sustainable Development Commission Panel, an ongoing forum to inform the SDC's policy and engagement work programmes'. Sustrans, another pressure group, is a common consultee for projects involving transport, while CoaST is increasingly recognised as a point of contact for sustainable tourism advice, especially within the region. As it points out on its web site:

> We get asked our opinion a lot. While we don't pretend to have the answer to everything, we are hugely passionate about what we do and have accumulated a lot of information, expertise and techniques for switching people on to behaviour change.

Clearly, consultation from the wider public community of practice is immensely helpful to organisations that interact with, and tap into funding from, public bodies like RDAs and RTBs that emphasise the value in knowledge creation and innovation within the region. Even two local authorities were identified as key consultees and sources of potential best practice and experience. The South Hams District Council was identified as a pioneer in green business schemes and as such it had much experience and knowledge to pass on to other local authorities. Caradon District Council was highlighted for a number of initiatives and projects, in which the principals of sustainable development were applied in a cross-cutting manner (cf. Vernon et al. 2005). This is interesting in so far as Caradon did not perform in relative terms as well as other districts in Cornwall in the allocation of Objective One funds (see Table 11.2). It does, though, contribute to the view that sustainable tourism is driven by projects of different scale and scope with a range of funding requirements and sources, both private and public.

Sometimes, key individuals were identified as repositories of knowledge about sustainable tourism and/or by their knowledge of how to access funding and construct applications. They were named first and the organisation they represent was almost a secondary feature. The identification of individuals was no accident, rather a function of the social networks that form in tourism governance in general and about sustainable tourism more specifically. One respondent noted that theirs was a small county, and that all the tourism officers knew one another personally pretty well, they met pretty frequently at functions and the like, and that they were all aware of what was going on the others' 'patches'. Knowledge of peers' plans and actions offered the potential for developing reciprocal support for each others' projects, as well as an opportunity for joint project bids. Clearly, in 'place wars' imitation can be the sincerest form of flattery and original, successful initiatives seldom remain such as ideas can be copied, reproduced, and refined by competitors to diminish a destination's competitive advantage (Coles 2003). The potential and—to a degree—the practice of poaching cannot be discounted. However, the social networks appear to have functioned to ensure buy-in for more discrete, even locally attuned projects and initiatives, widespread support for which would help to demonstrate that (public) funding would contribute to the wider public good and a more holistic regional response to the demands of sustainable development in tourism.

DISCUSSION: FACING UP TO THE NEOLIBERALISATION OF SUSTAINABLE TOURISM

Despite the considerable epistemological and methodological challenges in knowing and researching them as temporary organisational forms, projects are frequently used as vehicles to deliver sustainable tourism. The use of the word 'project' is no coincidence. As a widely used term in tourism governance around the world at a number of scales, it is emblematic of a series of disciplining principles and practices around which much contemporary economic and social activity is organised. A project-based reading emphasises not only the volume of 'work' purporting to advance sustainable tourism, but also the intricate inter-connectivities between projects, individuals, and organisations to this end. Far from single meta-initiatives, policies, or detailed planning, this chapter argues that sustainable tourism in South West England is actually advanced and delivered by a complex bricolage of projects and identifiable project ecologies involving sustainable tourism. As material in this chapter helps to demonstrate, it is no easy task to piece together the detail of the specific project ecologies from a methodological perspective. Some aspects of the social networks are not easily revealed without long-term, costly empirical research. Nevertheless, as the evidence presented here shows, the irregular patchwork of investment serves to reinforce or revise, not eliminate uneven patterns of development (see Table 11.2).

In fact, the inconvenient truth is that tourism governance structures in regions like the South West of England assist the projectification of sustainable tourism and will do so for some time to come. Among other things, they effectively devolve responsibility to deliver sustainable tourism to a fragmented array of public, voluntary and private sector actors and individuals at the local and regional levels. In general, neoliberal approaches to economic development encourage project teams in some cases to ally, in others to compete, with one another (as well as projects in other areas of interest) for limited resources (Harvey 2006). The situation is no different in the tourism sector. The technologies of project management offer regulatory safeguards for regional, national, and supranational (i.e., EU) bodies that they will receive acceptable outputs and outcomes from the projects they fund. Hence, complaints about the short-term nature of funding for sustainable tourism (Halme 2001, 113) rather fail to acknowledge the basic practicalities of contemporary regulatory frameworks. Neoliberal governance in tourism (like other sectors of economic activity) contributes to disruptive learning evidenced in such features as short-term views and intense competition among rivals who are vying for leadership of sustainable tourism and support for their projects. Funding frameworks may hint at co-ordinated efforts, however they encourage tactical team-building through know-whom, switching of ties, and personal links to build greater chances of success. There may appear to be a closer fit between the principles of sustainable tourism and cumulative learning regimes (Table 11.1) through the advocacy of such features as modularisation, recurring ties, repeatable solutions, conventions, and orchestration. Demonstration projects, the identification of best practice, award schemes, 'key ambassadors' and 'critical champions' would all appear to endorse this view. However, the sheer number of tourism projects being conducted (in the name of sustainable tourism) and professional rivalries contribute to a patchwork of praxis, selective 'organisational amnesia' to some initiatives, and remarkable consciousness of others.

CONCLUSION

Sustainable tourism is a regional policy priority in the South West of England and there have been notable investments towards its progression. Numerous epistemic communities have formed but closer coordination is important to ensure more effective generation, transfer, and sedimentation of the considerable volume of knowledge about sustainable tourism that has been created via, resides in, and is circulated among, separate projects and individual project participants. It is often dangerous to attempt to generalise from case studies. Nevertheless, there are strong reasons to suggest that the issues discussed in this chapter are emblematic of a more widespread trend in the management of sustainable tourism in one precise form or another. After all, the same EU apparatus that drives projectification in the South West of England is, broadly speaking, applied across all 25 member states. Furthermore, it helps to shape the functioning of member state and regional governance structures and frameworks. Similarly, the principles of project management would appear to underpin World Bank investment in tourism projects (Hawkins and Mann 2007), many of which are intended to contribute towards sustainable development. Finally, within academe, the activities of a community of practice like the Sustainable Tourism Co-operative Research Centre in Australia (ST CRC) demonstrate many hallmarks of projectification of knowledge production and application. Beyond the chapter's wider geographical resonances, there are important conceptual implications for scholars of tourism. The context of application is an increasingly important source of funding for tourism research (Tribe 2004; Cooper 2006; Coles 2008). Despite reservations in some quarters, Mode II knowledge production can progress theory and concept even though it may appear that there are more significant challenges in developing praxis.

REFERENCES

Burke, R. (2003) *Project Management. Planning and Control Techniques*. Chichester: Wiley.

Butler, R.W. (1999) 'Sustainable tourism: a state of the art review'. *Tourism Geographies* 1(1): 7–25.

Caffyn, A. (2000) 'Is there a tourism partnership life-cycle?' In Bramwell, B. and Lane, B. (eds.) *Tourism Collaboration and Partnerships: Politics, Practice and Sustainability*, 200–209. Clevedon: Channel View.

Coles, T.E. (2003) 'Urban tourism, place promotion and economic restructuring: the case of post-socialist Leipzig'. *Tourism Geographies* 5(2): 190–219.

Coles, T.E. (2008) 'The regulation of higher education in the United Kingdom and the spaces of tourism enquiry'. *Tourism Geographies* 11(1).

Coles, T.E., Liasidou, S., and Shaw, G. (in press) 'Tourism and new economic geography: Issues and challenges in moving from advocacy to adoption'. *Journal of Travel and Tourism Marketing* (in press).

Cooper, C. (2006) 'Knowledge management and tourism'. *Annals of Tourism Research* 33(1): 47–64.

Cornwall Sustainable Tourism Project (CoaST) (2007) *Who We Work With*. Available at: http://www.cstn.org.uk/Page3.asp?id=56 (14 September 2007).

DeFillippi, R. and Arthur, M.B. (1998) 'Paradox in project based enterprise: The case of film making'. *California Management Review* 40(2): 125–139.

Dornisch, D. (2002) 'The evolution of post-socialist projects: trajectory shift and transnational capacity in a Polish region'. *Regional Studies* 36(3): 307–321.

DTZ Pieda Consulting (DTZ) (2005) *Iconic Tourism Projects in the South West of England*. Bristol: DTZ Pieda Consulting.

Ekinsmyth, C. (2002) 'Project organization, embeddedness and risk in magazine publishing'. *Regional Studies* 36(3): 229–243.

Engwall, M. (2003) 'No project is an island: Linking projects to history and context'. *Research Policy* 32: 789–808.

Engwall, M. and Westling, G. (2004) 'Peripety in an R&D drama: Capturing turnaround in project dynamics'. *Organization Studies* 25(9): 1557–1578.

Future Footprints (FF) (2007) *Future Footprints—Keeping the South West Special!* Available at: http://www.sustainabilitysouthwest.org.uk/projects/future_footprints/ (14 September 2007).

Gaddis, P.O. (1959, May/June) 'The project manager'. *Harvard Business Review*, 89–97.

Gibbons, M., Limoges, C., Nowotny, H., Schwartzmann, S., Scott, P., and Trow, M. (1994) *The New Production of Knowledge. The Dynamics of Science and Research in Contemporary Societies.* London: Sage.

Goodwin, M., Jones, M., and Jones, R. (2005) 'Devolution, constitutional change and economic development: Explaining and understanding the new institutional geographies of the British state'. *Regional Studies* 39(4): 421–436.

Grabher, G. (2001) 'Ecologies of creativity: The Village, the Group, and the heterarchic organisation of the British advertising industry'. *Environment and Planning A* 33: 351–374.

Grabher, G. (2002a) 'Cool projects, boring institutions: Temporary collaboration in social context'. *Regional Studies* 36(3): 205–214.

Grabher, G. (2002b) 'The project ecology of advertising: Tasks, talents and teams'. *Regional Studies* 36(3): 245–262.

Grabher, G. (2004a) 'Learning in projects, remembering in networks? Communality, sociality, and connectivity in project ecologies'. *European Urban and Regional Studies* 11(2): 103–113.

Grabher, G. (2004b) 'Temporary architectures of learning: knowledge governance in project ecologies'. *Organization Studies* 25(9): 1491–1514.

Halme, M. (2001) 'Learning for sustainable development in tourism networks'. *Business Strategy and the Environment* 10: 100–114.

Hartman, F., Ashrafi, R., and Jergeas, G. (1998) 'Project management in the live entertainment industry: What is different?' *International Journal of Project Management* 16(5): 269–281.

Harvey, D. (2006) *Spaces of Global Capitalism: Towards a Theory of Uneven Geographical Development.* London: Verso.

Hawkins, D.E. and Mann, S. (2007) 'The World Bank's role in tourism development'. *Annals of Tourism Research* 34(2): 348–363.

Hobday, M. (2000) 'The project-based organisation: An ideal form for managing complex products and systems?' *Research Policy* 29: 871–893.

Hodgson, D. (2002) 'Disciplining the professional: the case of project management'. *Journal of Management Studies* 39(6): 803–821.

Hultman, J. and Gössling, S. (2008) 'Nature and environment as trans-boundary business strategies'. In Coles, T.E. and Hall, C.M. (eds.) *International Business and Tourism: Global Issues, Contemporary Interactions*, 70–83. London: Routledge.

Jackson, M. (2000) *Eden: the First Book.* Bodelva: The Eden Project.

Jamal, T. and Getz, D. (1995) 'Collaboration theory and community tourism planning'. *Annals of Tourism Research* 22(1): 186–204.

Lindgren, M. and Packendorff, J. (2007) 'Performing arts and the art of performing–on co-construction of project work and professional identities in theatres'. *International Journal of Project Management* 25: 354–364.

Lundin, R.A. and Söderholm, A. (1995) 'A theory of temporary organization'. *Scandinavian Journal of Management* 11(4): 437–455.

Midler, C. (1995) '"Projectification" of the firm: The Renault case'. *Scandinavian Management Journal* 11(4): 363–375.

Objective One (2005) *Programme for Cornwall and the Isles of Scilly, 2000–2006.* Available at: http://www.objectiveone.com/O1htm/01-whatis/publications.htm (7 September 2007).

Objective One (2007) *Objective One Innovates. The Objective One Partnership for Cornwall and the Isles of Scilly.* Available at: http://www.objectiveone.com/ (14 September 2007).

O'Sullivan, D., Stewart, E.J., Thomas, B., Sparkes, A., and Young, J. (2003) 'Evaluating European Union funding programmes for Tourism SMEs: A case from industrial South Wales'. *International Journal of Tourism Research* 5: 393–402.

Our South West (2007) *Our South West—Key Information & Guidance Towards a Sustainable Future for SW England.* Available at: www.oursouthwest.com/ (28 August 2007).

Paterson, M. (2006) *The Eden Project: an Overview.* Online Conference Proceedings for The Nature of Success: Success for Nature (2006). Available at: http://www.bgci.org/educationcongress/proceedings/Authors/Paterson%20 Mark%20-%20PA.pdf (10 May 2008).

Pollack, J. (2007) 'The changing paradigms of project management'. *International Journal of Project Management* 25: 266–274.

Prencipe, A. and Tell, F. (2001) 'Inter-project learning: Processes and outcomes of knowledge codification in project-based firms'. *Research Policy* 30: 1373–1394.

Smit, T. (2001) *Eden.* London: Corgi Books.

Smyth, H.J. and Morriss, P.W.G (2007) 'An epistemological evaluation of research into projects and their management: Methodological issues'. *International Journal of Project Management* 25: 423–436.

South West Tourism (2007) *Towards 2015. Shaping Tomorrow's Tourism.* Available at: http://www.towards2015.co.uk/downloads/vision_0105.pdf (28 August 2007).

Sustainability South West (SSW) (2007) *Sustainability South West. We are the Champions of Sustainability!* Available at: http://www.sustainabilitysouth-west.org.uk/ (14 September 2007).

Sydow, J. and Staber, U. (2002) 'The institutional embeddedness of project networks: The case of content production in German television'. *Regional Studies* 36(3): 215–227.

Sydow, J., Lindkvist, L., and DeFillippi, R. (2004) 'Project-based organizations, embeddedness and repositories of knowledge: Editorial'. *Organization Studies* 25(9): 1475–1489.

Tourism South East, South West Tourism and West Sussex County Council (2008) *Sustainable Tourism for Dummies.* Chichester: Wiley.

Treuren, G. and Lane, D. (2003) 'The tourism planning process in the context of organised interests, industry structure, state capacity, accumulation and sustainability'. *Current Issues in Tourism* 6(1): 1–21.

Tribe, J. (2004) 'Knowing about tourism: Epistemological issues'. In Phillimore, J. and Goodson, L. (eds.) *Qualitative Research in Tourism. Ontologies, Epistemologies and Methodologies.* London: Routledge.

Vernon, J., Essex, S., Pinder, D., and Curry, K. (2005) 'Collaborative policymaking. Local sustainable projects'. *Annals of Tourism Research* 32(2): 325–345.

Winter, M., Andersen, E.S., Elvin, R., and Leven, R. (2006) 'Focusing on business projects as an area for future research: An explanatory discussion of four different perspectives'. *International Journal of Project Management* 24: 699–709.

12 Carbon Labelling and Restructuring Travel Systems
Involving Travel Agencies in Climate Change Mitigation

Ghislain Dubois and Jean Paul Ceron

INTRODUCTION

Are there limits to tourism? Will climate change mitigation policies become so stringent that they limit the possibility of tourism mobility in the future? While the debate is still open as to how heavy the burden of climate change mitigation will be for the tourism industry (Dubois and Ceron 2007), there is now a broad consensus on the issues and the need for action that has been validated by the community of tourism stakeholders. Indeed, the Davos declaration, adopted at UNWTO-UNEP-WMO Davos Conference, 1–3 October 2007, unequivocally states that:

> The tourism sector must rapidly respond to climate change, within the evolving UN framework and progressively *reduce* its Greenhouse Gas (GHG) contribution if it is to grow in a sustainable manner.
>
> (UNWTO, UNEP, WMO 2007)

On a world scale, the target is not to exceed a long-term equilibrium global average temperature of between 2.0–2.4°C over pre-industrial levels. This corresponds to CO_2 concentrations at stabilisation of 350–400 ppm and CO_2-equivalent concentrations of 445–490 ppm. The IPCC Working Group III (2007) outlines that within that objective, global net emissions of CO_2 should peak no later than 2015, with a subsequent decline by -50 per cent to -85 per cent by 2050 (as a percentage of 2000 emissions). Tourism currently represents some 5 per cent of the global anthropogenic contribution to climate change. It is the fastest growing source of greenhouse gas emissions. In particular, aircraft manufacturers expect a continuing growth of air transport of +5 per cent per annum (Airbus 2006; Boeing 2006). UNWTO-UNEP-WMO (2008) conclude, in the technical report of the UNWTO-UNEP-WMO conference on tourism and climate change, that emissions from tourism would grow by more than 150 per cent by 2035 in a Business as Usual scenario (UNWTO-UNEP-WMO 2008). However, this forecasted growth would lead to unsustainable futures, for example with aviation exceeding the total CO_2 budget by 2050 for the UK (Bows,

Anderson, Peeters 2007), and shows that the inclusion of international air transport within global mitigation targets is a clear priority.

Facing this challenge, the tourism industry, as well as the consumers, is offered several options for reducing emissions. Four major mitigation strategies for addressing greenhouse gas emissions from tourism can be distinguished: reducing energy use, improving energy efficiency, increasing the use of renewable energy, and sequestering carbon through sinks. The UNWTO-UNEP-WMO expert team investigated these options, and concluded that:

> Reducing energy use is the most essential aspect of mitigation, which can be achieved by altering destination development and marketing (tour operators), destination choices (tourists), as well as shifts in transport use from car and aircraft to rail and coach. Changing management practices can be of importance for business tourism (videoconferencing). Tour operators play a key role in this process, as they bundle products into packages that are advertised to and purchased by tourists. Tour operators can also increase length of stay, which would very effectively reduce the carbon footprint per tourist day and increase economic opportunities for destinations. It has to be considered however that current tourism trends show an increase of short stays. *Overall, tour operators have a considerable influence on creating demand for less carbon intensive journeys* by creating attractive products that meet tourists' needs and desires.
>
> (UNWTO-UNEP-WMO 2008, 46)

Anticipating a carbon-constrained tourism economy thus appears as a vital challenge for the tourism industry, as well as a crucial priority for the well-being of citizens. Mitigation will likely involve a wide range of instruments: national or international caps on emissions, emissions trading schemes, taxation, infrastructure development, and behavioural changes. However, apart from this large scale and long term thinking, what can be done today, on an individual or corporate basis, given the state of travel systems? This chapter therefore elaborates on three statements:

1. It is possible to diminish drastically, often up to 50 per cent or more, the individual carbon footprint of mid-haul trips, through sound travel planning, combining measures such as modal shift to train for all of part of the trip, a choice of efficient aircrafts, and more direct routes, and provided the traveller is ready to accept (often but not always) constraints of time and price.
2. Despite the growth in consumers and corporate interest in more environmentally friendly travel, the difficulty to understand the above parameters, but all the more the impossibility to combine them on current reservation systems (GDS) limits the ability of individuals to reduce their emissions.

3. There is, therefore, a need to develop a set of calculation-optimisation-booking-selling tools for the travel industry, which will, in turn, offer a new market for travel agents. Some of these tools can take as a starting point emission calculators developed by some countries or private operators.

'Before travelling less, let's travel better'. From these statements, and from personal experience in reducing the carbon footprint of trips, TEC Consultants (www.tec-conseil.com), with Ghislain Dubois and Jean Paul Ceron as project managers, developed a project summarising this set of tools under the title 'Ecological travel agency (ECOTA)'. The aim of this chapter is to explore the rationale and main components of this project, which is still to be adopted and implemented.

PROJECT RATIONALE

Consumer and Corporate Growing Awareness of the Contribution of Travel to Climate Change

In 2003, the World Tourism Organization (UNWTO) held the first conference on Tourism and Climate Change, in Djerba, Tunisia (OMT 2003). This event provided one of the first opportunities to discuss the challenge of climate change at an international level, although the perspective was quite unbalanced with the issue of impacts of climate change on destinations and adaptation overwhelming that of the contribution of tourism to climate change and its mitigation. However, until 2003, the awareness of tourism GHG emissions remained confidential and limited to academic circles (Høyer 2000; Høyer and Næss 2001; Becken 2002; Becken and Simmons 2002; Gössling 2003). Since then there has been significant growth in awareness of the link between climate change and tourism, both for personal and for professional travel, which draws the boundaries of a market for ecologically-sound travel, distributed by 'Ecological travel agencies'.

With respect to *personal travel*, the environmental impact of aviation is increasingly mentioned in the media. In particular, the local sustainability of ecotourism is being balanced with its impact on the global environment, through long-haul travel (Figure 12.1). The individual carbon calculators (see, for example, www.bilancarbonepersonnel.org) have revealed the significance of air transport in contemporary ways of life, and the contradiction between tourism and transport on the one hand and sustainability on the other (Høyer 2000). The medium-term perspective of individual carbon budgets (Bows et al. 2006), conceived in terms of individual emission quotas, could lead households to operate some trade-off between different forms of mobility, between mobility and other uses of carbon, and to head for more efficient consumption patterns in general, in order to 'save' carbon credits.

Figure 12.1 Ecotourism is more and more related to the environmental impacts of travel than it was 1 or 2 years ago ('More Oxygen, Less CO_2': advertisement campaign for the publication of an ecotourism handbook).

Currently, a minority of travellers seem to be willing to pay for the mitigation of their tourism emissions through carbon offsets or carbon optimisation. Some tour operators (like the French 200 millions euros turnover Voyageurs du monde) have started implementing carbon offsetting for all of their products (on a 'mandatory' basis, i.e., included in the price), and are studying opportunities to reduce further their impact (Rial 2008, 22).

Business travel, in the framework of Corporate Social Responsibility (CSR), appears as another promising market for 'green' tourism mobility:

- Major companies are more and more engaged in sustainable development plans, with measurable objectives in terms of GHG emissions.
- Beyond voluntary initiatives, business travel could be, in a near future, included in carbon trading schemes, since they are a share of the company's carbon quota.

- For sectors such as consulting, business travel represents an important share of the company's environmental footprint.
- Conference tourism, incentives, and other professional events (fairs) are more and more often involved in carbon offsetting programmes.
- Contrary to individuals, large companies plan and optimise their travel policy through travel agents.

The Association for Corporate Travel Executives (ACTE) and KDS (a company located in Paris, specialised in 'Travel and Expenses' cost optimisation) have conducted an annual survey since 2005, which shows that CSR acceptance has grown exponentially in the business travel industry. The survey undertaken between November 2007 and January 2008, on a sample of 250 business travellers and 250 travel managers of large companies, showed that 45 per cent of respondents consider climate change has now a direct influence on individual decision-making for business travel. Compared to 2007 results, the share of companies which recommended diminishing business travel for environmental reasons grew from 25 per cent to 34 per cent. More corporate travel departments also reported GHG emissions of their trips to their general directorate. According to Susan Gurley, ACTE Manager:

> Travellers and their companies will incorporate more and more the environment in their decision, seeking for smarter, more productive and more ecological travel. Our industry will discover soon this new trend. (ACTE press release, 7 February 2008)

The Travel Agency Market Today: Opportunities and Constraints for Green Products

The Omnipresence of GDS

The travel business is more than ever shaped by Global Distribution Systems (GDS), whose comprehensive platforms and databases are unavoidable, given their rapidity and accessibility. Three companies running their own platform (Amadeus, Sabre, Galileo) share two-thirds of global air booking. The development of the Internet did not change this domination, and even reinforced it, since it shortcuts travel agents from the booking procedure. In recent years, only low cost carriers have been successful in implementing direct booking, without GDS.

In spite of this domination, the GDS still have shortcomings: booking complex trips is not always feasible, especially when an operator tries to combine several modes (train + plane + taxi for example). GDS focuses more on air transport than on other modes (coach is often ignored). It is therefore clearly difficult for travel agents to combine modes, to book international train circuits, or simply to book international train tickets.

The omnipresence of GDS presents several constraints and opportunities for the ECOTA project. Although trips can be sorted by date, duration, price, and company, they cannot currently be sorted by GHG emissions. However, adding a new criterion of choice could provide new business opportunities in that market, probably at an affordable price for GDS and travel agents.

Involving GDS into GHG calculation and optimisation would help generalise 'green' solutions in the travel sector. However, their reluctance to accept would force to develop more complicated, costly and probably less precise, customised tools for travel agents (e.g., databases and calculators). Nevertheless, recent experiments proved that the involvement of GDS is possible. UNITAID, the UN fund collecting the solidarity tax and voluntary contribution on air travel (financing health system in developing countries), reached an agreement with Amadeus, Galileo, and Sabre (to be implemented in 2009) so as to introduce an option in the booking process for the voluntary contribution.

The Need to Reposition Travel Agents' Activity

The business of travel agents has evolved considerably with the development of the Internet, which allows direct booking. The business model of travel agents, based on ticketing and commissioning on simple products, is no longer adapted to the new context. The sector needs to develop new markets, where it can value its expertise (i.e., become a 'travel advisor') in the organising of trips: customisation of package trips (mass customisation), and supplying more detailed options. Could environmentally friendly travel become part of this new offer? Travel agents have been relatively absent from the debate around sustainability, compared to other professional stakeholders like hoteliers or tour-operators. Their traditional role is that of simple intermediaries between a producer and a consumer. Will it change in this new context, 'travel advisors' offering proactively some low impact travel? For example, the UNWTO-UNEP-WMO Davos conference thematic sessions on Tour operators recommended to 'improving computer reservation systems and global distribution systems (CRS and GDS) in order to calculate GHG emissions, so as to allow travel agents to propose soft mobility products' (UNWTO-UNEP-WMO 2008, 13).

Towards a Carbon Labelling of Tourism Products?

Designing the ECOTA project can benefit from the current thinking on the carbon labelling of tourism products. Indeed, the minimum level of information needed for integrating climate change into the expression of consumers

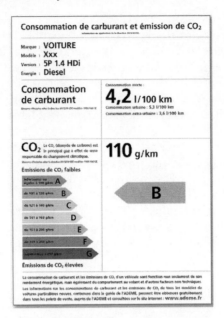

Figure 12.2 The French carbon labelling system for cars was implemented in 2006, as an operational measure of the National Mitigation Plan 2004–2012, 'Plan Climat'.

preferences, is the knowledge of emissions. Energy and carbon labelling have proved successful for household electrical appliances, and have been implemented in France for cars since 2006 (Figures 12.2, 12.3, and 12.4), with a ranking from A to G), and since 2008 are used as a baseline for financial incentives ('bonus-malus'). This debate on labelling is still an emerging one in the tourism sector. Some propose to label trips with ratios like 'CO_2-e per night stay', another approach could be to consider the eco-efficiency of tourism 'CO_2-e per euro spend' (Gössling et al. 2005).

A first step could be, before discussing a ranking system, to mention on each plane ticket the amount of GHG emissions, with the objective to raise customers' awareness and to provide a reference calculation procedure for any mitigation initiatives (see Figure 12.5). This would require the involvement of the International Air Transport Association (IATA) and of the International Civil Aviation Organisation (ICAO).

Climate Change Mitigation and the Travel Sector Today: Calculators, Offsetting and . . . Beyond?

The tour operator and travel agents sector initiated recently some initiatives on GHG calculation and mitigation. These are discussed later.

Figure 12.3 'Leclerc' supermarket started implementing, following the recommendations of the 'Grenelle de l'Environment', the calculation of the carbon footprint of shopping, based on the assessment of 250 products.

Figure 12.4 Audi's reluctance to carbon labelling 'Those who want to see our large cars disappear will be happy . . . to learn that 95 per cent of their material is recyclable' (Audi advertisement in French 'Cigare Magazine').

Figure 12.5   Towards a carbon labelling of air tickets?

Controversies in Carbon Calculation

In the current debate on climate change, figures are neither neutral nor reliable, especially in a context of scientific controversy. Statements—science-based or not—like 'new aircraft do not consume more fuel than a Prius' or 'air transport pollutes twenty times more than the train', will shape the perception of the issue of climate change by stakeholders, public authorities, and thus, future mitigation policies.

Given that climate change is still a recent topic, and that there are some still considerable scientific controversies on the impact of travel on climate change, especially with respect to the non-carbon impact of aviation, it is not surprising to notice debates on the various existing calculators, often pointing out their weaknesses. Non-carbon emissions are for example not accounted for, diverging ratios are used, neither the type of aircraft nor the detour, nor the transits of flight are taken into account. Initiatives are clearly competing to become the 'reference calculator'; ideally the science should provide a common ground of understanding.

In France, for example, Voyages-scnf.com (the French Railways travel agency), with the support of the national environment and energy agency (ADEME), launched an 'écocomparateur' (http://www.voyages-sncf.com, see 'ecocomparateur' section) comparing the time efficiency, the price, and the CO_2 emissions of different transport modes (but without taking into account intermodal solutions), much to the dissatisfaction of the *Air France* CEO. In response, *Air France* decided to launch its own calculator (http://developpement-durable.airfrance.com/FR/fr/local/calculateurCO2/calculateurAccueil.htm).

Similarly, in the UK, Carlson Wagonlit Travel started developing a calculator with the Department of Environment, Food and Rural Affairs (DEFRA). A few initiatives tried to establish standards for calculators. For example, ADEME, through its 'charter on carbon offsetting' is currently trying to implement its *Bilan Carbone* methodology for the four

French offsetting initiatives. For air transport, ICAO is apparently developing an international calculator.

Carbon Offsetting

Carbon offsetting is increasingly promoted within the tourism sector. Gössling et al. (2007) reported more than 60 offsetting initiatives applicable to aviation, worldwide. In a context of continuous, limited technological improvement of aircraft, and high income customers with a high desire to travel, the temptation is high, indeed, to refuse any cap or internal emission reductions for tourism, and to favour offsetting, e.g., financing in another sector and/or another country some energy efficiency or carbon storage projects (Caire 2008; Ceron and Dubois 2008). Clearly under pressure from the media and NGOs, the tourism industry has grasped this opportunity, and actively promoted carbon offsetting. To offset 10 per cent of global aviation emissions, however, the current amount of GHG offset by these initiatives should be multiplied by 400 (Gössling et al. 2007). Apart from a clear impact on awareness-raising, the effect on emission reductions is therefore likely to remain limited.

The Need to Go One Step Beyond

This overview of initiatives implemented within the travel market points at several serious limitations.

1. *Emissions calculators and comparators* are still to be improved. ADEME's 'Ecocomparateur' for example connects to a GDS (or to companies' databases and web sites), applies various emissions factors to the type of trips, and proposes one choice for each mode (with many blanks for international rail connections). Its limitations are therefore:
    a. no combination is possible between transport modes for a selected trip;
    b. only one trip per mode is proposed, instead of letting clients choose which is the most convenient for them;
    c. calculation methods are not harmonised at an international level (especially the uplift factor for air transport, accounting for non-carbon emissions);
    d. it does not handle international train trips, which are often missing from its comparison. This seems to be due to the software used by French railways (*SNCF*) for their booking, which does not manage complex trips (over two stopovers), while other companies like the German *Deutsche Bahn* perform this;
    e. most importantly, it is not integrated with the booking procedure. So far, 'Ecocomparateur' proved very successful to communicate on the issue of transport impacts on climate change

and was used by *SNCF* to lobby for the train. It might have influenced in some cases the final choice of consumers. However, it is neither a booking tool for individuals, nor a commercial tool, allowing a travel agent to develop a new offer. In short, its impact on emission reductions is limited. The traveller must first use a GDS, then use 'Ecocomparateur' to compare alternative trips in a long process, which does not allow optimisation.

2. *For mitigation, carbon offsetting is favoured against direct reduction measures*, which is highly questionable in the long-run, given the current and forecasted growth rate of the sector (Caire 2008; Ceron and Dubois 2008).

3. These two points justify our ECOTA project: The objective is to improve calculation and optimisation tools, and enhance the potential for emission reduction within the sector of travel.

PROJECT DESCRIPTION

The Concept of 'Carbon Optimisation of Trips'

Optimisation is a central concept for a travel agent's own vocabulary (see below), and travel agents can be defined as travel optimisers.

Carlson Wagonlit Travel (CWT) offers a variety of products and services to help clients *optimize their travel program and generate savings*. Generating and consolidating timely, relevant data is the foundation of program optimization. CWT also assesses clients' travel programs by *analyzing performance* for each of the eight key levers to effective travel management. This assessment includes helping clients identify successful practices, as well as opportunities for continuous improvement, and *benchmarking* against best-in-class practices. Together, CWT and its clients then define performance targets, priorities and action plans, which are translated into measurable results that are *continuously monitored to drive progress*.

(CWT Website http://www.carlsonwagonlit.com)

Optimising a trip regarding its contribution to climate change requires a combination of options.

- favouring direct flights, with no stopovers (low cost carriers multiply such opportunities, creating lines between small and medium airports)
- choosing more direct routes, avoiding detours (e.g., Marseilles to Athens by Milan, rather than by London)
- selecting more efficient aircraft: A350 or B787 instead of MD80, but also high occupancy flights, e.g., charter flights, flights with no business or 1st class seats, companies with better load factors

- substituting plane with train, for the whole trip, or for part of it, for example to reach the hub or the final destination
- at the end of the itinerary, proposing one or several offsetting options, for the remaining GHG emissions

Though possible in theory for any kind of trip, the notion of carbon optimisation is especially relevant for short and mid-haul trips, where possibilities of modal shift exist. Its potential is also greater in dense and populated regions of the world, where several companies operate, on different routes, with diversified transport infrastructures. Given its size, density, and its volume of internal air transport, the EU appears as the perfect market owing to its dense and high-speed railway network.

The following illustrate a few 'optimised trips', based on real experiments. Some alternatives appear as more accessible for motivated individuals (Bergen), given their time constraints, others could be attractive for business travellers as well (Helsingborg):

- Marseille (France) / Sevilla (Spain), in April 2008, for a personal trip: Marseille-Barcelona by plane, and Barcelona-Sevilla by AVE (high speed train) through Madrid. Eleven hours door to door instead of five, additional price of forty Euros, emissions halved
- two consecutives business trips merged in one, to Helsingborg (Sweden) and to Davos (Switzerland) starting from Marseille, in September 2007 adding a weekend in Copenhagen in between. Marseille to Zurich by train and back, Zurich to Copenhagen by plane and back, Copenhagen-Helsingborg by train. Large financial savings (one trip instead of two, since a 'normal' traveller would have come back home), emissions diminished by more than 66 per cent, but need to accept this additional weekend in Copenhagen (savings allow the company to offer this week-end to its employee)
- Marseille (France) to Ljubljana (Slovenia) in December 2007 for a one-day conference: Marseille to Zurich by train and Zurich to Ljubljana by plane, instead of a connecting flight in Munich. Emissions diminished by more than 50 per cent, large savings (200–400 euros depending on the reference trip). Given the limited distance, the short length of stay, and the efficient flight connection in Munich should it have been retained, the train option chosen involved a large loss of time (thirteen hours instead of five hours) for such a short stay. The train time, however, was devoted to (efficient) work, given the good travel conditions of TGV and Swiss trains
- Marseille (France) to Bergen (Norway; see Table 12.1), which is given the current state of high speed trains, at the highest range of distance acceptable (at least to TEC's consultant mind!), since the additional time spent in transport is large

Table 12.1   Marseille (FR)—Bergen (NW) Return Trip

	Proposed by a travel agent 1 month in advance	Optimized by TEC
Trip steps	Marseille—Munchen (plane) Munchen -Oslo (plane) Oslo—Bergen (plane)	Marseille—Paris (train) Paris CDG—Oslo (plane) Oslo—Bergen (train)
Duration	8h30 door-to-door	20h door-to-door
Price	420 €	460 €
GHG emissions	1,320 kg CO_2-e	740 kg CO_2-e
Avoided emissions		580 kg CO_2-e
Price of remaining compensation (Atmosfair—2007 price)	33 €	21 €
Additional cost		28 €
Additional price paid per avoided ton of CO_2-e		48€

Objectives

The fact that it is possible, in theory, to reduce the carbon footprint of trips, does not mean it is immediately achievable to generalise and standardise procedures, in short to achieve this at a large scale. The authors' experience of carbon optimisation proved the difficulty to understand all these parameters, but all the more the impossibility to combine them on current reservation systems (GDS). This clearly limits the ability of individuals to reduce their emissions. Travel agents are at the moment of no help, and the motivated traveller must open up (at least) a dozen of windows, on train companies websites, GDS, low cost carriers; refer to a map to choose the most direct routes; and use an Excel spreadsheet to calculate emissions.

To operationalise these ideas—adapting carbon optimisation to the constraints and expertise of travellers, as well as to those of travel agents—there is a clear need to develop a set of calculation-optimisation-booking-selling tools for the travel industry, which will, in turn, offer a new market for travel agents. Some of these tools can take as a starting point emission calculators developed by some countries or private operators.

Main Options

The main bifurcation of the ECOTA project is the involvement or not of a GDS, which would change the project objectives.

- A calculator will be far easier to develop, but above all to adapt to travel agents' needs, if it is integrated in their daily working environment. CO_2 could then become an additional sorting criterion. Without a GDS, the project should develop a customised tool for travel agents. A third alternative is to develop a tool technically compatible with GDS, even if not integrated in a the first step.
- The involvement of a GDS allows targeting all travel agents, developing, as said, a new criterion of choice, but not necessarily a new business model for a 'green' travel agency. In the 'no GDS' option, the agency, if it is to develop its own tools, will have to be specialised in greener mobility.

The pre-feasibility study will help selecting the best option.

Main steps

The project rationale defines a few steps, which can be developed sequentially, jointly, or even separately: Each 'work package' has its own outputs, partnership, and timing.

Work Package 1: Market Survey

Currently, the market for such an 'agency', or for green tourism mobility in general, is probably limited, but it is undoubtedly deemed to grow along the consumers' awareness and the development of corporate responsibility. One way to make this business model profitable could be to attract major companies involved in emissions reduction. Therefore, this travel agency should be a web agency, with call centres, and should have a large geographical coverage (national or European), to reach a market large enough to be profitable. A market survey should investigate personal and professional markets, groups and individuals, and should in particular study the willingness to pay of consumers, the acceptability of time constraints, and the psychological barriers.

Work Package 2: Developing a Reference Calculator

An improved calculator and booking tool must have the following characteristics:

- reliability and regular updating of emission factors
- a maximum number of parameters: load factors, type of plane, flight altitude, country (energy mix, electrification of the railway network)
- the possibility to combine modes (train, plane, coach, taxi), and include international travel by train
- a detailed set of sorting criteria
- fast and friendly use, so as to optimise the travel agents' time

Existing calculators, such as *Atmosfair's, DEFRA's,* or the *ADEME* 'Ecocomparateur' could be taken as starting points, at least for their emission factors.

Work Package 3: Developing Booking Tools and Procedures

The better the calculator, the more straightforward the booking and buying procedure will be. For example, if the project manages to integrate international trains, it will not be necessary to develop additional modules or training programmes for travel agents. Regardless, there is a need to develop, in close collaboration with travel agents, processes to answer 'ecotravellers' needs (how to sort results and present them, to propose alternative trips, and to monitor CO_2 savings, how to propose offsetting mechanisms for the remaining emissions). This requires some training tools to help travel agents using this new module/function.

Work Package 4: Developing Business Models

From the tools developed, the project will derive and design a few business models for green travel agencies, for example:

- an on-line agency devoted to soft mobility for individuals
- a travel agency or a group (like CWT) offering on its global platform these optional services for large companies buying business travel (with additional tools such as the initial emissions of the company, a setting of targets and action plans, an annual monitoring of CO_2 avoided. . . .)
- a new travel agency specialised in all 'green' markets, operating all over Europe, though its website and call centre

This work package will deal with the tricky issue of earnings for the travel agency. Generally, the process will result in a more expensive or longer travel. How can money be asked for such a 'service'? Given the additional time required to optimise the trip, what amount can the travel agency invoice on each trip? Beyond this ticketing service, can the travel agency charge other services, such as an annual report on corporate travel emissions, a corporate action plan for green mobility?

CONCLUSION

Implementing the ECOTA project means creating a technical and financial partnership (Table 12.2), defining a business model for the project itself, and finalising a work plan. Given the complexity of the project, and the current uncertainty on the involvement of partners, a prefeasibility

Figure 12.6   The French 'Ecocomparateur' links booking systems and emission factors, but is not integrated in the booking procedure.

step seems required, as a preliminary step to a full project, which duration could be two or three years. The ECOTA project itself still has to define its business model. In the short term, partnership and funding opportunities must be found, to help its implementation. In the mid term, does the project aim at developing free, open source tools, or to patent some innovations? What will be the ownership of the tool, depending on the scenarios of partnership? Is there a need to create a joint-venture, to develop the project?

Despite its feasibility, the financial costs, personal constraints, and efforts needed to mitigate individual carbon footprint of trips might appear high today, and could prevent developing the ECOTA project. However, there is undoubtedly a market for soft mobility products given contemporary concerns surrounding climate change and tourism, and thus a room for an 'ecological travel agency'. Furthermore, as the IPCC outlined (2007), the carbon price allowing significant emission reductions compared to baseline SRES scenarios would be between fifty and one hundred euros per ton. Compared to cheap current prices (around fifteen euros), this tends to favour carbon optimisation, against carbon offsetting, as mitigation measures. The trend, with the negotiation of post-Kyoto agreements, with the involvement of aviation in EU ETS, and with the growing awareness of consumers, is clearly towards more stringent climate policies. Thus, for the travel industry, the challenge is to be more involved in this debate than in the past, to anticipate future constraints, grasp opportunities, and to become prime movers in this new market.

Table 12.2   ECOTA Project Categories of Potential Partners

—Global distribution systems

—Travel agents (groups and independents)

—Corporate travel specialists

—Travel agents associations

—Transport databases holders

—Environmental partners

—NGOs

—University and researchers

—International organizations

The success of the ECOTA project appears, in that context, linked to the capacity to establish and animate a partnership around this challenging idea of carbon optimisation.

REFERENCES

ACTE (2008) *ACTE/KDS Survey Illustrates CSR Acceptance Has Grown Exponentially In The Business Travel Industry*. Press release, 7 Feb 2008. Available at: http://www.acte.org/resources/press_release.php?id=262 (30 August 2008).
Airbus (2006) *Global Market Forecast: The Future of Flying 2006–2025*. Toulouse, France: Airbus.
Becken, S. (2002) 'Analysing international tourist flows to estimate energy use associated with air travel'. *Journal of Sustainable Tourism* 10(2): 114–131.
Becken, S. and Simmons, D.G. (2002) 'Understanding energy consumption patterns of tourist attractions and activities in New Zealand'. *Tourism Management* 23: 343–354.
Boeing (2006) *Current Market Outlook*. Boeing. Available at: http://www.boeing.com/commercial/cmo/ (27 Feb 2006).
Bows, A., Anderson, K., and Peeters, P. (2007) 'Technologies, scenarios and uncertainties', E-CLAT Technical seminar, *Policy Dialogue on Tourism, Transport and Climate Change: Stakeholders Meet Researchers*. UNESCO, Paris, France.
Bows, A., Mander, S., Starkley, R., Bleda, M., and Anderson, K. (2006) *Living with a Carbon Budget*. Manchester, UK: Tyndall Centre.
Caire, G. (2008) 'La compensation CO2, une forme d'indulgence des temps modernes?' *Espaces* 237: 21–22.
Ceron, J.-P. and Dubois, G. (2008) 'Compensation volontaire des gaz à effet de serre. Enjeux et limites'. *Espaces* 257: 14–21.
CWT (2008) *Program optimization*. Available at: http://www.carlsonwagonlit.com/en/global/program_optimization/ (30 August 2008).
Dubois G. and Ceron, J.P. (2007) 'How heavy will the burden be? Using scenario analysis to assess future tourism greenhouse gas emissions'. In Peeters, P. (ed.) *Tourism and Climate Change Mitigation, Emissions*. Breda, The Netherlands, NHTV.

Gössling, S. (2003) *The Importance of Aviation for Tourism. Status and Trends*. European conference on aviation, atmosphere and climate, Friedrichshafen, 30 June to 3 July 2003.

Gössling, S., Broderick, J., Upham, P., Ceron, J.P., Dubois, G., and Stradas, W. (2007) 'Voluntary carbon offsetting schemes for aviation: Efficiency and credibility'. *Journal of Sustainable Tourism* 15(3): 223–248.

Gössling, S., Peeters, P., Ceron, J.P., Dubois, G., Patterson, T., and Richardson, R. (2005) 'The Eco-efficiency of tourism'. *Ecological Economics* 4(54): 417–434.

Høyer, K. and Næss, P. (2001) 'Conference tourism: A problem for the environment, as well as for research?' *Journal of Sustainable Tourism* 9(6): 451–470.

Høyer, K.G. (2000) 'Sustainable tourism or sustainable mobility'. *Journal of Sustainable Tourism* 8(2): 147–160.

IPCC (2007) 'Summary for policymakers'. In Metz, B., Davidson, O.R., Bosch, P.R., Dave, R., and Meyer, L.A. (eds.) *Climate Change 2007: Mitigation. Contribution of Working Group III to the Fourth Assessment Report of the Intergovernmental Panel on Climate Change*. Cambridge University Press, Cambridge, United Kingdom and New York, NY, USA.

OMT (2003) *Changement Climatique et Tourisme*. Madrid, OMT.

Rial J.F. (2008, March) 'La Compensation carbone est une nécessité'. *Espaces* 257: 22–25.

UNWTO-UNEP-WMO (*United Nations World Tourism Organization, United Nations Environment Programme, World Meteorological Organization*) (2008) 'Climate change and tourism: Responding to global challenges'. [Scott, D., Amelung, B., Becken, S., Ceron, J.-P., Dubois, G., Gössling, S., Peeters, P., and Simpson, M.] *United Nations World Tourism Organization (UNWTO), United Nations Environmental Programme (UNDP) and World Meteorological Organization (WMO)*. UNWTO, Madrid, Spain.

UNWTO-UNEP-WMO (2007) *Declaration. Climate Change and Tourism: Responding to Global Challenges*. Second International Conference on Climate Change and Tourism, Davos, Switzerland.

13 Moving Towards Low-Carbon Tourism
New Opportunities for Destinations and Tour Operators

Paul Peeters, Stefan Gössling, and Bernard Lane

INTRODUCTION

It is increasingly clear that tourism can be a very carbon-intense activity. While the sector's contribution to global emissions of carbon dioxide is only 5 per cent (UNWTO-UNEP-WMO 2008), its contribution is highly skewed. Aviation is particularly problematic, accounting for the major share of emissions from global domestic and international tourism; this is 40 per cent if expressed as CO_2, and considerably more if the radiative forcing of other greenhouse gases is taken into account as well[1], yet aviation serves only 17 per cent of all tourist trips. Further, it is estimated that only 2 to 3 per cent of humanity participates in international air travel on an annual basis (Peeters, Gössling, and Becken 2006). Emissions from air travel, therefore, are caused by a rather small part of the population. On a per capita basis, there is no other human activity that can increase individual carbon emissions as rapidly as air travel. A single journey from Europe to Australia and back will result in emissions in excess of four tons CO_2, equivalent to the average amount of CO_2 emitted by an average person's activities in a whole year. As global warming is arguably the sum of the individual actions of about 6.7 billion (in 2007) human beings, an increasing number of whom are becoming highly mobile, it is of great importance to address both individual tourism related carbon emissions and the business practices within tourism which are increasing greenhouse gas emissions.

Current debates on curtailing growing emissions from aviation and tourism more generally are technical in character. IATA (2007) and individual airlines acknowledge the contribution of aviation to climate change. Individual airlines such as Scandinavian Airlines (2008) have focused on air traffic management (ATM), new technology, emission trading, and biofuels as their main strategic pillars to achieve emission reductions. Technical options to reduce emissions from air travel are indeed valuable, but are outweighed by the rapid growth in air travel (Peeters 2007a; UNWTO-UNEP-WMO 2008). With new trends in international travel now including long-distance short-duration city breaks on a massive scale, it becomes increasingly important to consider behavioural mitigation strategies to achieve emission reductions. This poses both risks and opportunities for tourism. The risk is that certain tourism products could become nonviable from an emissions point of view, both because they may become unacceptable for environmentally aware travellers or more expensive as a result of mitigation policies such as those planned within the framework of the European Union Emission Trading System (Gössling, Peeters, and Scott 2008). Vice versa, low-carbon tourism products may become increasingly attractive for a growing number of travellers. For all these reasons, a closer investigation of the carbon intensity of their tourism products should be of interest to the tourism industry. This chapter focuses on tour operators, and discusses their potential to contribute to low-carbon tourism through new approaches to packaging, product development, and marketing.

TOUR OPERATORS

Tour operators are often seen as key players in the tourism industry that have considerable power to improve the sustainability of the sector, even though there appears to be a lack of studies investigating the role of tour operators in this context (Timothy and Ioannides 2002). Overall, the engagement of tour operators in contributing to sustainable tourism has remained very limited (van Wijk and Persson 2006).

As Timothy and Ioannides (2002) point out, tour operators may largely be responsible for the current situation in global tourism, as they have created the mechanisms for cheap mass holidays, and systematically developed and promoted new destinations with a focus on low price, through bulk purchasing of hotel and transport capacity, direct involvement in destination development, and the use of mass marketing techniques. With their power to influence tourist decision-making, tour operators have considerable power over travel flows. For instance, in early 2008, Swedish Fritidsresor (part of the large Touristik Union International [TUI] group), marketed the Maldives in billboard campaigns with the slogan '*the dream is just 9 hours away*'. Even more massive television-based advertisement campaigns were made by competitor Ving Resor a number of years before, announcing Borneo as a 'must' destination. As these examples show, and as is generally acknowledged, tour operators help determine market trends and influence the demand for certain destinations (e.g., Carey, Gountas, and Gilbert 1997). The number of tourists brought by a single tour operator to a destination can be massive, as are overall travel flows directed by tour operators. For instance, Carey et al. (1997) report that the thirty largest tour operators in the UK carried more than 17 million customers in 1995, and German TUI alone had over 7 million clients annually in the 1990s (Iwand 1995, cited in Timothy and Ioannides 2002). Most of the clients of these mass-tourism operators use aircraft to arrive at their destinations. With the emergence of a global leisure class soon involving a billion international travellers per year, tour operators have been one of the driving factors in the expansion of tourism, often in cooperation with international investors (for current developments in, for example, Cambodia, see The Guardian 2008). In contrast, their engagement in pro-sustainable action—or more specifically reduced air transport volumes—has been limited.

TUI may be the tour operator best recognised for its early involvement in environmental management, and can serve as an example of the role of tour operators in contributing to the development of sustainable tourism. In the mid-1990s, TUI got worldwide recognition for introducing environmental standards for many of its 7,000 hotel partners (Iwand 1995, cited in Timothy and Ioannides 2002). These standards addressed a wide range of issues, including waste and sewage disposal, water quality, energy consumption, preservation of natural resources, traffic congestion, beach erosion and cleanliness, as well as building ecological awareness among local leaders and residents (Iwand 1998, cited in Timothy and Ioannides 2002). As TUI's environmental manager Dr. Wolf Iwand put it:

> We insist on efficient practices in local waste and sewage management, whether in Kenya, on the Maldive Islands, On Fuerteventura, or in Bulgaria. We do interfere in the interest of our customers in the long-term preservation of the product we offer.
> (Iwand 1995, 9, cited in Timothy and Ioannides 2002, 188).

While this seemed proactive from the late 1990s point of view, it is clear today that most of the action taken by TUI was also economically viable— efficient resource use also means saving costs. Moreover, prior to this time TUI faced considerable criticism in Germany from various NGOs as well as travellers, and its engagement in pro-environmental action was not entirely voluntary. In the late 1990s, when TUI came under economic pressure, the company used its pro-environmental credentials to strategically create discourses on the positive effects of tourism, presenting tourism as a 'global strategy for sustainable development' (Iwand 1999). When it became clear in the early 2000s that tourism's most considerable environmental aspect may be its contribution to global warming, TUI dismissed the emissions problem (see TUI Environmental Report 2003/04, TUI 2004). In terms of pro-environmental action, the company then stagnated for several years, seemingly preoccupied with economic problems, the absorption of companies such as the major UK tour operator Thomson, and a restructuring process that gave greater operational freedom to its national branches. In recent years, several national dependencies of TUI have again become engaged in innovative environmental management. For instance, TUI Travel PLC in the UK, and Fritidsresor Sweden, have introduced measures to reduce emissions, ranging from the use of efficient aircraft with high occupancy rates to the use of trains on some routes, to carbon offsetting. It remains to be seen, however, if these measures will lead to absolute reductions in emissions, as travel to long-distance destinations remains heavily promoted.

Globally, serious action to achieve sustainable tourism, measurable with regard to climate change in terms of declining absolute emission levels, appears nonexistent. For instance, van der Duim and van Marwijk (2006) report that environmental actions by tour operators in the Netherlands include waste separation, reduction of paper use, power and energy, information on environmentally friendly behaviour, the use of environmentally friendly accommodation, and so on. But transport, in climate change terms the most pressing issue, remains one of the least addressed problems. Tour operators focus largely on using environmentally friendly transport to the airport or place of departure. Carey et al. (2007, 426) point out that '[tour] operators base their product development decisions largely on historic performances, and the competition's activities in following the market leaders', acknowledging that:

> The ever increasing choice of tourist destinations provides the tour operators with more choice of products to sell and therefore being less dependent on any one destination. The tour operators, like any other distributor, operate in a dynamic and price competitive market, selling a non-essential product, they try to cater to their potential clients' needs and not those of the destinations.
>
> (ibid.)

The 'ever increasing choice' reflects the expansion of packages to include the whole world, with mass destinations being created even in remote locations. This has partially been facilitated by low kerosene (and thus air transport) prices, and low infrastructure and labour costs in developing countries, making it attractive to move to ever more exotic and cheap locations. These developments have also led to increasing travel distances, as exemplified by Norwegians in the past twenty years (Table 13.1). As shown by Hille et al. (2008), distances travelled in international charter airlines have increased by 37 per cent, almost all of this in the past 10 years. In international scheduled air traffic, passenger kilometres (pkm) have increased by more than 500 per cent over the past 20 years. Emissions have increased concomitantly.

Table 13.1 Leisure and Business Travel by Norwegians

| Year | Air Travel (million pkm) | | |
	Domestic	International (charter)	International (scheduled)
1987	2,320	5,508	3,122
1997	3,747	5,600	7,417
2006	4,127	7,524	18,809
Change	+78 per cent	+37 per cent	+502 per cent

Source: Hille et al. 2008

In summary, the issue of transport and its emissions remains an urgent problem to be addressed (cf. Font et al. 2008; Peeters 2007b; Tepelus 2005; UNWTO-UNEP-WMO 2008; van Wijk and Persson 2006). Tour operators may, however, be able to influence tourist flows, and there may be many options towards the sustainable management of tourism.

AN ANALYSIS OF EMISSIONS FROM TOURISM

Opportunities for reducing tourism's contribution to climate change are significant, as indicated by the huge differences in the carbon footprint of different tourism products. For instance, while a holiday by train and bicycle may, in the best cases, entail emissions of only a few kg of CO_2, cruise ship tourism, in combination with a flight, may lead to emissions one hundred times larger (Figure 13.1). On global average, a tourist trip, including domestic and international trips, lasts 4.15 days and causes emissions of 0.25 t CO_2 (UNWTO-UNEP-WMO 2008). This average includes 4.8 billion tourist trips (in 2005), out of which 2 billion are domestic trips in developed and 2 billion domestic trips in other (developing) countries. However, international tourism may be more relevant in the context of this chapter. International trips last 8.3 days on global average and cause 0.66 tons of CO_2 emissions. It should be noted, however, that there are huge differences between international trips. As Figure 13.1 shows, a fourteen-day holiday from Europe to Thailand may cause emissions of 2.4 tonnes of CO_2, and a typical fly-cruise from the Netherlands to Antarctica produces some 9 t CO_2 (Lamers and Amelung 2007). These figures show that emissions caused by a single holiday can vastly exceed the annual per capita emissions of the average world citizen (4.3 t CO_2), or even the average EU citizen (9 t CO_2), while the radiative forcing of non-carbon emissions at flight altitude is not even considered here. Low-carbon trips, on the other hand, are mostly those involving trains or buses and to a lesser extent cars. As shown in Figure 13.1, even shorter car trips (if taking several passengers) can be relatively efficient, and increase annual per capita emissions only marginally. The main parameters determining emissions are the distance travelled and the mode chosen.

Another way of illustrating emissions is to calculate these on a per day basis (i.e., to compare average annual per capita per day emissions to those caused on a daily basis by various holidays). Figure 13.2 shows that per day emissions largely depend on the choice of transport mode, with air transport generally increasing emission levels substantially. Other factors influencing per day emissions are the distances travelled; average length of stay; the accommodation chosen, with luxury hotels generally having higher levels of resource use than more basic accommodation establishments; and the activities carried out. Destination choice (i.e., distance), the means of transport and average length of stay are however the main factors determining per trip per day emissions.

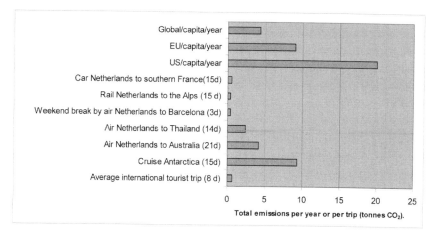

Figure 13.1   Annual per capita CO_2 emissions and emissions caused by various journeys.

Even though these figures are illustrative of the relative contribution of different journey types to emissions, they do not reveal another relevant insight, the fact that on a global scale, a minority of trips are responsible for a major share of emissions. It is the minority of long-distance trips that heavily influences tourism's overall contribution to climate change to the degree that there may be a Pareto distribution in emissions, in the sense that 20 per cent of the

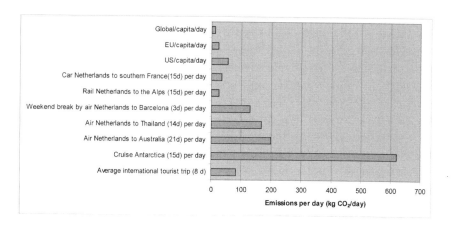

Figure 13.2   Daily average emissions per person and emissions per tourist-day for various journeys.

trips may be responsible for 80 per cent of overall emissions (see Figure 13.3, Peeters, Szimba, and Duijnisveld 2007; UNWTO-UNEP-WMO 2008).

The largest contribution to reducing emissions from tourism can thus be achieved by focusing on these trips, indicating a role for tour operators in creating demand for less distant destinations. Reductions of emissions in tourism will, therefore, largely depend on managing individual travel choices towards different destination choices and the use of low-carbon transport modes (cf. Gössling et al. 2005).

As differences in emissions caused by various journeys seem not directly related to the profitability of tourism products, there is also room for the optimisation of tourism products, i.e., to maintain or increase the profitability of tourism businesses while reducing their environmental impact (Becken and Simmons 2008; Gössling et al. 2005). For tourism more generally, there should be opportunities for restructuring products and travel flows, i.e., to maintain financial gains and economic growth while reducing emissions. One avenue could be to adjust marketing and sales strategies to promote economically profitable tourism products with a relatively low carbon footprint. Such strategies may also reduce vulnerability to climate policy or increasing oil prices (Gössling, Peeters, and

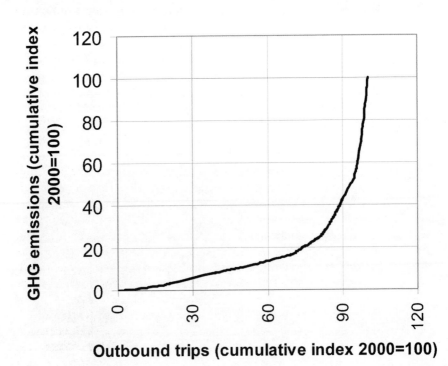

Figure 13.3 Lorentz graph of outbound trips and GHG emissions for the EU25 countries in 2000.

Scott 2008). An important question is, however, whether travellers will be willing to buy low-carbon products.

MARKETING AND BEHAVIOURAL CHANGE

Changes in tourist flows towards closer destinations and extended lengths of stay will require innovative marketing and better information on climate change issues. All forms of behavioural change, in any business and in any society, are difficult to achieve. They are particularly difficult in tourism, where powerful and emotive issues of pleasure, freedom, and status are involved. Three ways forward to achieve behavioural change in tourism are discussed below: the potential for de-marketing and social marketing techniques, and the role of government.

De-marketing

The concept of de-marketing was first floated as a technique by the American marketing expert Phillip Kotler in 1971 (Kotler and Levy 1971a). He discussed de-marketing in a commercial context, where a company wished to reduce demand, perhaps because of supply or quality problems, or to reduce demand from certain market segments, because of price or image issues. He also noted the dangers that de-marketing can, in some circumstances, bring increased demand by creating an elite concept of scarcity.

An early example of tourism management by de-marketing occurred in Cyprus in the 1980s to discourage young, noisy, drunken tourists, largely from the UK. It cut marketing to the unwanted market segments according to age and income, cut the promotion of certain types of activity, and avoided specific media outlets (Clements 1989). In the USA, Fort Lauderdale, Florida, discourages visits by spring break college and high school students for similar reasons. For different reasons the county of Somerset, UK, in an effort to retain traditional village life and landscapes, has cut all public sector marketing mention of a fragile range of beautiful hills, the Quantocks, and removed signage to the hills and through signing of villages on either side of the hills. In the USA, the Arches National Park in Utah carried out similar policies (Groff 1998). Beeton and Benfield (2002) provide a useful overview of demand management and de-marketing. It should be noted, however, that de-marketing is a *demand control* device rather than a *behavioural change* device. It is often heavily dependent on the public sector creating consensus and partnership approaches. It demands a high level of market knowledge. It can be useful at local and sub-regional level; but its economic and political consequences make it almost impossible to implement at regional or national levels.

Social Marketing

Social Marketing is an umbrella term for a range of techniques which have evolved over the last forty years to try to change behaviours. Social marketing can include de-marketing, but social marketing is a collection of more positive and proactive approaches towards encouraging behavioural change. Like de-marketing, social marketing owes much to Philip Kotler who first published on it in the *Journal of Marketing* in 1971. He described it as 'the use of marketing principles and techniques to advance a social cause, idea or behaviour' (Kotler and Zaltman 1971b quoted in Kotler, Roberto, and Lee 2002, 8). Unlike commercial marketing, where an owner or company profits from successful marketing, successful social marketing benefits society as a whole or the wider natural or cultural environment, as well as individuals. It aims to influence a target audience to voluntarily accept new, or reject old, or modify or abandon their behaviours for the benefit of individuals, groups, society, or the environment. Societal gain replaces to a greater or lesser extent financial gain—but there need not be a total switch away from financial gain. Commercial marketing sees competitors as those selling similar goods and services. Social marketing sees competitors as those partaking in current or 'problematic' behaviours, which opens up new opportunities in the context of low-carbon tourism.

Social marketing has a much more difficult task than commercial marketing. It tries to influence people to:

- give up often addictive behaviours.
- change often enjoyable lifestyles.
- overcome peer pressures to be like one's peers, risking ridicule.
- do perhaps uncomfortable and inconvenient things for a greater good.
- spend more money, often for intangible long term gains.
- reduce short term pleasures.
- think about the consequences arising from actions, invoking guilt.

None of the aforementioned tasks create instant pleasure, believed to be a powerful motivator to take holidays. Research has also uncovered a specific travel industry problem. Many people have powerful belief in their personal right and need to travel, coupled at the same time with a contradictory powerful belief that others should be denied that right for the good of the planet (Becken 2007; Høyer and Naess 2001; Shaw and Thomas 2006). To be successful in changing behaviours and consumption patterns, social marketing has to research and understand what motivates each market segment. To persuade the market to change destinations and travel mode it is necessary to learn both their current motivations and also to learn from the experience of tour operators: The process of changing behaviour is not a unidirectional one. Change needs an understanding of exchange theory and has to seek to show real benefits to the consumer from the promoted new behaviour. It needs to segment each of many audiences to find their specific drivers. It also must, while using the concept of social marketing, continue to understand and use traditional marketing techniques. It is particularly important in tourism that new behaviour has to give pleasure and be fashionable—it should be remembered that tourism is both a form of escapism and a fashion pursuit. Therefore it should fit within a social context of the desired behaviour. A recent in-depth report on using social marketing to encourage pro-environment behaviour change goes so far as to suggest that in many cases the best approach to achieving this would be to use non-environmental motivations (DEFRA 2008).

So far, social marketing techniques have remained rarely used in tourism; tourism marketing has concentrated on increasing visitor numbers and market share rather than the more difficult, demanding, and controversial tasks of changing behaviours. The rise of sustainable tourism, ecotourism, cultural tourism, and specifically the drive towards low-carbon tourism should, however, stimulate interest in the technique.

The Role of Governments

Tourism is largely a private sector activity, but it has very close relationships with the public sector at central, regional, and local government levels. Governments are involved in regulating the industry, and in creating infrastructure. They are also closely involved in 'boosterism', using image creation and marketing to create jobs and income flows. Governments are therefore essential—if sometimes unwitting—partners in behavioural change and in influencing visitor flows, travel modes, and activities.

Governments have powerful mechanisms that can be invoked to bring about change. On the negative side they can tax or prohibit certain activities. They can decide not to create infrastructure that could bring about higher emission levels. They can introduce quotas for using some areas—thus the New Zealand government issues only a limited number of permits for using some trails that are in high demand. Bhutan does not use quotas, but limits entry to travellers on pre-planned, prepaid programmes. More positively, governments could encourage the development of low emission infrastructure, including the retention and improvement of existing rail lines, and the building of new rail lines. And they could encourage successful destination image change through market research funding, and support for social marketing experiments. Tour operators work with governments through lobbying: Ideally it should be a two way process. Governments need to support tour operators who take the risk of trying to correct market failures which prevent more sustainable tourism development.

The next section outlines the central approaches to low-carbon, high eco-efficiency tourism. It is based on the insight that not only customer information to facilitate pro-environmental choices, an extended length of stay, the restructuring of destination portfolios towards closer destinations, and the choice of train or bus/coach are important in achieving more sustainable tourism, but also the development and promotion of new products that can reduce carbon emissions. Marketing should, however, eventually not focus on the carbon properties of such packages, rather it should stress aspects that provide a tangible value, such as their unique character or the experiences that can be gained through the consumption of the product. The marketing focus on enjoyable, unique experiences is probably the major reason for the success of ecotourism in Sweden (Gössling 2006).

OPPORTUNITIES FOR TOUR OPERATORS TO PROMOTE LOW-CARBON TOURISM

Various opportunities seem to exist to use social marketing techniques in developing more sustainable tourism. The following sections will present a discussion of three of them (i.e., carbon labelling, increasing length of stay, and changes in transport choices, and destination portfolios).

Pro-Environmental Choices: Carbon Labelling

Carbon labelling, i.e., the provision of information on the carbon intensity of consumption, could be relevant in building more sustainable tourism in at least two different ways. First, there are substantial differences in the carbon intensity of the same tourism product, such as a flight from A to B. As shown in Figure 13.4 for airlines operating in Sweden, the same distance flown by different airlines can entail energy use that is up to 50 per cent higher per seat kilometre. A traveller's choice to fly with My Travel Airways rather than Lufthansa could entail energy use (and corresponding emissions) that is more

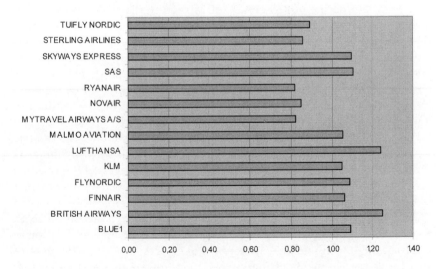

Figure 13.4   Fleet weighted and seat density corrected average energy consumption per seat-kilometre (MJ). Source: Peeters (2007c).

than 30 per cent lower. If two different airlines operate on the same route, passengers may make travel decisions in favour of the more environmentally friendly airline. This, however, requires the availability of information on environmental performance by various airlines, which may also depend on the aircraft type used on a particular route. Ideally, carbon information would thus have to be provided in the marketing and booking process, by displaying both the price for a certain trip as well as its carbon intensity.

Another area where carbon labelling could be used is for total trips, allowing for comparison of the carbon intensity of various destinations and packages, such as a train-based city break or a long-distance sun-sand-sea journey. German tour operator Forumandersreisen, an umbrella organisation for seventy-three small, environmentally and socially aware tour operators, provides for instance information on emissions of each of the packages offered (Forum Anders Reisen 2008). It is not clear, however, how such information will affect travel decisions, particularly if expressing the impact of a journey in terms of kg CO_2, a concept abstract in time and space (for discussion see Moisander 2007, 404; Budeanu 2007). A potentially valuable side effect might be that the information would alert tour operators themselves to the varying impacts of their products, and encourage them to use that information as a competitive marketing tool.

Increasing Length of Stay

Currently, packages are generally advertised by tour operators following the formula 'Destination x: from €y per person'. As the price of a package is largely the cost of transport plus the length of stay times the cost per night, this makes short trips generally more attractive, as the total price will be lower. However, as outlined earlier, increasing the average length of stay is a key to achieving low-carbon tourism (cf. UNWTO-UNEP-WMO 2008), given that most employed people have a fixed holiday entitlement. If tour operators marketed travel packages as 'Destination x: from €y per person *per day*', the customer's focus would change in favour of longer holidays, as longer stays would become comparatively cheaper, i.e., adding an additional day to the holiday would entail rather low costs.

For tour operators, such an approach would have a number of advantages. First, the average revenue per customer should increase because packages with a greater length of stay will yield a higher turnover per package sold. As the share of the yield that is an add-on is mostly accommodation and eventually activities-related, both of which are likely to have a higher profitability than air transport, this could be significant. Second, costs at the destination should decline with regard to cleaning, house-keeping, washing linen, check-in and check-out, welcome drinks, and other aspects. More generally, tour operators commencing this kind of advertisement should have a comparative advantage, as the prices in travel catalogues will appear to be considerably lower, given that packages are marketed on a per day basis. For tour operators, decreasing the demand for transport by promoting longer stays could lead to greater competition from transport undertakings, leading to lower prices and better service.

Changing Transport Choices and Restructuring of Destination Portfolios

The literature on tourism marketing largely ignores issues of sustainability, but sustainable tourism actors often seem to market their products based on their pro-environmental or pro-social properties. These products are generally aimed at niche markets of conscious tourists, who arguably are a minority (for a review of such studies see Simpson, Gössling, and Scott 2008). One key question is thus how to integrate commercial perspectives into low-carbon tourism products that could reach out to mass tourism. One idea is to simply enhance the eco-efficiency of the whole product portfolio, i.e., to choose an approach that combines economic and environmental goals (eco-efficiency is often defined as the amount of turnover divided by resource use or emissions, see Heijungs 2007). As has been shown by Gössling et al. (2005), there are huge differences in the eco-efficiency of various tourism products, defined as € turnover per kg of CO_2 emissions. Eco-efficiency offers opportunities for tour operators to reduce emissions step-by-step during product development and sales, and to market that efficiency effectively.

The development of the product has a direct impact on carbon intensity and eco-efficiency, which is mainly determined by the transport mode, distance to the destination, and to a lesser extent by the type of accommodation and activities offered. Consequently, a more favourable eco-efficiency, i.e., a higher turnover per kg of emissions, can be achieved with relatively minor changes to the product, such as the choice of more energy-efficient accommodation and transport. Packages in the short- to medium-range distance of up to 2000 kilometres can strongly benefit from a change of transport mode, i.e., from air or car to rail or bus/coach. This, however, will change the product significantly and make it less attractive to current customers, demanding social marketing approaches. For example, an aviation-based journey from Germany to Andalucía may become more interesting, if the train replacing the aircraft makes a stop-over in Barcelona, involving sightseeing or a restaurant visit, and thus enhancing the overall appeal of the trip from an experience point of view. From a logistical point of view, this may however be complicated to realise for tour operators. Much would also depend on the train operators offering more attractive on train experiences, and, in Europe, breaking down barriers between national train operators— and in some countries creating links between local operators.

The greatest amount of carbon can be saved by promoting closer destinations instead of remote ones. Emissions reductions will then be almost proportional to the reduction in flying distance (see Peeters et al. 2007). The core of the product, fast and easy access by air as well as the holiday type (for instance sun-sand-sea motives) can however be maintained in such an approach. The perceived attractiveness of for example a Mediterranean versus a Caribbean destination is the crucial point. Here, social marketing could focus on comfort advantages, such as shorter flight times and lower costs. If the destination choice then comes into the range of surface transport modes, the effects of modal shift and closer destination choice could eventually be combined, even though this is likely to require the development of a totally new product. Opportunities may nevertheless exist, as exemplified by German 'nature and outdoor' and 'fun and action' holidaymakers (see Rheinberger, Schmied, and Götz 2007).

New Marketing Concepts

In addition to considering different forms of marketing, much could be achieved by partnership promotion of lower carbon holidays. In many cases, the pioneers in lower carbon holidays are relatively smaller companies, often specifically set up with environmental and ethical approaches in mind. They have small marketing budgets, and can have limited marketing skills. Partnership marketing can pool resources, create greater impact and image promotion, and afford the services of skilled marketing personnel and the use of new technology. The same concept would have greater impact on the mass media of TV, radio, and national press outlets.

Two examples illustrate the point. In the UK, AITO (The Association of Independent Tour Operators) brings together over 150 smaller tour operators. In addition to working as a partnership lobby and marketing organisation, members pledge to work towards a mild form of responsible tourism. Responsible tourism has a headline place on AITO's web site (www.aito. co.uk). AITO has worked to promote sustainable and responsible tourism for over fifteen years. However, the tour operators within AITO cover long haul as well as short haul travel, thus ignoring the main sustainability issue of such travel. Even the tour operators offering European destinations like the Italian Alps, give most information about flights, less on driving yourself, and almost none on rail or coach travel. Climate change is almost exclusively addressed by proposing off-setting. More effective behavioural changes like choosing medium- or short-haul destinations, low carbon transport, or stay longer and fly less frequently are ignored.

A more environmentally effective approach is taken by Vertraeglich Reisen, a German company which produces a widely circulated (currently 250,000 copies) annual magazine, promoting sustainable tourism. The website puts great emphasis on avoiding travel that requires air transport over medium and long distances. It promotes train and bus as the most sustainable alternative and even gives detailed information about accessibility throughout Europe by these modes. The company has worked in this field since 1990. While concentrating on the German speaking market of over 100 million people, sections of its web site (www.vertraeglich-reisen.de) are available in English, including an especially informative reader survey and technical details of its operations. Vertraeglich Reisen provides a professional alternative platform for many small tour operators and destinations.

Most products to be found on the website are truly sustainable, in that they have a very small impact on the climate.

More generally, businesses may seek to reduce their carbon footprints through four measures (Lash and Wellington 2007, 112). In a first step, (1) they can seek to quantify the carbon footprint of their products; (2) then go on to assess carbon-related risks and opportunities; (3) restructure the business; and (4) 'do better then rivals' (i.e., have less CO_2 emissions for more profit). A market-based approach for tour operators and travel agencies may be developed building on these general principles, focusing on the reduction of average carbon footprints. Such an approach would seek to:

1. Determine the eco-efficiency of each product/product group by quantifying carbon footprints and comparing them with economic performance (turnover as well as profitability), also identifying the importance of each product group with regard to overall turnover and profitability.
2. Analyse current marketing efforts and product pricing and their consequences for eco-efficiency, i.e., the impact of marketing on steering customers towards certain products and product groups that may be more or less carbon intense.
3. Identify challenges and opportunities for shifting marketing budgets and redefining cross-subsidies between different product lines (for example reducing profit on short haul at the cost of long haul trips)
4. Implement changes on a level that allow control of overall turnover/profitability.

CONCLUDING REMARKS

Tourism becomes increasingly carbon intensive. Though accommodation and tourist leisure activities do contribute substantially to the CO_2 emissions of the sector, the main share (75 per cent) comes from transport, and within transport more than half is caused by aviation. Measured as radiative forcing, air-based tourism may stand for as much as 75 per cent of tourism's overall contribution to global warming, even though it is the means of transport used for only 17 per cent of all trips (domestic and international). Technological innovation in air transport will not be adequate to compensate for the growth of air transport, and a reduction of the number of kilometres travelled per trip is inevitable if tourism is to reduce its overall emissions. Tour operators and destinations will be key players in the restructuring process. However, there is not much evidence that tour operators will take this role seriously.

De-marketing and social marketing were identified in this chapter as little-explored means to achieve behavioural change in tourism. While de-marketing has been successfully applied in some tourism-related contexts, such as to reduce pressure on vulnerable nature reserves, or to avoid certain types of undesirable tourists, it may only be a concept of interest for tour operators if it at the same time has a strategy to boost low-carbon products. Social marketing, on the other hand, has a greater focus on commercial objectives, but it is inhibited by the difficulties inherent in any attempt to make people behave in ways that contradict addictive behaviour, or make less comfortable, or more altruistic holiday decisions.

Three options for tour operators to decarbonise their products were presented in this chapter. Carbon labelling may help customers to make pro-environmental choices. Changes in the packaging and pricing of products, as well as their marketing may be a second option. Changing the destination portfolio and the transport modes offered may be a third. None of these has been assessed scientifically, however, in the sense that tourist opinions on these issues have been evaluated.

From these considerations it becomes clear that both destinations and tour operators have various opportunities to restructure and change their product portfolios and marketing. Such changes should be possible even from an economic point of view, and they are certainly necessary from a climate change point of view.

NOTES

1. Radiative forcing is a measure of the overall contribution of various greenhouse gases to global warming. Emissions from aviation are particularly problematic as they are released in the higher troposphere and lower stratosphere, where nitrogen oxides and water emissions contribute to additional warming. As these emissions are short-lived, it is more complicated to compare these to CO_2, a long-lived greenhouse gas. For a recent discussion of the contribution to aviation to global warming and options for comparison with CO_2 see e.g., Graßl, H. and Brockhagen, D. (2007) Climate forcing of aviation emissions in high altitudes and comparison of metrics.

REFERENCES

Becken, S. (2007) 'Tourists' perception of international air travel's impact on global climate and potential climate change policies'. *Journal of Sustainable Tourism* 15(4): 351–368.

Becken, S. and Simmons, D.G. (in press) 'Using the concept of yield to assess the sustainability of different tourist types'. *Ecological Economics* (in press).

Beeton, S. and Benfield, R. (2002) 'Demand control: The case for demarketing as a visitor and environmental management tool'. *Journal of Sustainable Tourism* 10(6): 497–513.

Budeanu, A. (2007) 'Sustainable tourist behaviour—a discussion of opportunities for change'. *International Journal of Consumer Studies* 31(5): 499–508.

Carey, J.M., Beilin, R., Boxshall, A., Burgman, M.A., and Flander, L. (2007) 'Risk-based approaches to deal with uncertainty in a data-poor system: Stakeholder involvement in hazard identification for marine national parks and marine sanctuaries in Victoria, Australia'. *An International Journal* 27(1): 271–281.

Carey, S., Gountas, Y., and Gilbert, D. (1997) 'Tour operators and destination sustainability'. *Tourism Management* 18(7): 425–431.

Clements, M.A. (1989) 'Selecting tourist traffic by demarketing'. *Tourism Management* 10(2): 89–94.

DEFRA (Department of Environment, Food and Rural Affairs) (2008) *A Framework for Pro-Environmental Behaviours.* London, DEFRA. Available at www. defra.gov.uk/evidence/social/behaviour/index.htm

Font, X., Tapper, R., Schwartz, K., and Kornilaki, M. (2008) 'Sustainable supply chain management in tourism'. *Business Strategy and The Environment* 17: 260–271.

Forum Anders Reisen (2008) *Reiseperlen 2008.* Available at: http://forum andersreisen.de/downloads/reiseperlen_2008.pdf (14 May 2008).

Gössling, S. (2006) 'Ecotourism as experience-tourism'. In Gössling, S. and Hultman, J. (eds.) *Ecotourism in Scandinavia. Lessons in Theory and Practice,* 89–97. Wallingford: CABI Publishing.

Gössling, S., Peeters, P., and Scott, D. (2008) 'Consequences of climate policy for international tourist arrivals in developing countries'. *Third World Quarterly* 29(5): 869–897.

Gössling, S., Peeters, P.M., Ceron, J.-P., Dubois, G., Patterson, T., and Richardson, R.B. (2005) 'The eco-efficiency of tourism'. *Ecological Economics* 54(4): 417–434.

Graßl, H. and Brockhagen, D. (2007) *'Climate Forcing of Aviation Emissions in High Altitudes and Comparison of Metrics'.* Max-Planck-Institut fuer Meteorologie, Hamburg. Available at http://www.mpimet.mpg.de/wissenschaft/ publikationen/papers/climate-forcing-of-aviation-emissions-in-high-altitudes-and-comparison-of-metrics.html (31 May 2008).

Groff, C. (1998) 'Demarketing in park and recreation management'. *Managing Leisure* 3: 128–135.

Heijungs, R. (2007) 'From thermodynamic efficiency to eco-efficiency'. In Huppes, G. and Ishikawa, M. (eds.) *Quantified eco-efficiency. An introduction with applications,* 79–103. Dordrecht, Netherlands: Springer.

Hille, J., Storm, H., Sataøen, H., Aall, C., and Holden, E. (2008) *Miljøbelastningen fra norsk forbruk og norsk produksjon 1987–2007. En analyse i forbindelse med 20-årsjubileet for utgivelse av rapporten 'Vår Felles Framtid' og lanseringen av målet om en bærekraftig utvikling.* (Environmental impacts of Norwegian consumption and production 1987–2007. An analysis in connection with the 20th jubilee of 'Our Common Future' and the launching of the goal of sustainable development). VF-rapport 2/08 Sogndal: Vestlandsforsking.

Høyer, K.G. and Naess, P. (2001) 'Conference tourism: A problem for the environment as well as for research?' *Journal of Sustainable Tourism* 9(6): 451–470.

International Air Transport Association (IATA) (2007) *IATA Calls for a zero emissions future.* Available at: http://www.iata.org/pressroom/pr/2007-06-04-02 (27 September 2008).

Iwand, W.M. (1995) 'Instruments, procedures, and experiences of integrating the environment into tourism development by a major tour operator'. *World Conference on Sustainable Tourism,* 24–29. April 1995, Lanzarote.

Iwand, W.M. (1998) 'Supporting biodiversity: Taking the benefit of tourism into account', *Biodiversity—Treasures in the World's Forests Forum.* 4 July 1998, Schneverdingen, Germany.

Iwand, W.M. (1999) 'Tourismus als globale Strategie für nachhaltige Entwicklung'. *International Tourism Exchange (ITB).* March 7, 1999, Berlin.

Kotler, P. and Levy, S.J. (1971a) 'Demarketing? Yes, demarketing!' *Harvard Business Review* 49(6): 74–80.

Kotler, P., Roberto, N., and Lee, N. (2002) *Social marketing: Improving the Quality of Life.* London: Sage Publications Ltd.

Kotler, P. and Zaltman, G. (1971b) 'Social Marketing: An approach to planned change'. *Journal of Marketing* 35: 3–12.

Lamers, M. and Amelung, B. (2007) 'The environmental impacts of tourism, to Antarctica. A global perspective'. In Peeters, P.M. (ed.) *Tourism and Climate Change Mitigation. Methods, Greenhouse Gas Reductions and Policies*, 51–62. Breda: NHTV.

Lash, J. and Wellington, F. (2007) 'Competitive advantage on a warming planet'. *Harvard Business Review* 11, 85 (3): 95–102.

Moisander, J. (2007) 'Motivational complexity of green consumerism'. *International Journal of Consumer Studies* 31(4): 404–409.

Peeters, P. (2007a, April 15) 'The impact of tourism on climate change'. *Policy Dialogue on Tourism, Transport and Climate Change: Stakeholders Meet Researchers*. Paris: Eclat.

Peeters, P. (2007b) 'Mitigating tourism's contribution to climate change—an introduction'. In Peeters, P.M. (ed.) *Tourism and Climate Change Mitigation. Methods, Greenhouse Gas Reductions and Policies*, 11–26. Breda: NHTV.

Peeters, P.M. (2007c) *Report on the Environmental Performance Class of Airlines Flying at Swedish Airports*. Breda: NHTV CSTT.

Peeters, P.M., Gössling, S., and Becken, S. (2006) 'Innovation towards tourism sustainability: Climate change and aviation'. *Journal of Innovation and Sustainable Development* 1(3): 184–200.

Peeters, P., Szimba, E., and Duijnisveld, M. (2007) 'Major environmental impacts of European tourist transport'. *Journal of Transport Geography* 15: 83–93.

Rheinberger, U., Schmied, M., and Götz, K. (2007) 'Greenhouse gas emissions reduction by target-group tailored holidays'. In Peeters, P.M. (ed.) *Tourism and Climate Change Mitigation. Methods, Greenhouse Gas Reductions and Policies*, 143–155. Breda: NHTV.

Scandinavian Airlines (2008) *Your Choice for Air Travel. SAS Group Annual Report and Sustainability Report 2007*. Stockholm, Sweden: SAS AB.

Shaw, S. and Thomas, C. (2006) 'Social and cultural dimensions of air travel demand'. *Journal of Sustainable Tourism* 14(2): 209–215.

Simpson, M.C., Gössling, S., and Scott, D. (2008) *Report on the International Policy and Market Response to Global Warming and the Challenges and Opportunities that Climate Change Issues Present for the Caribbean Tourism Sector*. Barbados: Caribbean Tourism Organization.

Tepelus, C.M. (2005) 'Aiming for sustainability in the tour operating business'. *Journal of Cleaner Production* 13: 99–107.

The Guardian (2008) *Country for Sale, 26-April-2008*. Available at: http://www.guardian.co.uk/world/2008/apr/26/cambodia (7 May 2008).

Timothy, D.J. and Ioannides, D. (2002) 'Tour operator hegemony: Dependency and oligopoly in insular destinations'. In Apostolopoulos, Y. and Gayle, D.J. (eds.) *Island Tourism and Sustainable Development: Caribbean, Pacific, and Mediterranean Experiences*. Westport: Preager.

TUI (2004) *TUI Environmental Report (2003/04). Managing Sustainably at World of TUI. Group Environmental Reporting 2003/2004*. Available at: http://www.econsense.de/_CSR_MITGLIEDER/_CSR_NACHHALTIGKEITSBERICHTE/images/TUI/TUI_EnvRep_03_4.pdf (7 July 2008).

UNWTO-UNEP-WMO (2008) *Climate Change and Tourism: Responding to Global Challenges*. Madrid: UNWTO.

van der Duim, R. and van Marwijk, R. (2006) 'The implementation of an environmental management system for Dutch tour operators: An actor-network perspective'. *Journal of Sustainable Tourism* 14(5): 449–472.

van Wijk, J. and Persson, W. (2006) 'A long-haul destination: Sustainability reporting among tour operators'. *European Management Journal* 24(6): 381–395.

14 Sustainable Transportation Guidelines for Nature-Based Tour Operators

Wolfgang Strasdas

WHAT IS THE ISSUE?

The Fourth Assessment Report of the Intergovernmental Panel on Climate Change (IPCC), published in the course of 2007, has made a compelling case that global warming is a reality and that much of it is man-made. What has ensued is an almost frantic discussion on what to do to avert further damage. Unfortunately, there is a tendency to see others as the main culprit for climate change and thus as the ones most obligated to take counter-measures. This is not only true for countries, but also for economic sectors. Tourism is no exception. A recent study conducted on behalf of the World Tourism Organization (UNWTO) has estimated tourism's contribution to the man-made greenhouse effect to be around five per cent, putting it as an industry well ahead of major polluting countries such as Germany or Japan (UNWTO-UNEP-WMO 2008).[1] This means that tourism is not only affected by climate change, but also significantly contributing to it.

The industry has been somewhat slow to acknowledge this fact. Although the UNWTO already conducted an international conference on the issue in 2003 in Djerba (Tunisia), this did not lead to any specific action. It took the 2007 IPCC report with its tremendous impact on the public, professionals and politics worldwide to instigate widespread concern among the tourism sector. For example, global warming was one of the dominating themes of the 2007 International Tourism Exchange (ITB) in Berlin leading to a vague feeling of responsibility among tourism professionals, but without much of a clue of what exactly could be done to reduce tourism's impact on the world climate (Lund-Durlacher, Strasdas, and Seltmann 2007). The 2007 UNWTO conference on climate change in Davos (Switzerland) produced a more specific declaration than the one in Djerba, but fell short of any concrete action. Much of the conference (which was attended by the author of this chapter) was dominated by trying to shift the burden of mitigation to other sectors or by pointing out past achievements concerning the sustainability of tourism.

It is conspicuous that sustainable tourism concepts and guidelines, including formal certification schemes, so far have been centred on the destination and its facilities. The trip between the destination and the place or country of tourists' origin has been largely disregarded. However, several studies have shown that transportation is the major variable when it comes to tourism's impacts on the global climate. Sixty-two per cent for domestic tourism in Germany (with its easily accessible tourist attractions and excellent public transportation network), 78 per cent within New Zealand, over 90 per cent for long-haul holidays taken by German tourists, and an extreme of 97 per cent for the Seychelles (a typical long-haul resort destination), of the emissions of a holiday trip can be attributed to transportation (UBA/Öko-Institut 2002; Simmons and Becken 2004; Gössling 2000). Yet, most sustainable tourism concepts have focussed on energy-saving light bulbs in hotels and reducing other facility-related emissions while disregarding the bulk of emissions caused by transport. Furthermore, from a consumer's perspective a lifestyle of frequent flying is extremely unsustainable. One intercontinental trip may emit as much greenhouse gases as driving a car for several years (VCD 2006; for a calculator see www.atmosfair.de).

This is of particular concern since transportation is a *conditio sine qua non* for tourism. Unfortunately, nature-based tourism is not an exception in this respect. In fact, the opposite is usually the case. Since nature travel often involves 'exotic' countries or remote areas with little infrastructure, getting there by public transportation or even tour buses is logistically difficult if not impossible. In addition, in contrast to the typical beach resort vacation, nature-based travel usually combines visiting several sites during one trip, which means that transportation *within* the destination becomes an important factor as well. Simmons and Becken (2004) have studied this phenomenon in New Zealand, a tourism destination known for its natural attractions. They found that the average New Zealand traveler consumes three times as much energy as the typical resort vacationer in the Caribbean.

Thus, strategies to mitigate tourism's contribution to climate change impact must put more emphasis on transport than on local facilities, especially with regard to long-haul tourism. Taking active steps in this direction must become an indispensable component of any sustainable tourism concept or certification scheme. However, this does not mean that hotels and other tourism facilities should be disregarded. In the case of Germany with its relatively tight regulations on energy efficiency for buildings, they still represent a sizeable twenty-eight per cent of the overall energy consumption of domestic tourism (UBA/Öko-Institut 2002).

This chapter, which is the result of a six-months research project conducted by the author in cooperation with The International Ecotourism Society (TIES) and Stanford University/USA in 2007, has a certain, but not exclusive focus on nature-based tourism or ecotourism.[2] Although eco tourism represents a comparatively small segment of the overall tourism market, businesses and travelers in this field have traditionally been at the forefront of sustainable tourism development and may be expected to play this role once again with respect to climate change mitigation. Furthermore, in order to cover the carbon footprint of an entire trip, the project has emphasised the key role of tour operators since they are usually the ones who design trips, market them and contract other suppliers involved. While outbound operators play a larger role in the international arena, inbound and local operators are to be primarily addressed when national and local transport is being considered.

This chapter will first outline the general principles and requirements of sustainable transportation as derived from the existing literature on the subject. It will then analyse sustainable tourism certification schemes and codes of conduct that are relevant for tour operators with regard to sustainable transportation principles. As a third step, it will specifically examine individual companies' perceptions, attitudes, and actions in relation to transportation. Based on the results, a framework as well as a set of concrete criteria for sustainable transportation to be implemented by tour operators will be developed. The intention here is to lay the foundation for a set of guidelines for voluntary measures to be integrated into existing certification schemes or codes of conduct. Although there is a special consideration of nature-based tourism, many of the proposed measures are also applicable to mainstream tour operators, especially in relation to international transport.

SUSTAINABLE TRANSPORTATION PRINCIPLES

Reducing tourism's impact on the world climate and local environments through sustainable transportation is by no means a new agenda. Environmental groups, particularly in Europe, have been lobbying for a different travel behaviour, the use of public transportation, and strict government regulation since the 1980s. And although some success has been achieved on the local level (for example, by creating automobile-free tourist resort towns in the Alps (among others, see Zeppenfeld 2004) the opposite has happened in relation to the trips taken to and from the destinations. For instance, in Germany the combined share of vacationers using buses or trains—the more environmentally friendly modes of travel—has constantly decreased to about 15 per cent in 2006, whereas the use of the airplane has risen to over 35 per cent (F.U.R. 2000, 2007), an increase which has recently been exacerbated by the proliferation of low-cost airlines.

With the now more imminent threat posed by global warming and a growing awareness of tourism's role in it, sustainable transportation concepts, especially in relation to air travel, have taken on a new urgency. A variety of tools or actions have been proposed by different authors, environmental groups, and government agencies (such as the German *Umweltbundesamt*, the Federal Environmental Agency) targeting different stakeholders within tourism including consumers and the tourism industry (for a more detailed discussion, see Strasdas 2007):

• use of regulatory instruments (such as compulsory emissions standards, fees and taxes, cap-and-trade systems for transportation emissions). Government regulation is probably inevitable to achieve significant results. Tour operators should prepare for this and actively shape and support the process rather than obstructing it. If implemented on a global scale, regulatory instruments will create equal conditions for all competitors and may even favor early adaptors
• technological solutions (improved energy efficiency, decreased emissions, use of renewable energies): Tour operators cannot directly influence the development of new travel technologies, but they can favor them by appropriate supply chain management or by purchasing fuel-efficient and otherwise climate-friendly vehicles
• operational solutions (air traffic management, increased occupation rates, vehicle logistics and maintenance, training of drivers, etc). Again, this is more of an issue for transport companies or ground operators who are the suppliers of outbound or inbound operators
• modal shift (switching to more energy-efficient means of transportation such as railways and tour buses). Provided that appropriate suppliers (e.g., public transit systems) are available, tour operators can contribute to modal shift by integrating it into their tour product design and marketing
• change of travel patterns in relation to space and time (increased length of stay in combination with fewer trips, preference of nearby destinations for short trips, less multiple-destination trips). This can be a powerful way of reducing transport energy without decreasing the overall time spent in destinations (see UNWTO-UNEP-WMO 2008). While it is ultimately up to consumers to change their travel behaviour, tour operators can support this process by creating and promoting attractive 'slow travel' products allowing for more quality time spent at selected sites
• voluntary compensation of greenhouse gas emissions (carbon offsets). Since transportation is an integral part of tourism and associated emissions cannot be completely avoided, especially on long-haul trips, carbon-offsetting becomes a prime climate protection measure for tour operators *in addition* to (often limited) direct emissions reduction as described earlier. Compensation costs may be internalised, i.e., integrated into a trip's price calculation, and/or recommended to the customer. Carbon offsetting may also be required from suppliers (for standards of carbon-offsetting, see Strasdas 2007)

In view of the massive cuts in emissions that industrialised countries will have to achieve in order to keep climate change within supposedly manageable limits (eighty per cent by 2050 according to Gössling et al. 2007), a combination of different measures is probably necessary. Tour operators can implement such measures to different degrees and in various fields of their management. By using the five business management fields of tour operators as laid out by the Tour Operators' Initiative's Sustainability Reporting Guidelines (see in next section), the following observations can be made in relation to sustainable transportation:

1. Product management and development is considered to be a core area where trips are being designed. This area is directly under the control of the (outbound) tour operator, but is of course dependent on demand and the feasibility of certain travel ideas. There are different opinions as to what degree tourism companies are able to *create* or at least influence demand by offering climate-friendly products. However, because the travelling public in many countries is acutely aware of tourism's role in global warming (see Eijgelaar 2007 for an excellent overview of studies of this issue), one might expect that there are market opportunities for such products.

2. Supply chain management: In this field it needs to be made sure that the designed trips can be carried out using the highest available standards. Tour operators have less influence here since they are dependent on what their suppliers can offer, which may even force them to modify their trip designs, for example, if public transport is not available. On the other hand, tour operators may influence suppliers, for instance by creating a demand for energy-efficient buses or carbon-neutral airlines.

3. Customer relations management: This is another key area because sustainable transport needs to be accepted by the demand. In addition to simply offering climate-friendly trips or mentioning the possibility of carbon-offsetting, consumers may need to be informed and educated about climate-friendly travelling and such offers may have to be specially promoted. Different communication strategies may be necessary for different segments of the demand. As with supply chain management, there is limited control by the tour operator.

4. Cooperation with the destination: This is a secondary field not belonging to the core areas of business management. It is defined as any initiative going beyond immediate supply chain management, for example philanthropic activities, support to protected areas, the implementation of carbon-offset projects, or helping destinations to cope with the consequences of global warming.

5. Internal management: This is an important area in that it forms the basis for the above activities (company's goals, definition of procedures and management systems, staff responsibilities, etc.). However, in terms of emissions, trips taken by staff and 'office ecology' are less important than the trips taken by the customers themselves.

The second structural element that needs to be taken into account when contemplating sustainable transportation is a physical one related to travel phases and the spatial level where transportation takes place:

- International transportation: This is obviously the most important part of a trip in terms of greenhouse gas emissions, mostly referring to travel between the country of origin and the destination. However, multiple-destination trips involving several countries may entail additional international transport. Unless neighbouring countries are visited, the use of airplanes is usually inevitable.
- National transportation: This can either relate to the second leg of a trip between the international gateway and the final destination, ranging from a simple hotel transfer from the airport to several domestic flights in the case of multiple-destination trips. Travel emissions from domestic tourism in large countries may equal or surpass international transport emissions.
- Local transportation: Although relatively insignificant in terms of greenhouse gas emissions, the local impact on both people and fragile natural areas (e.g., through air/water pollution, noise, wildlife disturbance) may be considerable. Certain motorised leisure activities that are typical for nature-based tourism (e.g., the use of power boats for whale-watching, helicopters, or scenic flights) are both energy-intensive and have a strong local impact.

An overriding goal of sustainable transportation concepts must be to reduce the amount of greenhouse gas emissions per day of travel or per unit of economic value of a given trip through the different measures described in this chapter. Quantifying this amount (for example, measured as CO_2 equivalents) is a difficult task in view of the multitude of components of a package tour or an individual trip. There have been a number of attempts for such a calculation, but so far none of them has been turned into a practicable system to be used by tour operators. However, a very useful, if basic instrument for this purpose has been developed by the University of Breda (Netherlands) and *De Kleine Aarde* ('The Small Earth', a Dutch advocacy group). Its 'Holiday Footprint' allows individual travelers to estimate their emissions based on the types of transport, accommodation, and activities chosen for their vacation (see www.vakantievoetafdruk.nl).

SUSTAINABLE TOURISM CERTIFICATION SCHEMES AND CODES OF CONDUCT FOR TOUR OPERATORS

The project's initial hypothesis according to which sustainable tourism concepts so far have been focused on destinations rather than on how to get there and back was to be verified by analysing existing certification schemes and codes of conduct. These can be seen as the most concrete and sometimes measurable outcome of sustainable tourism concepts. Second, they also appear to be the most logical place to integrate sustainable transportation principles into overall sustainable practice. In spite of some criticism as to the effectiveness of such schemes they nevertheless offer guidance and orientation to the tourism sector.

There are currently close to eighty sustainable tourism certification schemes worldwide, most of them in Europe (TIES 2007). More than half of them relate to accommodations only. The majority of the remainder certifies other service providers in the destinations (such as marinas) or spatial entities within destinations (such as beaches or protected areas). Formal, externally verified certification schemes specifically for outbound tour operators do not currently exist. Green Globe and Sustainable Travel International's STEP Programme (Sustainable Tourism Eco-Certification Programme) are intended for a variety of tourism companies including tour operators, but at the time of the research (May 2007) did not list any outbound tour operators as being certified.

ISO 14.001 and EMAS II (Eco-Management and Audit Scheme of the European Union) are process-oriented environmental management schemes that can be applied to any kind of company. At present, only a few larger tour operators and transport companies are certified under these schemes. These include TUI (Touristik Union International), Europe's biggest tour operator, and Studiosus (Germany's leading outbound operator in the cultural and nature tourism segment). In addition there is a small number of nonbinding codes of conduct or reporting schemes for outbound tour operators.

By way of contrast, a few formal certification schemes specifically for inbound or local operators do exist, most notably in Australia and Sweden. Following, the most important schemes and codes relating to tour operators are being analysed, mostly based on information provided by their respective websites. However, this is not necessarily an inclusive selection since not all existing schemes or codes could be specifically examined. Others may as well contain certain elements that could be relevant to tour operators.

Outbound Operators

Tour Operators Initiative (TOI; www.toinitiative.org)

The TOI, a loose network of mostly European large-scale outbound operators, has developed a framework (Sustainability Reporting Guidelines) specifically for tour operators relating to the five fields of business management mentioned in Section 1. This is a process-oriented tool which makes no specific mention of transportation. It is not known to what degree this framework has been implemented by individual companies associated with the TOI. However, the basic structure of the framework is very useful for the management practice of tour operators and will therefore be used for the Sustainable Transportation Guidelines (see Section 4).

Forum anders reisen[3]

Forum anders reisen (FAR; www.forumandersreisen.de) is an association of over one hundred German outbound tour operators who have developed an extensive catalogue of mostly performance-oriented sustainability criteria. The catalogue has to be underwritten by the association's members, but there is no external verification of compliance. The criteria are structured according to travel phases (trip to/from destination, type of destination, accommodation, leisure activities) as well as internal management.

In terms of sustainable transportation the FAR criteria are unique and can be regarded as exemplary concerning transport to and from the destination. FAR operators do not leave sustainability to their suppliers, but see it as part of their own product development. This includes the use of environmentally friendly means of transportation to the destination wherever reasonably possible; the exclusion of plane trips under 700 km as well as a minimum length of stay in mid-range or long-haul destinations. However, the criteria are much less strict for the destinations themselves where domestic flights are common components of FAR members' travel packages. On the local level, motorised leisure activities (such as snow scooters or helicopter flights) are excluded unless unavoidable for transferring guests to remote sites. All of these principles are to be clearly communicated to customers by each individual company. FAR members also ask their clients to offset their flight emissions by paying into certified compensation projects managed by their partner agency *atmosfair.*

However, as exemplary as FAR's sustainable transport criteria may be in relation to public transit and the time–distance ratio of a trip, one has to keep in mind that these are based on the German outbound market, where there is an excellent railroad network and travelers have several weeks of paid vacation per year. In spite of this there is increasing evidence of noncompliance by member companies with those rules. The association is now intending to increase transparency by introducing a Corporate Social Responsibility reporting system (based on its catalogue of criteria) which will be compulsory for each member company.

The Travel Foundation

The Travel Foundation (TF; www.thetravelfoundation.org.uk) is a non-profit organisation established by major British tourism associations and airlines. In its 'Insider Guide for Managers—Climate Change and Tourism' TF outlines a number of principles and recommendations relevant to tour operators who would like to reduce their energy consumption and thus their contribution to climate change. TF emphasises supply chain management, where its guidelines are particularly strong (relating to 'Sustainable Aviation' principles and providers as well as to local transport), customer information, and internal management. The two latter are more general and less transport-focused. In terms of product development TF does not advocate different travel patterns (such as using public ground transport or longer stays at destinations). Instead it urges companies to use carbon offsetting. The three major British outbound tour operator associations (who are also TF members) have recently established a special compensation scheme, the Tourism Industry Carbon Offset Service (TICOS; www.ticos.co.uk).

It is not known to what degree individual British tour operators actually apply TF's guidelines or use TICOS to compensate their greenhouse gas emissions. However, in the United Kingdom climate change and aviation's and tourism's contribution to it is an issue which is probably more controversially discussed than in any other country, thus creating both political and public pressure on the tourism industry to reduce their climate footprint (see Eijgelaar 2007).

Travelife[4]

Travelife is a common sustainability management system for European outbound tour operators which is presently being developed by the Dutch ANVR, the British FTO, the Belgium ABTO, and *forum anders reisen* of Germany, all of whom are predominantly outbound tour operator associations. The project is being funded by the European Union. According to the project manager, all members of those associations will have to implement this management system in the future. The system originally focused on supply chain management, but has been expanded to all other fields of business management, including product development and customer relations. An essential part of the system is the evaluation of suppliers through a common online tool based on social and environmental sustainability criteria. Standards for transportation are under development. Noteworthy are also *Travelife*'s Best Practice Standards for local transport, for example on bus driving and boating. In relation to product development *Travelife* suggests a shift to more environmentally friendly modes of transport and the adoption of some of the FAR criteria. Carbon-offsetting is expected from suppliers, and customers need to be informed about it.

Outbound and Inbound Operators

Green Globe (www.greenglobe.com)

Green Globe, one of the first international certification schemes for sustainable tourism, is exemplary in that it has put the reduction of greenhouse gas emissions at the top of its agenda according to information provided on their website. Apart from tour operators, the programme also certifies transportation companies, but so far has only done so to a very limited degree. Green Globe requires certified companies to reduce their energy use as much as possible, mostly by technological and operational measures, including renewable energy sources. Second, carbon offsetting is recommended, with an emphasis on sequestration through tree planting. There is no mentioning of more climate-friendly travel patterns and products.

Sustainable Tourism Eco-Certification Programme (STEP)

Sustainable Travel International's STEP programme (www.sustainabletravelinternational.org)[5] is similar to Green Globe in that it offers certification to a variety of companies including outbound and inbound tour operators and transportation service providers. The programme also explicitly lists global warming and climate change and the associated minimisation of greenhouse gas emissions, specifically in relation to transportation, as one of eleven fields of action. There is a strong emphasis on carbon-offsetting, which is being offered through STI itself, as well as on the use of environmentally friendly technology and logistics for local transportation. The programme's second emphasis is on companies' internal management (including staff's commuting and avoiding business trips through tele-conferencing, among others). There is no information available on how many tour operators have been STEP-certified, but several of STI's members have signed up for compensating their emissions.

Inbound Operators

Nature's Best (NB; www.naturensbasta.se)

Nature's Best is a joint project by the Swedish Ecotourism Association, the Swedish Travel & Tourism Council, and the Swedish Conservation Society. It is a formal certification scheme for inbound/domestic and (mostly) local nature tour operators. Nature's Best has very strong sustainable transport principles for the local level and, to a more limited degree, even for the national level (shift to environmentally friendly modes of transport, mostly). According to Nature's Best, 'alternative modes of transport that saves energy, provides less emissions and noise, are sought after, encouraged and preferred.' And:

> The transportation to the destination in itself is a polluting factor, which increases with the distance covered. The tour operator tries to minimize this in different ways. In part through using transport alternatives using the best available technology, and in part by making it possible for the customer to choose means of transportation with less environmental impact.

Criteria for local transport are differentiated and include shuttles to meet train arrivals (to be actively communicated to the customer), the use of environmentally friendly technology and optimised logistics to reduce energy consumption. Motorised leisure activities are excluded for NB-certified products: 'The basic rule is that motorized vehicles can be used for transfers to and from the attraction, but should not be the attraction itself.' There is a bonus for companies who offer products without any motorised transport.

Ecotourism Australia's EcoCertification Programme

Similar to Nature's Best, Ecotourism Australia (EA; www.ecotourism. org.au) is an association that includes local and inbound tour operators and tour guides. The organisation's certification programme, which is one of the oldest in the field of sustainable tourism, has a highly differentiated section on local transport, but is much weaker with regard to transportation at the national level. Apart from fuel efficiency, local traffic impacts (noise, air pollution, physical damage, disturbance of wildlife) are a major area of concern for EA. There are specific guidelines for land-based vehicles and for power boats relating to environmentally friendly technologies, logistics, and driving techniques, which are also to be applied to suppliers. Customers should be encouraged to do the same and to minimise motorised transport, but energy-intensive product components (such as scenic flights) are not excluded from certification. Ecotourism Australia is presently in the process of having its sustainability criteria amended to include broader sustainable transportation principles (Pahl, personal communication, March 2008).

Sustainable Tourism Certification of the Americas

This network of tourism certification programmes in Latin America has been derived from Costa Rica's pioneering Certification for Sustainable Tourism (CST) programme (www.rainforest-alliance.org/tourism). Although these schemes are centered on accommodation businesses, CST has recently been applied to inbound tour operators. However, although energy saving and the use of renewable energy sources are generally being advocated, there are very few criteria explicitly relating to climate change and sustainable transportation. These are limited to requiring certified businesses to use fuel-efficient vehicles and to make sure that legal emission requirements are being met (including by suppliers). There is no mentioning of climate-friendly product design.

Smart Voyager, the certification scheme for boat operators in the Galapagos Islands and a member of the certification network, has some very specific criteria to minimise the local impact of boat operations (e.g., the use of four-stroke external engines) and to increase fuel efficiency. Again, there is a provision requiring certified companies to comply with national and international emission standards.

'Soft Mobility' in Europe

Sustainable tourism transportation has been at the center of several initiatives in Central Europe, including the 'Alpine Pearls' (a network of vacation resorts—sometimes car-free—and regions in the Alps advocating the use of public transit; see www.alpine-pearls.com) and the German *Reiselust* (www.vcd.org)[6] projects. These projects have promoted the use of public transportation at the international, national and local levels by creating an uninterrupted 'travel chain' from tourists' homes to their accommodation, thus connecting the source country with the destination. In addition, a local and regional public transit system with a focus on reaching typical tourist attractions is being developed to guarantee tourists' mobility during their stay. The prime target groups of these projects have been destination management organisations (as the suppliers) and individual travelers, but the former have also acted as incoming tour operators by offering full packages including public transportation.

Conclusion

The original hypothesis according to which sustainable transportation principles do not yet play an important role in sustainable tourism concepts could be affirmed. Although several certification programmes and guidelines for tour operators already include soft mobility elements, none of them offers a comprehensive set of criteria, and some aspects miss entirely, especially in relation to product development. There appears to be a certain tendency to shift responsibility for climate-friendly travel behaviour away from the operator to its suppliers and customers. Furthermore, none of the schemes for outbound operators is binding; in fact, most of them are mere recommendations (*forum anders reisen* is partially an exception here), and there seems to be relatively little concern for this issue outside of Europe. Clearly, in relation to transportation, even the sustainable tourism community has been taken by surprise by the current climate change controversy. This includes carbon-offsetting which is still in its infancy, but appears to play an increasingly important role in the schemes analysed here.

In general, there is a stronger focus on local transportation rather than on the national and international levels where the biggest amount of energy consumption and greenhouse gas emissions occurs. Those schemes that have a section on local transportation are usually more concerned about the local impact of emissions. This is also in line with some schemes' emphasis on internal management, often concentrating on 'office ecology'. Nevertheless, on the local level sustainable transportation principles are more developed, more measurable and more often subject to a formal assessment process than on the international level and for outbound tour operators.

TOUR OPERATORS' PERCEPTIONS, ATTITUDES, AND ACTIONS REGARDING SUSTAINABLE TRANSPORTATION

The third step of the research project consisted in exploring present perceptions, attitudes, and actions concerning global warming and sustainable transport strategies among nature-based tour operators. As a complementary to the previous analysis which was mostly based on websites of formal schemes and guidelines, this part was about individual companies and included an in-depth survey. Thus, it allowed an understanding of tour operators' actual business management practice in relation to global warming and transportation issues. Furthermore, this analysis specifically examined a selected segment of the overall tourism market—nature-based or adventure tourism.

The surveyed companies belong to three organisations which advocate sustainable tourism in this segment to varying degrees, but are usually not formally certified. These were The International Ecotourism Society (TIES), Sustainable Travel International (STI), and the Adventure Travel Trade Association (ATTA). Sustainability is a primary goal of TIES and STI and is more generally advocated by the ATTA. Membership in any of the three organisations does not require any formal statements of adherence to sustainability principles, but one may expect that members, especially of TIES and STI, at least support those principles and the philosophy behind them.

All three organisations are based in the United States and their members include mostly North American outbound tour operators and incoming operators in developing countries, many of them in Latin America. Therefore, this part of the research also introduced a different geographic focus compared to the previous one which was incidentally centered on Europe (because of the concentration of sustainable tourism certification schemes and guidelines there). European members are underrepresented in all three organisations, which may be due to the fact that (Western) Europe has its own tradition of sustainable tourism organisations and that Europe as a destination is more culture- than nature-oriented.

The analysis was implemented in two steps during the first quarter of 2007:

(a) a website analysis of all outbound and inbound tour operators (N = 248) who are members of TIES, STI, and the ATTA
(b) an in-depth online survey of those same operators as well as of those who are members of national ecotourism organisations associated with TIES based on self selection after having contacted all companies by e-mail with the support of the three organisations (N = 67).

The results of the study can be summarised as follows (Driscoll, Mansfield, and Strasdas 2007):

- About 15 per cent of all 248 tour operators mentioned climate change mitigation or sustainable transportation activities on their websites.

Considering the fact that this a sample of companies who are members of organisations committed to sustainable travel this figure is low. Furthermore, the related information showed different degrees of precision and was presented at different levels of companies' websites, not necessarily at prominent places. This confirms that sustainable transport has not really found its way into ecotourism practice yet.

- However, the survey shows that there is a very high level of awareness of climate change. A clear majority of those surveyed see tourism as both affected by, and contributing to global warming. In their view, the latter statement includes nature-based tourism, although to a lesser degree than tourism in general. It has to be kept in mind, though, that participation in the survey was based on self-selection, therefore this is not representative, but rather the prevailing opinion of a small elite of operators. Furthermore, there was sometimes limited knowledge about the exact interrelationship and suitable ways to counteract global warming, especially in developing countries.

- About two thirds of those surveyed have already started to take action to mitigate their impact. Two patterns can be observed here: Outbound operators usually favor emissions compensation of the flights they offer as part of their packages. In most cases they ask their customers to do so, but several have gone one or two steps further by including their internal management into the process (including the compensation of business trips) and by partially or entirely paying for offsetting the emissions caused by the trips booked by their customers. This result shows that outbound operators tend to assume more of a global responsibility for the trips they offer.

- Inbound operators are more concerned about the impacts of local transportation, including its immediate local impacts (such as air pollution). Technical solutions are usually preferred (increased energy efficiency, use of renewable fuels), but in some cases efforts were reported to minimise motorised local transportation altogether. This focus of concern makes sense in that it relates to the immediate sphere of influence of those operators. However, neither outbound nor inbound tour operators seem to feel responsible for the sustainability of the national transportation level (frequent domestic flights while in the destination country).

- When asked about future activities it becomes clear that the present state of mitigation is only the beginning. Many operators want to do even more if they can. This includes sustainable supply chain management (working with climate-friendly carriers and transport companies) and a stronger integration of carbon offsetting into the booking process and even the entire company's operations. Educating the consumer also ranged high on many operators' agendas.

- The absence of climate-friendly travel patterns, a substantial modal shift to public transport and adapted product development (longer stays, less distances covered on a given trip) is conspicuous, even where future plans are concerned. Although those measures are very effective ways to reduce tourism's energy consumption per day of travel, they are hardly considered by anyone. This may be due to customers' time constraints and the nonavailability of suitable public transit systems in most countries of the world.

- Most of those surveyed mentioned one or several barriers that keep them from implementing far-reaching mitigation measures. The most frequently cited concern is related to increasing costs, which consumers may not be willing to bear, along with a perceived unwillingness of the latter to adapt their travel behaviour accordingly. Other barriers mentioned were a lack of suitable suppliers, a nonsupportive political framework, and in some cases lack of access to alternative technologies and know-how.

CONCLUSION

The study shows that there is not yet a broad discussion about climate change mitigation strategies in the nature-based tourism industry, at least not at the time when the research was carried out in early 2007. Ecotourism may have lost its cutting edge position in relation to climate change mitigation which has also become a major issue within mainstream tourism—and perhaps even more so as the activities of the UNWTO and ongoing discussions at the ITB Berlin 2008 have shown.[7] The website analysis is more telling in this respect than the survey, which most likely represents a nonresponse bias.

However, the level of awareness seems to be high, and a number of operators have begun to take action. It can be expected that the nature tourism sector will be very open to implementing more encompassing sustainable transportation principles in the near future, especially in view of continuous scientific and media reports which point at tourism's vulnerability and responsibility and also at consumers' awareness of this. However, there are barriers to voluntary mitigation which need to be taken seriously, particularly for small companies who may be less capable of coping with increased costs. Apart from knowledge and technology transfer to tour operators, educating the consumer will be crucial to gain support and marketability of adapted travel products and full emissions compensation.

SUSTAINABLE TRANSPORTATION GUIDELINES

The results of the two studies presented in the previous sections show that encompassing sustainable transportation principles have only partially found their way into sustainable tourism practice of tour operators. Especially for national and international transportation there is a need for action, whereas local transport has received more attention in some certification programmes and regional projects. The tour operator survey shows that there are practical barriers to sustainable transportation, especially in developing countries and outside of Europe (lack of suitable suppliers, lack of adequate public transit systems, etc.) which need to be taken into account by the Sustainable Transportation Guidelines to be developed so as not to overtax the willingness and capability of tour operators to voluntarily adhere to those guidelines. Nevertheless Sustainable Transportation Guidelines do not need to be created 'from scratch' since many elements have already been covered by different schemes (see Table 14.1), although mostly in the form of nonbinding recommendations rather than formal certification programmes.

Based on these considerations, it is suggested to set up an overall structure that provides:

(a) a process-oriented framework,
(b) a set of performance-oriented criteria and indicators, which may be divided into core (= compulsory) and advanced (= optional) criteria.

As pointed out earlier, it appears to be most appropriate to structure the framework according to the five business management fields of tour operators as laid out by the TOI Sustainability Reporting Guidelines (see Section 1). Second, because of their different greenhouse effect, there should be a differentiation between the international, national, and local transport levels. Putting these two structural elements together results in the following framework (see Table 14.1).

Table 14.1 Conceptual Framework for Sustainable Transportation Guidelines

Spatial level → *Business management field*	*1. International Transportation*	*2. National Transportation*	*3. Local Transportation*
Product development and management	*forum anders reisen*	Alpine Pearls, *Reiselust*	Nature's Best, Ecotourism Australia
Supply chain management	Travel Foundation	Travelife	Nature's Best, Ecotourism Australia, Travelife
Customer relations management	*forum anders reisen*	Alpine Pearls, *Reiselust*	Nature's Best, Ecotourism Australia
Cooperation with destinations		Alpine Pearls, *Reiselust*	
Internal management			STEP, Travel Foundation

Note: Colors measure the relative importance of each field of action (dark gray = very important, light gray = important, white = less important). Names indicate certification schemes with some exemplary criteria in the respective field.

The relative importance of each field of action varies as already discussed earlier. Concerning climate protection, taking action in relation to international transportation must have the highest priority, especially when designing and marketing products (customer relations). However, supply chain management is a less effective tool in this respect since most international airlines already have quite similar standards concerning fuel efficiency and emissions. National and local transportation have been assigned the same importance, although the former is more significant for its greenhouse effect. However, local transportation gains similar importance in the context of nature-based tourism because of its potential impacts on fragile local ecosystems.

When comparing the different fields of business management, product development, and customer relations should have the highest priority. This is the level where substantial savings of greenhouse gas emissions can be achieved, for example by designing and marketing trips with increased lengths of stay. Supply chain management is also an important field, especially when working with national and local transport companies which can have significant differences in terms of energy efficiency and emissions. When implementing the Guidelines through a formal certification scheme, a consideration may be to assign different values to criteria in each of these fields.

A tentative compilation of criteria for the Sustainable Transportation Guideline are to be found in the annex.[8] More stakeholder input is needed to assess the feasibility of the criteria in different environments. For instance, tour products based on public transportation are much easier to implement in Europe than in North America or in many developing countries where such infrastructure is deficient or even nonexistent. Also, in order to be integrated into existing certification schemes, the Guidelines may have to be modified.

A further development of the Guidelines could consist of a methodology to calculate the amount of greenhouse gas emissions per day of travel, following the example of the University of Breda's 'Holiday Footprint'. Similar to refrigerators and other electric devices, energy consumption categories could be established, perhaps differentiated for domestic, intra-continental and intercontinental trips. This may be a powerful instrument to encourage customers to choose more climate-friendly products and/or to compensate their emissions accordingly.

It is important to note that implementing climate protection strategies may have additional benefits not relating to climate change mitigation to both operators and consumers. For example, the minimisation or better coordination of transport will reduce operating costs and prices. In developing countries and peripheral areas (which are becoming increasingly popular destinations for adventure and nature tourists) motorised transport also leads to a high degree of leakage from local or national economies since vehicles and fossil fuels usually have to be imported from abroad. By contrast, nonmotorised forms of local transport may often result in more benefits to local people offering guiding services or providing traditional forms of transport (including pack animals). Finally, such forms of transport and more quality time spent at a destination also enhance the travelers' experience of an adventure or nature trip.

Advocating voluntary Sustainable Transportation Guidelines does not mean that the reduction of tourism's impact on the world climate should be left to voluntary measures alone. On the contrary, since enormous overall reductions of greenhouse gas emissions will be necessary to keep global warming within manageable limits, government intervention, and binding regulations will be inevitable. However, due to the difficulties to reach global intergovernmental agreements on effective climate protection within a reasonable timeframe, the implementation of voluntary changes by environmentally conscious businesses and consumers may be an important strategy to either precede and/or supplement government regulation (see Strasdas 2007 for an in-depth discussion of this issue).

Last but not least, it should be noted that while it would certainly be a relief to the world climate to prefer short-haul over long-haul destinations, a widespread implementation of such a strategy would have devastating economic consequences for typical long-haul tourist destinations with a small or nonexistent domestic market or with few other (sustainable) economic alternatives. Therefore, it is not the intention of these guidelines to generally 'penalise' operators offering trips to long-haul destinations as long as the Sustainable Transportation Guidelines are taken into account. Fewer trips, but with extended stays, in such destinations would not hurt them economically, and carbon-offsetting, if implemented in those destinations, may even provide additional benefits.

PROPOSED SET OF CRITERIA[9]

Product Development and Management

International Transportation

- If feasible and in line with your profile, diversify your range of travel products by increasing the percentage of domestic trips and to nearby destinations as well as to destinations that can be reached through ground or maritime transportation. Offer a minimum number of such trips.
- Maintain a reasonable ratio between trip length and the distance of the destination. Offer a minimum number of trips following this ratio (perhaps following the provisions required by *forum anders reisen*).

Advanced operators will apply this rule to all of their trips. Any trips not following this rule must include compulsory emission compensation.

- The time–distance ratio should be reflected in the amount of energy consumed and/or of greenhouse gases produced per day of travel (see discussion earlier on). Alternatively, a steady reduction may be required here.
- If direct flights to the destination are available, these should be preferred over nondirect routes involving stopovers.
- Use public transportation (trains, buses, ferries) or chartered buses/trains to get to a destination wherever reasonably possible and environmentally sustainable. Offer a minimum number of trips following this rule, if possible.
- Organise such 'slow' trips in a way that they become part of the experience of travelling, for example by including stopovers at attractive sites or by providing entertainment/interpretation during the journey.
- Do not offer short-haul trips by plane unless this is the only way to access a destination. Explanation is needed why this is not otherwise possible. If so, such trips must include compulsory emission compensation.
- If reasonably possible, include public transportation to the international departure airport as part of the package. Do not offer domestic connecting flights to the international airport, if these can be avoided. Explanation is needed if this is not possible. If so, such flights must include compulsory emission compensation, if booked through the tour operator.
- Packages should include a complete travel chain management from the customer's home (e.g., pick-up service) all the way through to the destination.
- Each trip must include a calculation of greenhouse gas emissions and state the compensation amount due by using an appropriate, high-quality carbon calculator. Payment may be left to customers, but advanced operators include compensation costs into the package price.
- The tour operator should partner with a distinguished carbon offset provider offering high-quality offsets. Alternatively, a tour operator may implement its own compensation project provided it follows the same strict rules.

National Transportation

For domestic tourism (especially in big countries), the criteria just listed are to be applied accordingly, including carbon-offsetting. In addition, the following rules should apply:

- Offer a minimum number of trips following this rule: reduce the number of sites visited during a multiple-destination trip with the aim to spend

more quality time at each site. Offer a minimum number of multiple-destination trips spending quality time at a reduced number of sites.
- Once at the destination, strive to minimise the amount of energy consumed and of greenhouse gases produced per day of travel.

Local Transportation

- Minimise local transportation wherever possible. Trip duration should be in reasonable proportion to the attraction of the site visited and the length of stay there.
- If available and convenient, prefer public transportation or chartered buses/ trains over individual motorised traffic.
- If visiting highly frequented sites, prefer those with a good public transport system or those striving to develop such systems.
- Make nonmotorised modes of transportation and physical activities (bicycles, canoes, horseback riding, walking safaris, etc.) part of the experience, provided that customers are physically able to comfortably participate in those activities.
- Exclude energy-intensive motorised leisure activities, such as helicopter skiing, scenic flights, power boating, snow scooters. Exceptions are possible where such modes of transportation are needed for transfer.
- Organise programmes and local transfers in a way that makes it possible for guests to arrive by means of public transportation (if available) and be met by the local operator at the time of arrival.

Supply Chain Management

International Transportation

- Preferably use airlines that are certified for their environmental management or are otherwise climate-friendly. This includes certification through ISO 14.001 or EMAS II or airlines that are fully or partially carbon-neutral by offsetting their emissions through high-quality compensation projects. Other criteria may be that those airlines support binding regulation or substantially invest into energy efficiency.[10]

National Transportation

For domestic tourism (especially in big countries) and for travel within the destination the previous criterion is to be applied accordingly.

- Depending on local conditions and companies' standards, fuel-efficient domestic airlines may in some cases be more environmentally sustainable than also existing ground transportation. Operators preferring air transport over ground transport even though the latter would be a viable option must show that this is, in fact, the more climate-friendly alternative.

- Preferably work with railroad, bus/coach, or car rental companies that are certified for their environmental management or are otherwise climate-friendly, for example by using renewable fuel sources or hybrid vehicles.

Local Transportation

- If possible and convenient, use local services providers offering traditional nonmotorised means of transportation, such as pack animals, canoes, rickshaws, or porters.
- Preferably work with bus/coach companies that are certified for their environmental management (for example, optimisation of logistics, increased seat occupancy, trained drivers) or can show that they have such a system in place.
- Preferably work with motorboat companies that are certified for their environmental management or otherwise strive to maximise fuel-efficiency and reduce emissions (for example, by using four-stroke engines).[11]
- Preferably work with hotels and restaurants that save energy and sell local (preferably certified) products.
- Generally work with suppliers that buy locally, thus reducing the amount of energy needed for the transport of the goods purchased.

Customer Relations

International Transportation

- Generally educate customers about tourism's (especially aviation's) contribution to global warming.
- Calculate the 'climate footprint' (amount of greenhouse gas emissions) of each trip and per trip day and include it in the travel description.
- Specially promote climate-friendly products (see first section of chapter) over others, for example in catalogs, on websites, and in newsletters.
- Calculate and display costs per trip day to show the relative decrease of travel costs for trips with extended lengths of stay.
- Explain carbon-offsetting to customers, ask them to make their contribution and integrate it into the booking process (for example in the form of an opt-in or opt-out box).

National Transportation

For domestic tourism (especially in big countries) and for travel within the destination the criteria listed above are to be applied accordingly, especially in relation to multiple-destination trips, where longer quality-time experiences should be emphasised.

Local Transportation

- Educate customers about their local climate footprint in relation to different means of local transportation.
- Encourage physically able customers to use traditional nonmotorised and muscle-powered forms of local transport.

Cooperation with Destination[12]

General

- Support destinations, especially long-haul destinations in developing countries, by channeling compensation payments from international flights into carbon-offset projects located in those countries. Those projects should have additional benefits such as community development or biodiversity conservation.
- Support sites visited during a trip (for example a forest area rich in biodiversity) through compensation payments.
- Support local communities through technology transfer and/or compensation payments into innovative technologies (renewable and energy efficiency) to be used by those communities.

Internal Management

General

- Make climate protection an integral element of the company's sustainability goals and environmental management system. Emissions reduction should have priority over emissions compensation.
- Have an environmental manager who is in charge of, and familiar with, climate change issues.

International Transportation

- Minimise international business travel through video/phone conferences and by delegating as much as possible to partners (inbound operators) in the destinations.
- Offset inevitable emissions caused by business trips.

National Transportation

- Minimise national business travel through video/phone conferences.
- Use public transportation for business trips wherever possible and convenient.
- Offset inevitable emissions, at least from air travel.

Local Transportation

- Encourage and reward employees to travel to work on foot, by bicycle, or by using public transit.
- Use fuel-efficient or alternative energy vehicles for business purposes.

NOTES

1. Please note that the 5 per cent figure used here does not include non-CO_2 emissions.
2. Defined as sustainable nature-based tourism 'that conserves the environment and sustains the well-being of local people' (The International Ecotourism Society, www.ecotourism.org). In this chapter the term *nature tourism* or *nature-based tourism* is used in a more factual sense in that it simply describes trips taken to natural areas, whereas ecotourism is seen as a concept aspiring to achieve sustainable impacts which in reality may or may not be the case.
3. In German, *anders reisen* means 'travel differently'.
4. Based on information provided by Naut Kusters (personal communication April 2007 and www.travelife.eu).
5. STI is a small nonprofit organisation based in the United States whose constituency are mostly nature-based tourism companies and organisations.
6. In German, *Reiselust* means 'joy of travelling'.
7. University of Eberswalde/MODUL University Vienna 2008.
8. Comments on this compilation were received (in alphabetical order) by: Susanne Becken (Lincoln University, New Zealand), Megan Epler Wood (Native Energy Travel Offsets, USA), Stefan Gössling (Lund University, Sweden), Naut Kusters (ECEAT/Travelife, Netherlands), Paul Peeters (NHTV Centre for Sustainable Tourism and Transport, Breda/Netherlands), Rolf Pfeifer (forum anders reisen, Germany), Jan Wigsten (TIES director, Nomadic Journeys, Sweden/Mongolia).
9. Please note that the following list of criteria is not complete. It needs more specifications of several aspects and a broader discussion of potentially controversial requirements.
10. An excellent reference for this is are the guidelines of 'Sustainable Aviation', an initiative of the British aviation industry (www.sustainableaviation.co.uk).
11. Excellent criteria for this can be found in 'Smart Voyager', the certification programme for the Galápagos and Ecotourism Australia's EcoCertification programme.
12. Criteria in this section are optional. Their compliance reflects an advanced state of sustainability.

REFERENCES

Driscoll, L., Mansfield, C., and W. Strasdas (2007) '*Nature Tour Operators' Attitudes and Actions Concerning Travel Related Greenhouse Gas Emissions*. Website analysis and survey of members of The International Ecotourism Society, Sustainable Travel International and the Adventure Travel Trade Association'. Stanford University/USA, May 2007.

Eijgelaar, E. (2007, October) *Voluntary Carbon-offset Schemes and Tourism Emissions—Assessment of Mitigation Potential Through Analysis of Online*

Provider Communication on and Review of Public and Industry Attitudes to Climate Change and Air Travel. Master thesis, University of Eberswalde.

Forschungsgemeinschaft Urlaub und Reisen (F.U.R.) (2000) *German Travel Analysis*. Kiel: Forschungsgemeinschaft Urlaub und Reisen.

Forschungsgemeinschaft Urlaub und Reisen (F.U.R.) (2007) *German Travel Analysis*. Kiel: Forschungsgemeinschaft Urlaub und Reisen. Available at: www.reise-analyse.de (12 September 2008).

Gössling, S. (2000) 'Sustainable tourism development in developing countries: Some aspects of energy-use'. *Journal of Sustainable Tourism* 8(5): 410–425.

Gössling, S., Broderick, J., Upham, P., Ceron, J.P., Dubois, G., Peeters, P., and Strasdas, W. (2007) 'Voluntary carbon offsetting schemes for aviation—Efficiency and credibility'. *Journal of Sustainable Tourism* 15(3): 223–248.

Intergovernmental Panel on Climate Change (IPCC) (2007, February 2) *Climate Change 2007—The Physical Science Basis*. Working Group I Report of the IPCC Fourth Assessment Report. Paris.

Lund-Durlacher, D., Strasdas, W., and Seltmann, R. (2007) *Climate Change and Tourism—Perception of Problems and Solutions Proposed by Tourism Professionals*. Results of a survey among exhibitors at the ITB Berlin 2007 (*in German*): oikos Verlag, FH Eberswalde. Available at: www.fh-eberswalde.de/tour (30 May 2008).

Simmons, D. and Becken, S. (2004) 'The cost of getting there—impacts of travel to ecotourism destinations'. In Buckley, R. (ed.) *Environmental Impacts of Ecotourism*, 15–23. CABI Publishing.

Strasdas, W. (2007) 'Voluntary offsetting of flight emissions—an effective way to mitigate the environmental impacts of long-haul tourism?' (*in German*). In Egger, R., Herdin, T. (ed.) *Tourismus—Herausforderung Zukunft*, Wien/Berlin: Lit-Verlag (English translation available at: www.fh-eberswalde.de/tour) (30 May 2008).

The International Ecotourism Society (TIES) (2007) *Ecotourism and Certification*. Available at: www.ecotourism.org (30 May 2008).

Umweltbundesamt/Öko-Institut (2002) *Tourism and the Environment—Data, Facts, Perspectives* (in German). Berlin: Erich Schmidt Verlag.

University of Eberswalde / MODUL University Vienna (2008, March) *Climate Change—Responsibility is only hesitantly Assumed by the tourism Industry* (in German). Press release on the preliminary results of a survey among ITB exhibitors, Eberswalde, 9.

UNWTO-UNEP-WMO (*United Nations World Tourism Organization, United Nations Environment Programme, World Meteorological Organization*) (2008) 'Climate change and tourism: Responding to global challenges'. [Scott, D., Amelung, B., Becken, S., Ceron, J.-P., Dubois, G., Gössling, S., Peeters, P., and Simpson, M.] *United Nations World Tourism Organization (UNWTO), United Nations Environmental Programme (UNDP) and World Meteorological Organization (WMO)*. UNWTO, Madrid, Spain.

Verkehrsclub Deutschland (VCD) (2006) *Reducing the Environmental Effects of Air traffic* (in German). Berlin: VCD Fakten.

Zeppenfeld, R. (2004) *Automobile-Free Holidays—Product and marketing concept Developed for the VCD Reiselust Project* (in German). Master Thesis, University of Eberswalde.

15 Tourism Firm Innovation and Sustainability

C. Michael Hall

Innovation is an activity that occurs at various scales of organisation ranging through various scales of governance from the global to the national and down to the regional and local, but also being enacted by firms and nongovernment public interest organisations with various levels of 'reach'. As noted in Chapter 1, the very concept of sustainability itself can be regarded as an innovation in terms of thinking with respect to environmental stewardship and the development of an understanding of the extent to which ecological, social, environmental, and political issues are interrelated. Rather than focus on the changing ideas of sustainability at a meta-theoretical level, this chapter focuses on innovation as an integral component of how organisations and places can respond to the challenges of environmental change in a sustainable fashion. It is divided into three sections. The first briefly discusses the surprising lack of interplay between the sustainability and innovation literatures. The second notes the potential significance of an innovation system's approach to identifying the ways in which industries, regions, and countries seek to develop sustainable tourism. Third, the chapter utilises a model of the tourist firm to illustrate responses to different innovation strategies for sustainability at an organisational level.

BRINGING SUSTAINABILITY AND INNOVATION TOGETHER

'Innovation' has been at core of ideas of competition, firms and entrepreneurship since the pioneering work of Schumpeter (1934, 1942). However, like sustainability, there is no single accepted definition of the concept of innovation. To some users it implies newness (Nowotny, Gibbons, and Scott 2001) while others it suggests the entire process by which a firm adapts and responds to its environment. The latter is well illustrated in Kanter's (1983, 20–21) definition of innovation.

> Innovation refers to the process of bringing any new, problem solving idea into use. Ideas for reorganizing, cutting costs, putting in new

budgetary systems, improving communication or assembling products in teams are also innovations. Innovation is the generation, acceptance and implementation of new ideas, processes, products or services. . . . Acceptance and implementation are central to this definition; it involves the capacity to change and adapt.

Clearly, a critical issue is what is actually new particularly when innovation can take a wide variety of forms. There, an innovation needs to be understood in terms of what Hall and Williams (2008) describe as its 'focus', that is the form of the innovation, as well as its 'impact range', meaning that for something to be considered an innovation it does not have to be considered new at a world or national level, only in a particular market segment or regional/destination context (Sundbo 1998).

The dimensions of focus and range are combined in Abernathy and Clark's (1988) widely cited classification of innovation as:

- *niche* (opening new market opportunities via the use of existing technologies)
- *regular* (incremental)
- *revolutionary* (involving significant new technologies but whose impact is not industry wide)
- *architectural* (which can change the entire industry)

Hjalager (2002) noted the potential utility of this model for understanding tourism innovation, although there has been little follow-up to this useful observation. For example, such categories can be usefully applied with understanding the impact of new information and transport technologies in tourism with respect to such things as capacity for e-bookings (an *architectural* change which has established an entirely new distribution channel for tourism); download of interpretive material on mobile phones (a *revolutionary* tool for interpretation in some natural and cultural heritage settings); development and maintenance of new customer and business networks via e-mail (a *regular* incremental innovation); and the extension of submarine technology to tourist marine dives (a *niche* innovation using long-standing technology). Nevertheless, a ready criticism of such an approach, as well as of much innovation research in general, is that it is often focussed on manufacturing and the innovation of physical products rather than being readily extend to service industries, such as tourism, which have different characteristics in terms of products and the processes by which they are developed.

Hall and Williams (2008) drawing particularly on Sirilli and Evangelista (1998), Tether (2004), and Van der Aa and Elfring (2002), identified four distinctive features of service innovation: the coterminality of service production and consumption; information-intensity; the importance of the human factor; and the critical role of organisational factors, that all apply, to some degree, to tourism.

Coterminality implies the close interaction between consumers and producers of services to the extent that service products are often described as 'cocreated' or 'coproduced'. This means that the customers are themselves an important source of innovation while the degree of interaction can also imply a substantial blurring between product and process innovation. Information-intensity is significant for services because firms often have to deal with large numbers of separate interactions with their customers. This means that it not only provides an incentive to utilise information technologies but also the means to manage the information and knowledge flows that can result from them. Quality management plays a major role in service forms that is often played out in terms of customer satisfaction. This has meant that quality criteria are often set for various dimensions of producer–customer interaction including not only the appearance and social attributes of staff, but also the physical design of the space in which the service encounter takes place. Finally, organisational innovations are exemplified by multi-unit organisational forms and new combinations of services (Van der Aa and Elfring 2002).

To the characteristics of innovation in service sectors can also be added some distinctive features of tourism (Shaw and Williams 2002; Hall and Page 2006). These were identified by Hall and Williams (2008) as consisting of the clustering of a complex set of complementary activities and experiences; temporality in terms of time specificity; spatiality with respect to the relative spatial fixity of tourist consumption; tourist–tourism industry encounters; and tourist–host community encounters and environmental encounters that delimit the tourism experience.

Such dimensions clearly serve to influence the capacity of tourism firms to innovate but there is surprisingly little literature on tourism innovation in general (see Hall and Williams 2008 for a review) and even less so specifically with respect to the environment (see Hjalager 1996, 1997 as notable exceptions). Such a situation is quite remarkable when considering the extent to which tourist firms in general can be understood as adapting to their business environment and perhaps even more so given the emphasis in studies of sustainable tourism towards mitigation and adaptation of undesirable aspects of tourism development (e.g., Gössling and Hall 2006). Although, arguably the notion of adaptation has only come into popular usage in the tourism sphere following recognition of tourism's role in climate change and the subsequent importation of the discourse of climate change studies into tourism (see Chapter 1). Nevertheless, it should also be noted that there has also been relatively little interplay between the innovation literature and the climate change literature. Primarily this can be regarded as a result of climate change research historically focussing on more system wide adaptations to climate change rather than individual firm response, whereas innovation studies have historically focussed on the firm although more recently there has been much greater attention given to the embeddedness of firms in systems (e.g., Cooke et al. 1997; Carlson et al.

2002; Edquist 2005) including tourism systems (Hall and Williams 2008; Hjalager et al. 2008).

The bringing together of understanding of innovation in terms of both spatial scales of governance and firms and organisations potentially has great implications for trying to develop more sustainable forms of tourism as well as sustainable development overall. This is because while the shortcomings of tourism with respect to sustainability is reasonably well recognised, being able to change firm behaviours in terms of adopting more sustainable business paths is not.

It is possible of course that firms may be intrinsically unable to develop sustainable business forms because of the way in which they are designed to function within a cultural-economic system. This is because with many national economic systems and associated models of management the maximisation of shareholder value has emerged as the dominant goal and metric of corporate performance. Within such a setting shareholder value is operationalised in terms of future cash flows and return on investment. Neoconservative and neoliberal economists justify lack of attention to social issues by arguing that social welfare is maximised when organisations exclusively pursue profits (e.g., Friedman 1970). As a result management theorists argue that in order to maximise social welfare via the pursuit of profits then firms need to focus on a single objective function, which is shareholder wealth creation (e.g., Jensen 2002; Sundaram and Inkpen 2004). Such a perspective stands in contrast to much of the way in which sustainable tourism and sustainable development in general is understood, given its emphasis on equity, collaboration, and understanding the business environment in terms of stakeholders rather than shareholders (e.g., Bramwell and Lane 2000; Weaver 2006). However, it may well help us better understand the reason behind the substantial gap that exists between sustainable tourism theory and practice, at least from the perspective of commercial organisations and the limited progress that has been made with self-regulatory and market driven approaches towards sustainable business development in the tourism industry. In short, research on sustainable tourism, and possibly tourism in general, has an extremely limited understanding of the tourism firm as well as the systems within which it is located (Hall and Williams 2008).

INNOVATIVE TOURISM SYSTEMS

Tourism firms are embedded in various innovation systems. This observation reflects the fact that innovation is 'an intrinsically territorial, localised phenomenon, which is highly dependent on resources which are location specific, linked to specific places and impossible to reproduce elsewhere' (Longhi and Keeble 2000, 27). These include broader national and regional innovation systems as well as sectoral specific systems. Innovation systems are constituted by 'interconnected agents' that interact in influencing the execution of innovation at a particular economic spatial scale. For example, national systems of innovation may be styled as segmented layers of institutions and production modes that integrate national, regional and local ensembles of actors, institutions, and resources that pose particular issues of governance and the role of the state (Hall and Williams 2008). The key characteristics of an innovation system are usually summarised as:

- firms which are part of a network of public and private sector institutions whose activities and interactions, initiate, import, modify, and diffuse knowledge, including new technologies.
- linkages (both formal and informal) between institutions.
- flows of intellectual resources between institutions.
- learning as a key economic resource.
- geography and location matter (Holbrook and Wolfe 2000).

Figure 15.1 illustrates the various components of innovation systems at various scales. A critical observation is that firm innovation along with innovation policies are embedded in a broader socioeconomic context for which national and regional governance structures provide a permeable boundary. Furthermore, local innovation systems, such as that which occur at tourism destinations (Hjalager et al. 2008), along with sectoral systems, such as those of the tourism industry or its constituent sectors, must also be understood as being co-evolutionary with national systems of innovation. Nevertheless, recognition of scale of analysis along with the scale of innovation system is important as it will reflect different institutional settings and innovation needs.

Any reading of sustainable development highlights that some destinations appear better able to engage in capacity building than others and to effectively respond to external change. Destinations can therefore be regarded as a form of regional innovation system, 'the set of economic, political and institutional relationships occurring in a given geographical area which generates a collective learning process leading to the rapid diffusion of knowledge and best practice' (Nauwelaers and Reid 1995, 13). Different destinations will have different regional innovation systems that will vary with respect to:

- the ability of firms and other relevant nonprofit and public organisations to innovate due to their specialisations, as well as their functional and organisational characteristics.
- their propensity to interact depending on the existence of clusters, networks, and the attitude of actors towards cooperation.
- their capacity to construct relevant institutions (for example, in research, education, knowledge transfer) and in their 'governance model', which is dependent on their decision making powers, financial resources, and their policy orientation.

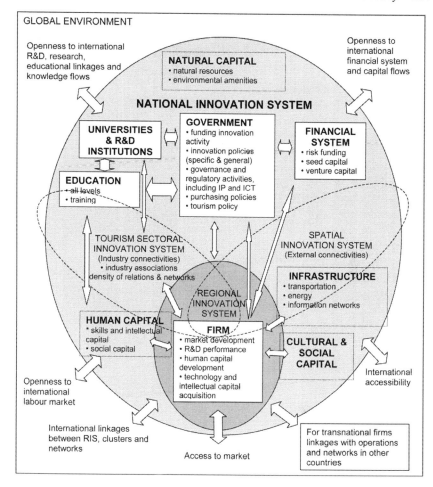

Figure 15.1 Elements of national innovation systems. From Hall and Williams 2008: 113.

As a result of these differences, it can be expected that some destinations will have a weak innovation system while others will have systemic interaction to a much higher degree (Tödtling and Kaufmann 1998). This observation is clearly important in understanding how information with respect to knowledge of sustainable tourism practices can be effectively disseminated and implemented (Hall 2008). Unfortunately, much research on tourism planning and sustainability has failed to recognise the role of blockages in policy and planning systems and that the reasons for nonadoption of sustainable practices lie in the realm of the social construction of sustainability; local cultures of innovation and policy; as well as power relationships, including corruption (Hall and Coles 2008). This is not the prescriptive policy realm of most tourism academic or UNWTO pronouncements but it is the reality of late or nonadoption of innovation, or the lack of effective indigenous innovation.

The issue of course arises as to how can systems be modified in order to encourage innovation and the development of improved adaptation and mitigation measures? The answer, perhaps unfortunately, is slowly. The development of appropriate governance, institutions, and knowledge transfer measures usually takes time. Radical system change, such as those encouraged by a crisis, can sometimes occur but crisis can just as easily provide for continuation of business as usual and a sense of denial. However, one clear finding is that an improved understanding of innovation systems may help locate some of the better points of intervention and leverage, but it will require a much more sophisticated approach to tourism destination systems than what has hitherto been the case which will also require better knowledge of the behaviours of actors and agents within the system with respect to adaptation and innovation. Therefore, the next section provides a model of the tourism firm that may help better understand the potential adaptive behaviour of firms.

THE TOURISM FIRM

Sustainable development, and especially climate change adaptation and mitigation, occurs at a number of different scales of governance and organisation. Tourism firms themselves also have a major role to play in adaptation to climate change as part of systems of innovation. Yet the notion of adaptation as a form of innovation that is understandable in the context of tourism business practice has not been well articulated but it remains an essential component of understanding the capacities of destinations to adapt and respond to the challenges of climate change (Hall 2007). Especially as the capacity of businesses to survive and grow is a measure of their individual resilience as well as a potential indicator of system resilience at a destination level (see also Carpenter et al. 2001).

Much of the focus of adaptation to climatic and environmental change is on technical responses to climate change. However, the development and transfer of innovative technology, for example with respect to improved fuel efficiencies or emissions reductions, is only a small element of what constitutes innovative tourism business practice. Indeed, research on innovation in tourism and similar service firms indicate that there are a range of other measures that firms can adopt to respond to external stimuli and stresses, such as those brought about directly and indirectly by climate change, in order to survive and, ideally, maintain or even increase margins (Hall and Williams 2008).

In order to understand how tourism firms may adapt to an element of sustainable development, such as climate change for example, it is essential to understand the main drivers of business change. These include:

- political drivers, i.e., changed government regulations, for example with respect to emissions
- economic drivers, i.e., impositions of taxes or new costs of inputs such as fuel
- social drivers, i.e., changed motivations of small or family business owners, or changes in consumer demand
- environmental drivers, i.e., changes in the resources on which tourism firms may depend including the quality of the natural environment or even resources such as water

Yet fundamentally these drivers must be considered within the desire of firms to (a) ensure their survival, and (b) make a profit. These are clearly essential elements of firm sustainability and its function as an actor in the ecology of business. This understanding is essential if we are to understand the reasons why businesses from micro-entrepreneurs, such as self-employed tour guides or tourist taxi drivers, through to family businesses, such as a family run hotel, and large corporations, such as an integrated resort company, may seek to operate and therefore innovate in response to challenges such as climate change. Unless tourist firms of all sizes can be persuaded that adaptation will positively influence their survival and returns they will not act in the short term in response to the threats of climate change no matter how concerned they may also be about climate change in the longer term (see Hall 2006). Therefore, in order to understand and identify the capacities of firms to innovate in relation to climate change it is necessary to utilise a model that conveys how value is created in a tourism business.

Value creation can occur in different areas of firm activity (Kim and Mauborgne 1999). Ideally, these elements are mapped by firms in the development of particular products, with the different elements then becoming part of a service-profit chain (SPC) which is a framework for linking service operations to a firm's profitability. A number of variants of SPCs have been produced that emphasise particular dimensions such as service satisfaction or environmental management practices (Kassinis and Soterlou 2003). However, the reality is that many businesses, especially small-scale entrepreneurs, do not map how they create value. Therefore, for the purposes of identifying potential adaptations at the firm level it is possible to utilise a model of how tourism businesses may generate value.

A business model is a conceptual tool that contains a set of elements and their relationships that allows the expression of the business logic of a specific firm (Hall and Williams 2008). It is a description of the value a company offers to one or several segments of customers and of the architecture of the firm and its network of partners for creating, marketing, and delivering this value and relationship capital, to generate profitable and sustainable revenue streams (Osterwalder, Pigneur, and Tucci 2005, 17–18).

Most literature on the conceptual dimensions of business models cover nine 'building blocks' that relate to four different dimensions of business models. The dimensions are 'product/service', 'customer interface', 'infrastructure management', and 'financial aspects'. The building blocks are the 'value proposition', 'target customer', 'distribution channel', 'customer relationship', 'value configuration', 'capability', 'partnership', 'cost structure', and the 'revenue model' (Osterwalder, Pigneur, and Tucci 2005). Several of these elements have been incorporated into Figure 15.2 which presents a business model of the innovation value creation points in the tourism firm (Hall and Williams 2008). The main idea of creating a reference model is that it identifies the domains, concepts, and relationships shared among tourism firms no matter where they are located. Nevertheless, the service aspects of tourism allow for the consideration of a number of common points. Eleven points are identified: Business model (referring to the overall approach), networks and alliances, enabling process, service design and development, service value, distribution, brand, servicescape, customer service experience, customer satisfaction, and customer loyalty although it should be noted that the elements of the model are not causally related. In addition, a further dimension exists which is the way in which all the elements of the business model are put together—what would commonly be described as strategy. Critically for present considerations, each of these points or elements are opportunities for innovation and adaptation at the firm level.

Tourism firms tend to cluster into two groups with distinctly different innovation strategies (Hall and Williams 2008):

- When tourism services are inseparable and difficult to standardise, there exists a fundamental problem of information asymmetry. In such situations, customers will have difficulties, or not be able to evaluate the quality of a potential service supplier before-hand, since alternative offerings are hard to compare. Innovation among firms and organisations in this group often tends to focus on reducing the risk for clients (ECON Analysis 2006). Examples of this type of operations include concerts and events, museums, art galleries, and tours. This type of firm has a strong external focus.

- When services are characterised by a stronger degree of separation and standardisation, innovations are more geared towards process improvements (Econ Analysis 2006). Airline and transport operations and highly standardised forms of accommodation or food provision, for example, are part of chains and franchises. This type of firm has a strong internal focus.

However, each of the value creation points has innovation potential with respect to sustainability and climate change.

Business Model

Different components of the model can be combined in different ways to create new ways of doing business thereby potentially creating new sources of competitive advantage. For example, community owned businesses may be more resilient to change than those of nonlocally owned businesses because their business model is influenced by social as well as economic concerns. One important component of the business model is the selection of business location as this can clearly have implications for the environmental impact of any development both directly, via its immediate site, and indirectly via the environmental costs of the firm's supply chain and the firm's customers.

Networks and Alliances

The development of networks and alliances and associated areas of actor cooperation and trust has been a significant theme in tourism research since the early 1990s. The creation of networks allows for the development of mutual learning and knowledge spillovers among actors as well as for more pragmatic actions with respect to cost sharing and cooperative action. The development of new sets of relations over time and the maintenance of valued existing relationships can provide a means of innovation on an individual and shared level (Medina, Lavado, and Cabrera 2005). For example, networks can share the cost of new energy saving and low-emissions technologies which otherwise may have been prohibitive to members at an individual level. Advanced marketing networks can also share resources to undertake innovative market segmentation and product matching exercises, for example with respect to the promotion of low energy accommodation and holidays, such as using beach huts rather than air-conditioned hotel rooms in tropical coastal and island destinations. Similarly, for many tourist hotels and restaurants in developing countries, ensuring the development of a local food supply chain not only helps increase the resilience of the local economy to global change but can also reduce the energy costs of food as well as provide a more authentic experience to the consumer.

Enabling Process

The enabling process refers to the way in which organisations give support to core business processes and the maintenance and enhancement of human resources. In some tourism firms one of the most basic means of innovation is for managers to listen to staff feedback so as to be able to respond to information and issues. This can be a very important source of new ideas in terms of things such as simple energy saving measures.

Service Design and Development

Service design and development is the process by which new services are developed, designed, and introduced to the market. An important innovation in new service design for externally focussed firms has been the increasing focus on customers as a source of learning for new service development, with customers often being regarded as more innovative than professional service developers (Matthing, Sandén, and Edvardsson 2004). For internally focussed firms, such as the transport and mass accommodation sector, physical design change may be more significant with respect to lowering emissions although changes in processes, such as encouraging greater reuse and recycling, can also have significant energy implications and cost savings.

Service Value

Service value refers to the value that customers gain from products both with respect to the basic product as well as value-added services which is where innovations tend to be made. Importantly, value-adding does not necessarily mean increasing physical consumption, rather value-adding will often come from innovation in terms of experiences. For example, the authenticity of many accommodation products developed as part of culturally and environmentally friendly tourism initiatives comes from their relative simplicity.

Distribution

Distribution is the means by which product offerings get to the market and the customer. Often the innovation focus is on the distribution channel, the chain of intermediaries, between the producer and the final consumers. For most tourism businesses this aspect refers to flows of marketing information as well as booking. E-business, for example, has meant the development of significant innovations in the flow of product information and capacity to purchase and therefore potentially allow for direct sales to customers and greater returns. In some cases innovative 'alternative' distribution strategies may actually signal a return to 'old fashioned' distribution channels. For example, many small agricultural producers may sell their products directly to visitors at a market or direct to small hotels, rather than through mainstream retail channels that often have much more energy costs with respect to transport.

INCREASING CUSTOMER FOCUS →

Innovation points in value chain	Business model	Networks and alliances	Enabling process	Service design and development	Service value	Distribution	Brand	Servicescape	Customer's service experience	Customer satisfaction	Customer loyalty
Intellectual capital	How the firm makes its money and how it is structured	Economic and knowledge relationships with other firms and actors	Support firm's core processes and employees	Service development architecture and systems. Matching of service to market segments	Provision of value to customers and consumers with respect to products	How the product offerings reach the market	The symbolic embodiment of all the information connected to the product, values and expectations	The physical evidence of service and manifestation of brand	The experiences that customers have at point of co-creation / production	Levels of customer satisfaction with experience over time	Long term relationships with customers
Innovative capacities and potential	New business models, including ownership structure, payment and cash flow system, and location	New sets of actor relations create new economic and social capital. New supply chains also affect product offerings	Improving employee satisfaction, retention and productivity	New service design, including service blueprints, assessment of internal service quality processes, and market segmentation	New services and means of adding value for customers	Utilisation of new channels, combinations of channels and intermediaries	Utilisation of new ways to communicate and create brand value	Development and design of new servicescapes, including the virtual world	Providing new experiences	Tracking customer satisfaction	Customer relationship management strategies
Examples of potential sustainability related firm innovations	Place-based ownership reinforces local supply chains	Developing local supply chain minimises emission costs of supply	Improvements in productivity through reducing energy consumption and waste	Services can be designed so as to reduce emissions and be more energy efficient	New green services can attract new customers as well as generate new firm product	Developing distribution channels with low emissions and energy demands	Utilising the brand to emphasise positive environmental values at a firm and destination scale	Design can incorporate measures to minimise emissions and energy use as well as enhance experiences	Knowledge of experience can help make environmental measures acceptable to customers	Better knowledge can lead to increased yield per customer at lower environmental cost	Loyal customers become advocates for firm's green business activities

Source: After Hall & Williams 2008

Figure 15.2 Innovation value creation points in the tourism firm in relation to sustainability.

Brands

Brands are the symbolic embodiment of all information connected to a product and are evidenced via names, slogans, logos, symbols, designs, sounds, smells, and colours. Brand innovation can refer to new ways of reinforcing existing brands, the development of new ones and new means to protect the values and designs of existing brands, including the extent to which they are seen as being environmentally friendly. Increasingly, destinations and tourism businesses are keen to ensure that their brand has positive environmental perceptions in order to sell to an environmentally conscious market.

Servicescape

A servicescape is the physical evidence of a service and refers to physical environment dimensions such as ambient conditions, space/function and signs, symbols and artifacts as customer perceptions of the servicescape. Servicescape innovations are one of the most common areas of innovation in tourism given the relative ease of changing physical surroundings and designs, i.e., in terms of using more environmentally friendly products, ensuring new structures are appropriate in terms of emissions and energy requirements.

Customer's Service Experience

Customer's service experience refers to innovation with respect to the experience the customer has with a particular organisation as well as that organisation's understanding of the experience. In a number of ecotourism-type operations, customer experiences can help influence environmental attitudes and behaviours following the trip.

Customer Satisfaction

Customer satisfaction refers to the satisfaction a customer has with a service encounter as well as a tourist firm or destination overall. In addition, customer satisfaction is also a focus for innovation with respect to tracking satisfaction over time. This can be extremely important for firms to increasing the returns they get from their existing visitor base rather than necessarily expanding the numbers of tourists they receive.

Customer Loyalty

Customer loyalty refers to the maintenance of customer relationships over time so they continue to purchase firm products, often as repeat visitors, as well as become advocates for that business, including its environmental and social performance.

Maintaining and innovating along the various value points of the tourism business will potentially bring greater returns to the firm and therefore enhance the likelihood that it will survive and hence contribute to the resilience of the destination as whole. As Baumol (2002, 1) noted, innovative activity is 'mandatory, a life-and-death matter for the firm and innovation has replaced price as the name of the game in a number of important industries'. Moreover each innovation point represents part of a firm's capacity to adapt and attract new markets in light of the turbulence that challenges such as climate change will bring to tourism consumption. Significantly, the model highlights that innovative and adaptation is not just a case of the development or adoption of technological solutions at the firm level. Instead, value for a tourism firm is derived from a number of different creation points. The model therefore suggests that there are a wide range of behavioural, organisational, and marketing responses that can all help firms remain in business and operate at a profit while contributing to sustainable development at regional, national, or global scales.

CONCLUSIONS

This chapter has highlighted the potential value of the innovation literature for understanding the possibilities of sustainable tourism development. It initially noted the ways in which the innovation, adaptation, and resilience, and sustainable tourism literatures had converged with respect to the importance of understanding system-wide effects and the role of agents, such as firms, within them.

The chapter then went on to discuss the concept of innovation systems that operate at various scales. Destination innovation systems may be regarded as a particular form of regional innovation system. The lack of innovation with respect to sustainable development within a destination or region can be usefully explained by reference to an innovation system that highlights the importance of the 'soft' characteristics of innovation with respect to culture, institutional and firm networks, and knowledge transfer. Such elements cannot be changed overnight but specific intervention points may be identified so as to increase the adaptive capacity of destinations.

Finally, the chapter emphasised the importance of an improved understanding of the firm as an element of sustainable tourism. Firms and the innovation systems within which they are embedded co-evolve in response to externally induced change, including some of the pressures arising from climate change. Sustainable firms are therefore extremely important with respect to the resilience and hence sustainability of destinations although this observation is not the same as Friedmannite economics that the social responsibility of business is to increase profits (Friedman 1970). Nevertheless, as business ecological actors, sustainable firms should be understood in terms of survivability and profit, as clearly they are not sustainable if they cease to exist. However, the notion of survivability also suggests that the conceptualisation of firm behaviour needs to be extended beyond that of being solely responsible to shareholders with respect to the provision of immediate returns and instead indicates that return needs to be understood over extended time periods, in a sense approximating the concept of sustained yield. The incorporation of survivability of the firm as an indicator of sustainability also provides support for the extension of the concept of responsibility of managers from being 'beholden to shareholders' to one that is more stakeholder based.

Finally the capacity of tourism firm to adapt and innovate was also placed in the context of a model of the firm that allowed for recognition of some of the value creation points by which firms may differentiate themselves, develop appropriate business strategies, and generate profit. Each of these value points also provided opportunities for adaptive business practices that also encouraged environmentally positive behaviours. But perhaps most importantly the model indicated the need to understand innovation for sustainable tourism as being primarily based in business organisation and behaviour—the social dimensions of innovation—rather than in the utilisation of technology. Sustainable tourism, like innovation itself, is socially constructed.

ACKNOWLEDGMENTS

The author would like to acknowledge the contribution and insights of Allan Williams to the work on which a number of the insights of this chapter is based.

REFERENCES

Abernathy, W.J. and Clark, K.B. (1988) 'Innovation: Mapping the winds of creative destruction'. In Tushman, M.L. and Moore, W.L. (eds.) *Readings in the Management of Innovation*, 2nd ed., 55–78. Cambridge, MA: Ballinger.

Baumol, W. (2002) *The Free Market Innovation Machine: Analyzing the Growth Miracle of Capitalism*. Princeton: Princeton University Press.

Bramwell, B. and Lane, B. (eds.) (2000) *Tourism Collaboration and Partnership: Politics, Practice and Sustainability*. Clevedon: Channelview.

Carlsson, B., Jacobsson, S., Holmén, M., and Rickne, A. (2002) 'Innovation systems: Analytical and methodological issues'. *Research Policy* 31: 233–245.

Carpenter, S., Walker, B., Anderies, M.J., and Abel, N. (2001) 'From metaphor to measurement: The resilience of what to what?' *Ecosystems* 4: 765–781.

Cooke, P., Gomez, O.M., and Etxebarria, G. (1997) 'Regional innovation systems: Institutional and organizational dimensions'. *Research Policy* 26: 475–491.

ECON Analysis (2006) *Innovation in Services: Typology, Case Studies and Policy Implications.* Commissioned by the Norwegian Ministry of Industry and Trade, ECIN-Report no. 2006–025, Project no. 44720, AHA/LAG/pil, TVA, 28. February 2006. Oslo: Ministry of Industry and Trade.

Edquist, C. (2005) 'Systems of innovation: Perspectives and challenges'. In Fagerberg, J. , Mowery, D., and Nelson, R.R. (eds.) *The Oxford Handbook of Innovation*, 181–208. Norfolk: Oxford University Press.

Friedman, M. (1970) 'The social responsibility of business is to increase its profits'. *New York Times Magazine*, 13 September: 32–33, 122, 124, 126.

Gössling, S. and Hall, C.M. (eds.) (2006) *Tourism and Global Environmental Change.* London: Routledge.

Hall, C.M. (2006) 'New Zealand tourism entrepreneur attitudes and behaviours with respect to climate change adaption and mitigation'. *International Journal of Innovation and Sustainable Development* 1(3): 229–237.

Hall, C.M. (2007, September 11–14) 'Key issues in achieving sustainable tourism: Tourism, innovation systems and sustainability'. Paper presented at Achieving Sustainable Tourism Conference, Helsingborg, Sweden, 2007.

Hall, C.M. (2008) *Tourism Planning*, 2nd ed. Harlow: Pearson Education.

Hall, C.M. and Coles, T. (2008) 'Conclusion: Mobilities of commerce'. In Coles, T. and Hall, C.M. (eds.) *International Business and Tourism: Global Issues, Contemporary Interactions*, 273–283. London: Routledge.

Hall, C.M. and Page, S. (2006) *The Geography of Tourism and Recreation: Environment, Place and Space,* 3rd ed. London: Routledge.

Hall, C.M. and Williams, A.M. (2008) *Tourism and Innovation.* London: Routledge.

Hjalager, A.-M. (1996) 'Tourism and the environment: The innovation connection'. *Journal of Sustainable Tourism* 4(4): 201–218.

Hjalager, A.-M. (1997), 'Innovation patterns in sustainable tourism: An analytical typology'. *Tourism Management* 18(1): 35–41.

Hjalager, A.-M. (2002) 'Repairing innovation defectiveness in tourism'. *Tourism Management* 23: 465–474.

Hjalager, A-M., Huijbens, E.H., Björk, P., Nordin, S., Flagestad, A., and Knútsson, Ö. (2008) *Innovation Systems in Nordic Tourism.* Oslo: Nordic Innovation Centre.

Holbrook, J.A. and Wolfe, D.A. (eds.) (2000) *Knowledge, Clusters and Regional Innovation: Economic Development in Canada, Montreal and Kingston.* McGill-Queen's University Press for the School of Policy Studies, Queen's University.

Jensen, M.C. (2002) 'Value maximization, stake-holder theory, and the corporate objective function'. *Business Ethics Quarterly* 12: 235–256.

Kanter, R.M. (1983) *The Change Masters.* London: Unwin.

Kassinis, G.I. and Soterlou, A.C. (2003) 'Greening the service profit chain: The impact of environmental management practices'. *Productions and Operations Management* 12(3): 386–403.

Kim, W.C. and Mauborgne, R. (1999) 'Strategy, value innovation and the knowledge economy'. *MIT Sloan Management Review* Spring: 41–54.

Longhi, C. and Keeble, D. (2000) 'High technology clusers and evolutionary trends in the 1990s'. In Keeble, D. and Wilkinson, F. (eds.) *High Technology Clusters, Networking and Collective Learning in Europe*, 21–56. Aldershot: Ashgate.

Matthing, J., Sandén, B., and Edvardsson, B. (2004) 'New service development: Learning from and with customers'. *International Journal of Service Industry Management* 15(5): 479–498.

Medina, C.C., Lavado, A.C., and Cabrera, R.V. (2005) 'Characteristics of innovative companies: A case study of sectors'. *Creativity and Innovation Management* 14(3): 272–287.

Nauwelaers, C. and Reid, A. (1995) *Innovative Regions? A Comparative Review of Methods of Evaluating Regional Innovation Potential.* European Innovation Monitoring System (EIMS) Publication No. 21. Luxembourg: European Commission, Directorate General XIII.

Nowotny, H., Gibbons, M., and Scott, P. (2001) *Rethinking Science: Knowledge and the Public.* Cambridge: Polity Press.

Osterwalder, A., Pigneur, Y., and Tucci, C.L. (2005) 'Clarifying business models: Origins, present, and future of the concept'. *Communications of the Association for Information Systems* 15(preprint): 1–40.

Schumpeter, J.A. (1934) *The Theory of Economic Development.* Cambridge, MA: Harvard University Press.

Schumpeter, J.A. (1942) *Capitalism, Socialism and Democracy.* New York: Harper & Row.

Shaw, G. and Williams, A.M. (2002) *Critical Issues in Tourism: A Geographical Perspective,* 2nd ed. Oxford: Blackwell.

Sirilli, G. and Evangelista, R. (1998) 'Technological innovation in services and manufacturing: Results from Italian survey'. *Research Policy* 27(9): 881–899.

Sundaram, A.K. and Inkpen, A.C. (2004) 'The corporate objective revisited'. *Organization Science* 15(3): 350–363.

Sundbo, J. (1998) *The Theory of Innovation: Entrepreneurs, Technology and Strategy.* Cheltenham: Edward Elgar.

Tether, B. (2004) *Do Services Innovate Differently?* Discussion paper 66, Manchester: University of Manchester, Centre for Innovation and Competition.

Tödtling, F. and Kaufmann, A. (1998, August 28–September 1) 'Innovation systems in regions of Europe—a comparative perspective'. Paper presented to the 38th Congress of the European Regional Science Association, Vienna.

Van der Aa, W. and Elfring, J. (2002) 'Realizing innovation in services'. *Scandinavian Journal of Management* 18: 155–171.

Weaver, D. (2006) *Sustainable Tourism: Theory and Practice.* Oxford: Butterworth-Heinemann.

16 Synthesis and Conclusions

Stefan Gössling, C. Michael Hall,
and David B. Weaver

This collection of chapters effectively exposes the progress and the problems associated with the tourism sector's pursuit of a sustainability-focused agenda. Awareness of associated ecological and social issues, as Lane (Chapter 2) reminds us, is now increasingly evident among tourism stakeholders, as is awareness of the need to do something about them. However, rhetorical enthusiasm aside, the emerging landscape of actual sustainable tourism practice reveals isolated pockets of incipient innovation driven by embryonic and often ephemeral stakeholder partnerships. A number of contributors have sought to provide reasons as to why the implementation of sustainable tourism practices has been so poor. Adoption of sustainable practices according to Weaver (Chapter 3) is still neither broad nor deep and is perhaps indicative of an adaptive 'paradigm nudge' rather than a transformational 'paradigm shift'. Weaver argues that in large part this is because consumers remain only conditionally committed to altering their lifestyles and travel behaviour, thereby denying industry and government a necessary incentive to pursue transformational change. Hall (Chapter 15) also argues that behavioural change is required, but with respect to the role of the producers of tourism, especially firms, as well as consumers. In particular he argues that current management focus on providing shareholder value in the immediate and short-term does not sit easily with the longer-time horizons of sustainability, particularly when it is argued that social welfare is maximised when organisations exclusively pursue profits (e.g., Friedman 1970). Nevertheless, profit along with survival does remain a key indicator of firm sustainability, particularly as Hall suggests that the capacity of tourism firms to survive and grow is a measure of their individual resilience as well as a potential indicator of system resilience at the destination scale. Therefore, Hall argues that in order to understand the lack of adoption of sustainable practices by firms we therefore need much more sophisticated understandings of how firms create value via innovation as well as the way in which firms are embedded within regional and national innovation systems. Indeed, an implicit theme of a number of chapters is that there is a need for a far greater overall understanding of how tourism systems operate and the co-evolutionary relationships between consumers, firms, and destinations with respect to their responses to drivers for change.

CLIMATE CHANGE AS CATALYST

Climate change has now emerged as a justifiably powerful catalyst for the appearance of new participants and innovations (as evidenced by the subject matter of most of the chapters), but whether it will break the apparent impasse of paradigm nudge is questionable. Indeed, what Lane describes as the narrowly focused 'politics of the carbon footprint' may inhibit the embrace of a more holistic and integrative approach toward sustainable tourism implementation. Broderick (Chapter 10), for example, expresses concerns over the credibility and effectiveness of politically motivated voluntary carbon offsetting schemes within the travel industry, and cites the possibility of their consequent dissuasive effects on more constructive avenues of innovation. Nature-based tour operators, as described by Strasdas (Chapter 14), further illustrate this conundrum in their strong focus on global warming and its transportation implications as they move to develop appropriate sector-specific sustainability guidelines and indicators. At a planning and policy level, it is notable that the scenario planning undertaken by VisitScotland and described by Page, Yeoman, and Greenwood in Chapter 5 is almost completely focused on transportation in its bold if quixotic attempt to outline relevant travel parameters for Scotland in the year 2025. The chapters thus show a somewhat paradoxical situation: While climate change and emissions of greenhouse gases have for almost two decades been nearly completely ignored as issues central to the development of sustainable tourism, there may now be a tendency by many researchers and parts of the industry and its organisations to focus almost exclusively on same. It is hoped, however, that some normalisation will occur and that climate change will receive the place it deserves in sustainable tourism while not ignoring other issues of relevance.

THE ISSUE ATTENTION CYCLE OF SUSTAINABILITY

The current focus on climate change rather than seeking to provide a more integrative perspective on the broader challenges of sustainability raise a number of issues with respect to how policy concerns rise and fall on the political agenda as well as the academic agenda. A number of the current concerns with respect to the sustainability of tourism, such as the cost of fuel, and broader concerns over resource availability, and the quality of the environment have been here before but then drifted off the policy map (see Chapter 1). For example, it is salutary to read the opening page of the now unfashionable report of The Club of Rome (Meadows et al. 1972, 17) which commented on a 1969 speech by the then UN Secretary-General U. Thant with respect to the state of the world.

The problems that U. Thant mentions—the arms race, environmental deterioration, the population explosion, and economic stagnation—are often cited as the central, long-term problems of modern man. Many people believe that the future course of human society, perhaps even the survival of human society, depends on the speed and effectiveness with which the world responds to these issues. And yet only a small fraction of the world's population is actively concerned with understanding these problems or seeking their solutions (Meadows et al. 1972, 17).

Indeed, the goal of the 'model output' for Meadows et al. (1972, 158), 'a world system that is:

1. sustainable without sudden and uncontrollable collapse; and
2. capable of satisfying the basic material requirements of all its people.'

Little is different from the basic principle of sustainable development enunciated in the present-day. Indeed, strong parallels exist between the environmental and energy crisis of the 1970s and the present-day. For example, in May 2008 the price of oil reached the cost of oil at the time of the 1979 oil spike. There were increasing protests in Europe by fisherman and truck drivers with respect to the price of fuel, while riots were occurring over the increasing cost of food in the developing world. In the UK Prime Minister Gordon Brown warned that the world was facing the 'third great oil shock of recent decades' that could only be addressed by urgent action on a global scale (Brown 2008). However, in the same month *The Guardian* also reported that 'a fifth of the UK bee colonies had been killed', 'supermarkets were failing in reducing packaging', 'Exxon investors rejected environmental initiatives', 'more than one in four of all individual animals, birds and fish on the planet have disappeared in just over thirty years', and that an interim report of the economics of ecosystems and biodiversity, was suggesting that '11% of the world's untouched forests and 60% of its coral reefs could be lost by 2030. About 60% of the Earth's ecosystem, examined by the researchers, has been degraded in the past 50 years'. All this at a time when many of the world's developed economies were experiencing a 'credit crunch' and increasing concerns over the economy. Nevertheless, *The Guardian* carried a leader on the environment that concluded, 'We cannot abandon our commitment to green policies in the face of economic difficulty. Just as the need to control inflation is still critical at a time of economic challenge, so, too, is the need to reduce our carbon dependency' (King 2008).

For some readers the response to the latest environmental crisis (in which we can include energy concerns) will likely be that we have been here before and such crises have been 'solved'. Nevertheless, it is arguably the case that the underlying issues with respect to population growth, resource use and scarcity, and distribution of material well-remains the same but that the intensity of concern is increasing. Moreover, tourism is now directly becoming a focal point of environmental concern in a way that it has not in previous environmental crises. This may be partly because it is such a visible element of conspicuous environmental consumption but it is also likely because it is one of the industry sectors that despite per capita improvement in energy efficiency and emissions the absolute amount of energy consumption and emissions from the sector will likely continue to grow globally in the foreseeable future (Gössling and Hall 2006; UNWTO-UNEP-WMO 2008). For example, climate change, increasing noise pollution, and congestion have united middle class Londoners, environmental groups, local communities, and more than 20 councils in public rallies against the UK government-backed British Aviation Authority plan for a third runway and a sixth terminal leading at Heathrow Airport (Vidal 2008). Indeed, tourism may well become an easy target for environmental and human rights groups in the future if the aviation and tourism industry's use of energy and biofuel comes to be regarded as a contributing factor to increasing costs of food and loss of biodiversity in the developing world. A campaign based on the leisure of the wealthy leading to the starvation of the poor would obviously not be a positive impression for an industry trying increasingly hard to state its credentials as a sustainable industry. For example, in opening the 2008 WTTC Global Travel and Tourism Summit in Dubai His Excellency Khalid bin Sulayem, Director General, Dubai Government Department of Tourism and Commerce Marketing, stated that 'The development of sustainable tourism is our mission in achieving our goals. Green initiatives are being unveiled by the government to ensure eco-friendly development projects across the emirate, positioning Dubai as the first city in the region to adopt a green building strategy, as well as innovating new eco-developments' (Global Travel and Tourism Summit 2008a). Similarly, organisations such as IATA and the UNWTO have also promoted the environmental claims of the tourism and travel industry. Undoubtedly, the tourism industry has made positive steps towards improving its environmental record since the 1970s, the problem is that many of these improvements have been lost in the vast growth of domestic and international tourism within that time.

Incremental Innovation

The contemporary landscape of sustainable tourism indicates substantial progress over what existed in the early 1990s, and may well constitute the basis for subsequent exponential and more holistic change should a parallel critical mass of consumers become sufficiently compelled to advocate for a higher order of action from industry and government. A promising trend in the incremental accumulation of innovation is the extension of sustainability considerations and applications to sectors not usually associated with same. Upham (Chapter 9), for example, demonstrates an emergent methodology for calculating the carbon footprint of arts festivals, while Dubois and Ceron (Chapter 12) describe how such outcomes can be disseminated to consumers through travel agencies to influence the travel planning process. Likewise, Peeters et al. (Chapter 13) demonstrate that there is a huge potential for social marketing to influence travellers' decision making that has largely remained unexplored in the context of tourism. It is not difficult in all three cases to imagine extensions of these applications that take into account broader environmental and social considerations. Other sectors of the tourism industry have already entered this new territory through the efforts of various corporate and governmental leaders. For example, Bohdanowicz (Chapter 7) describes how the accommodation sector is experimenting with an unprecedented array of benchmarking and reporting tools and how the Scandic chain functions as an exemplar in moving this process beyond conventional financial reporting protocols. Similarly, Del Matto and Scott (Chapter 8) reveal Whistler Blackcomb in British Columbia as being high on the 'ideal type' scale of integrated sustainable ski resort principles, though not yet 'fully realised' on any of the latter.

Partnerships

A thread which pervades all the contributions in this text is the crucial role of stakeholder partnerships and the development of networks for innovation development and knowledge transfer. Scandic and Whistler Blackcomb are unable to achieve their transformational aspirations in isolation, but rather are shown to rely on the cooperation of industry partners, various layers of government, and environmental and other nongovernmental organisations. Coles (Chapter 11) in particular reminds us that the practitioner of sustainable tourism must also be a master of acronyms, as demonstrated by the complex reticulate network of organisations that has formed to pursue sustainable outcomes in the southwestern region of England. This represents an opportunity in so far as it indicates consensus on the broad aspiration to achieve sustainable outcomes as well as the mobilisation of a critical mass of resources toward this end. Yet it is also a challenge to the extent that each new partner brings to bear an additional agenda and an increased likelihood of intra-network competition and conflict and, as a result, systemic inertia. The challenge is all the greater if we logically allow that these emergent reticulate networks must additionally include resource stakeholders outside of the tourism system *per se*.

Challenging the Growth Paradigm

Ultimately, a major consideration in the pursuit of sustainable tourism outcomes, as elaborated by Dodds and Butler in Chapter 4, is stakeholder perception of the growth paradigm that has dominated the tourism sector during its modern era of exponential expansion. International tourist arrivals and annual growth rates have been seen to represent the overall benefit brought by tourism to a country, with high growth rates being implicitly understood as a sign of progress. By extension any climate change mitigation policy or other sustainability strategy that adversely impact arrivals represents an inherently negative impact. Exacerbating this dilemma are generic policy formation constraints such as power politics and difficulties in attaining stakeholder integration and coordination (Dodds and Butler, Chapter 4). Moreover, as described by Ceron and Dubois (Chapter 6), data currently available to facilitate policy and management decisions remain inadequate. Yet, destinations are well advised to reconsider their tourism marketing and development strategies in the light of ever more compelling evidence of climate and environmental change and its detrimental impacts for the tourism industry, consumer markets that are increasingly aware of this and other contemporary environmental and social concerns, and governments that may be more willing to aggressively exercise their regulatory imperatives in light of the first two of these considerations. Hence, future viability and indeed, profitability, may be contingent upon industry's proactive reconsideration of the growth paradigm in a way that could indicate a transformational paradigm shift rather than an adaptive paradigm nudge.

THE END OR A NEW BEGINNING FOR TOURISM AS WE KNOW IT

This book has clearly highlighted that mass tourism is facing a high challenge with respect to its sustainability. It has also been suggested that the focus of tourism research has sought to move beyond issues of definition to ensure that practical steps and measures can be taken to ensure sustainability. Ironically, such steps may not have necessarily been recognised by the tourism industry. For example, Jean-Claude Baumgarten, President & CEO, WTTC in his Opening Remarks to the 2008 Global Travel and Tourism Summit:

> Against the background of consistent growth trends for Travel & Tourism in our rapidly-evolving world, this Summit is designed to facilitate an open exchange of ideas among industry and government on the responsibilities that the Travel & Tourism sector should exercise as its global influence increases.
> Since the definition of responsible Travel & Tourism is still a matter for discussion, the Summit should make a crucial contribution to refining the thinking on what this should mean.
> (Global Travel and Tourism Summit 2008b)

This is not to suggest that tourism researchers need to revisit old ground. Sustainable tourism is like an elephant, hard to describe but you know it when you see one. Effective knowledge transfer therefore remains crucial if the concerns of tourism academics and researchers will be translated in concrete actions by the tourism industry. It has been often noted in this book that consumers, by their preferences and behaviours, and governments, by their capacities to regulate, will play an essential role in determining the future sustainability of tourism. However, perhaps what is not so well recognised, and which will likely be part of the future agenda for sustainable tourism research, is the extent to which business, and big business in particular, influences the preferences of consumers and the policies of government.

Sustainable tourism innovation, including the adaptation and mitigation of climate and other negative environmental change, therefore needs to be better understood within what tourism academics describe as the tourism or destination system, and researchers on innovation, refers to as an innovation system (Hall and Williams 2008). Either way, governance, institutions, cultures, and resources all serve to influence the actions of actors such as tourism firms and destination management organisations. Tourism researchers are also embedded in this system the reality of which is that it is currently the market and hence profit which is regarded as the best means by which sustainability of tourism may be gained. Although it is significant that organisations such as the WTTC recognise the need to extend their focus. As Jean-Claude Baumgarten stated in the final session of the 2008 Global Travel and Tourism Summit:

> Over the last 20 years our message was very economics focused. We now have to move beyond it. We have to explain to the world that, since Travel & Tourism is one of the major sectors, we have obligations that go beyond our economic impact—such as human resources and infrastructure. So let's adapt our message.
>
> (Global Travel and Tourism Summit 2008c)

Such a statement suggests that the industry message, or at least the corporate message of the tourism industry, may well go little beyond business as usual. Unfortunately, as this and other chapters in the book have argued,'business as usual' may well in the future mean dramatic changes to tourism including potentially 'no business' for some firms and destinations. It may therefore be ironic that many members of an industry whose promotion often promises the advantage and value of change to its consumers may not be able to change quickly enough itself to meet the demands of their consumers to live and travel sustainably.

REFERENCES

Brown, G. (2008) 'Gordon Brown: We must all act together. The oil crisis is a global problem requiring global solutions. And the Opec cartel has to play its part'. *The Guardian* 28 May. Available at: http://www.guardian.co.uk/commentisfree/2008/may/28/gordonbrown.oil (29 May 2008).

Friedman, M. (1970) 'The social responsibility of business is to increase its profits'. *New York Times Magazine* 13 September: 32–33, 122, 124, 126.

Global Travel and Tourism Summit (2008a) *Opening Ceremony, Quotes and Speeches, HE Khalid bin Sulayem, Director General, Dubai Government*

Department of Tourism and Commerce Marketing. Available at: http://www. globaltraveltourism.com/media_detail.asp?m_gallery_id=42&ss=0&id=243 (29 May 2008).

Global Travel and Tourism Summit (2008b) *Opening Ceremony, Quotes and Speeches, Jean-Claude Baumgarten, President & CEO, WTTC—Opening Remarks*. Available at: http://www.globaltraveltourism.com/media_detail. asp?m_gallery_id=42&ss=0&id=208 (29 May 2008).

Global Travel and Tourism Summit (2008c) *Session 6: We've always looked ahead—now we're looking further, Jean-Claude Baumgarten, President & CEO, WTTC*. Available at: http://www.globaltraveltourism.com/media_detail. asp?m_gallery_id=40&ss=&id=437 (29 May 2008).

Gössling, S. and Hall, C.M. (eds.) (2006) *Tourism and Global Environmental Change*. London: Routledge.

Hall, C.M. and Williams, A. (2008) *Tourism and Innovation*. London: Routledge.

King, D. (2008) 'Now is not the time to abandon our ambition to be green', *The Guardian* 1 June. Available at: http://www.guardian.co.uk/commentis-free/2008/jun/01/carbonemissions.greenpolitics (29 May 2008).

Meadows, D.H., Meadows, D.L., Randers, J., and Behrens III, W.W. (1972) *The Limits to Growth. A Report for The Club of Rome's Project on the Predicament of Mankind*. London: Pan Books.

UNWTO-UNEP-WMO *(United Nations World Tourism Organization, United Nations Environment Programme, World Meteorological Organization)* (2008) 'Climate change and tourism: Responding to global challenges'. [Scott, D., Amelung, B., Becken, S., Ceron, J.-P., Dubois, G., Gössling, S., Peeters, P., and Simpson, M.] *United Nations World Tourism Organization (UNWTO), United Nations Environmental Programme (UNDP) and World Meteorological Organization (WMO). UNWTO, Madrid, Spain.*

Vidal, J. (2008) 'Thousands expected at "carnival" to fight Heathrow expansion'. *The Guardian* 31 May. Available at: http://www.guardian.co.uk/environment/2008/may/31/travelandtransport.communities (31 May 2008).

Glossary

AAU
Compliance permit traded and surrendered by states under the Kyoto Protocol.

Additionality
The requirement that an emissions reduction project would not have occurred without support from credit finance. Definition is difficult in practice and always subjective.

Afforestation
Planting of new forests on lands that historically have not contained forests.

Annex 1 Countries
Signatories to the UNFCCC who were members of the OECD in 1992 plus those with 'economies in transition' in Central and Eastern Europe.

Annex B Countries
Those countries that have binding GHG emissions targets under the 1997 Kyoto Protocol.

Baseline
A projected level of emissions against which to define reductions.

CDM
Provision within Kyoto Protocol to create a market in emissions reductions made in non-Annex 1 nations.

CDM Pipeline
The sum of all projects within the CDM from validation to registration and issuance of credits.

CER
Credit issued by the Clean Development Mechanism equivalent to 1 tonne (metric ton) of CO_2 emissions, calculated using Global Warming Potentials.

DOE
A third party organisation approved by the Executive Board of the CDM, providing technical and audit services.

DNA
Official state body created to review and approve CDM and JI projects within a party.

EUA
Allowance permits traded and surrendered by entities in the EU ETS.

Executive Board	Senior regulatory authority of the CDM ultimately responsible for overseeing methodologies, DOEs, and issuance of CERs.
Global Warming Potential	An index used to calculate equivalent warming effect of a unit mass of a given greenhouse gas integrated over a specified time period, usually 100 years, relative to CO_2.
Host country	Country in which credit project is located.
Joint Implementation	Provision within Kyoto Protocol to create a market in emissions reductions within Annex 1 nations. Emissions reductions units are transferred between parties so unlike CDM no new commodity is generate.
Linking Directive	Amendment to EU ETS to allow surrender of CERs and ERUs in place of EUAs.
LULUCF	Catch-all term for changes in terrestrial biological systems, including agriculture and forestry.
Methodology	In CDM terminology a quantification and monitoring method, including a baseline, which is used to define credits. Each must be approved by the EB but shared by projects of a similar type.
NAP	National Allocation Plan submitted by each state to the EU ETS to assign permits to each emitting installation within its jurisdiction.
PDD	Document specifying the technical details, participants, management, and organisation of an emissions reduction project including baseline and monitoring plan.
Radiative forcing	Measure of the change in vertical irradiance (in Wm^{-2}) at the tropopause as a result of a change in composition of the atmosphere.
Radiative Forcing Index	Ratio of total radiative forcing of aircraft emissions to that of CO_2 emissions alone over a defined time period.
Reforestation	Planting of forests on lands that had previously contained forests but have subsequently been converted to some other use.
Registration	Acceptance of a project to the CDM by the EB.
Registry	Electronic database recording provenance of all credits issued by a particular regime and subsequent trades or transfers.

Retirement	Cancellation of credits or permits. Permanent removal from circulation.
Sequestration	Removal of carbon dioxide from the atmosphere to another reservoir either biological or geological.
Sink	A natural or man-made process that sequesters carbon.
Validation	In CDM, evaluation of PDD by a DOE prior to registration.
Verification	Periodic review of project monitoring data against methodology to quantify reductions achieved.

Contributors

Dr. Paulina Bohdanowicz has a PhD in energy technology (Royal Institute of Technology, Stockholm, Sweden) and a PhD in social science (University of Gdansk, Poland). She is currently a lecturer in sustainable tourism and hospitality at Gdansk Academy of Sports and Physical Education and a sustainability consultant for hospitality companies (www.greenthehotels.com).

Philip Boucher is currently undertaking a PhD with The Tyndall Centre for Climate Change Research and The University of Manchester. His work, funded by Supergen's Bioenergy Consortium, focuses on institutional communications on biofuels and their associated social and environmental impacts.

John Broderick is a doctoral researcher with the Tyndall Centre for Climate Change Research at Manchester Business School. His work on emissions trading in the aviation industry is funded by the UK Energy Research Centre (UKERC).

Professor Richard W. Butler was trained as a geographer in the UK, and spent thirty years in Canada at the University of Western Ontario. He returned to the UK in 1997 to the University of Surrey and is currently at the University of Strathclyde. His main interests are in destination development and sustainability.

Dr. Jean-Paul Ceron is a social scientist who has been working for three decades on environmental issues first within CIRED, a team which is specialised on climate change issues and now at CRIDEAU (University of Limoges). The relationship between tourism and climate change is now his main fields of interest. He has been a lead author within the Fourth Assessment Report of the IPCC, dealing there mainly with tourism.

Dr. Tim Coles is a Senior Lecturer in Management and Co-Director of the Centre for Tourism Studies in the School of Business and Economics at the University of Exeter in the United Kingdom.

Tania Del Matto is co-founder of My Sustainable Canada, a national organisation that advocates for sustainable consumption. Since 2003, Tania has helped Canadian ski resorts become more sustainable. She has a Master of Environmental Studies degree from the University of Waterloo.

Dr. Rachel Dodds is an assistant professor at the Ted Rogers School of Hospitality and Tourism Management at Ryerson University in Toronto, Canada. Her research interests include sustainable tourism, corporate social responsibility and climate change.

Ghislain Dubois is an associate professor at the University of Versailles Saint-Quentin-en-Yvelines. He is a consultant for several international organisations and a researcher involved in the field of tourism and climate change for ten years, through contracts and peer-reviewed research. He also contributed to the IPCC 4th assessment report (WG2).

Dr. Stefan Gössling is research coordinator at the Centre for Sustainable and Geotourism, Western Norway Research Institute, and an Professor at the Service Management programme, Lund University/Sweden. He has been contributing author to the IPCC's 4th Assessment Report and has recently contributed to Climate Change and Tourism: Responding to Global Challenges (published by UNWTO-UNEP-WMO).

Chris Greenwood is a research analyst with VisitScotland in Edinburgh and has degrees in Geography.

Dr. C. Michael Hall is a professor in the Department of Management, University of Canterbury, New Zealand and Docent in the Department of Geography, University of Oulu, Finland. Co-editor of *Current Issues in Tourism*, he has published widely in the fields of tourism, environmental history, and gastronomy, including research on social marketing, climate change, and sustainable consumption.

Dr. Drew Hemment is associate director of ImaginationLancaster. He explores the connections between people, emerging technologies, and possible futures. His work focuses on Art and Social Technologies, and he was instrumental in the emergence of the field of Locative Media. Areas of research interest include Social Technologies; Art and Technology; Sustainability in Urban Environments; Locative Media; The City and Technology; Everyday Creativity; Collaborative Art; Open Source Culture and the Public Sphere; Ethics, Privacy, and Control.

Dr. Bernard Lane founded and co-edits the *Journal of Sustainable Tourism*. He is an Associate at Red Kite Environment, a consultancy specialising in sustainable tourism and heritage management, and Visiting

Research Fellow at the Centre for International Tourism Research, Sheffield Hallam University.

Professor Stephen Page is the Scottish Enterprise Professor of Tourism Management at the University of Stirling and has worked extensively with the tourism industry on scenario planning and consultancy projects to assist in the economic development of tourism in different localities. He is the Reviews Editor for the top tourism journal— *Tourism Management,* published by Elsevier and co-author of the leading introductory text— *Tourism: A Modern Synthesis.*

Paul Peeters is an associate professor and studies the relations between tourism transport and the environment, with a focus on climate change. Since 2002 he is a professor at the Center for Sustainable Tourism and Transport, NHTV University for Applied Science, Breda, the Netherlands.

Dr. Daniel Scott is a Canada Research Chair in Global Change and Tourism at The University of Waterloo. He is the Chair of the World Meteorological Organisation's Expert Team on climate and tourism and is Co-chair of The International Society of Biometeorology's Commission on Climate and Tourism.

Professor Wolfgang Strasdas is the managing director of the "Sustainable Tourism Management" Master programme at the University of Eberswalde, Germany. He is also a board member of The International Ecotourism Society and has extensive experience as a consultant and trainer in developing countries.

Dr. Paul Upham is an environmental social scientist at the Tyndall Centre Manchester, University of Manchester, with research interests in public and stakeholder perceptions of low carbon energy systems and climate change policy for aviation.

Dr. David Weaver is Professor of Tourism Research at Griffith University, Australia. He is author or co-author of over eighty refereed journal articles and book chapters as well as ten books on tourism management.

Dr. Ian Yeoman is the world's only professional crystal ball gazer or futurologist specialising in travel and tourism. Ian is an associate professor of tourism at Victoria University, New Zealand. Further details can be found at www.tomorrowstourist.com.

Index